Recent Titles in This Series

144 **O. V. Troshkin,** Nontraditional methods in mathematical hydrodynamics, 1995
143 **V. A. Malyshev and R. A. Minlos,** Linear infinite-particle operators, 1995
142 **N. V. Krylov,** Introduction to the theory of diffusion processes, 1995
141 **A. A. Davydov,** Qualitative theory of control systems, 1994
140 **Aizik I. Volpert, Vitaly A. Volpert, and Vladimir A. Volpert,** Traveling wave solutions of parabolic systems, 1994
139 **I. V. Skrypnik,** Methods for analysis of nonlinear elliptic boundary value problems, 1994
138 **Yu. P. Razmyslov,** Identities of algebras and their representations, 1994
137 **F. I. Karpelevich and A. Ya. Kreinin,** Heavy traffic limits for multiphase queues, 1994
136 **Masayoshi Miyanishi,** Algebraic geometry, 1994
135 **Masaru Takeuchi,** Modern spherical functions, 1994
134 **V. V. Prasolov,** Problems and theorems in linear algebra, 1994
133 **P. I. Naumkin and I. A. Shishmarev,** Nonlinear nonlocal equations in the theory of waves, 1994
132 **Hajime Urakawa,** Calculus of variations and harmonic maps, 1993
131 **V. V. Sharko,** Functions on manifolds: Algebraic and topological aspects, 1993
130 **V. V. Vershinin,** Cobordisms and spectral sequences, 1993
129 **Mitsuo Morimoto,** An introduction to Sato's hyperfunctions, 1993
128 **V. P. Orevkov,** Complexity of proofs and their transformations in axiomatic theories, 1993
127 **F. L. Zak,** Tangents and secants of algebraic varieties, 1993
126 **M. L. Agranovskiĭ,** Invariant function spaces on homogeneous manifolds of Lie groups and applications, 1993
125 **Masayoshi Nagata,** Theory of commutative fields, 1993
124 **Masahisa Adachi,** Embeddings and immersions, 1993
123 **M. A. Akivis and B. A. Rosenfeld,** Élie Cartan (1869–1951), 1993
122 **Zhang Guan-Hou,** Theory of entire and meromorphic functions: Deficient and asymptotic values and singular directions, 1993
121 **I. B. Fesenko and S. V. Vostokov,** Local fields and their extensions: A constructive approach, 1993
120 **Takeyuki Hida and Masuyuki Hitsuda,** Gaussian processes, 1993
119 **M. V. Karasev and V. P. Maslov,** Nonlinear Poisson brackets. Geometry and quantization, 1993
118 **Kenkichi Iwasawa,** Algebraic functions, 1993
117 **Boris Zilber,** Uncountably categorical theories, 1993
116 **G. M. Fel′dman,** Arithmetic of probability distributions, and characterization problems on abelian groups, 1993
115 **Nikolai V. Ivanov,** Subgroups of Teichmüller modular groups, 1992
114 **Seizô Itô,** Diffusion equations, 1992
113 **Michail Zhitomirskiĭ,** Typical singularities of differential 1-forms and Pfaffian equations, 1992
112 **S. A. Lomov,** Introduction to the general theory of singular perturbations, 1992
111 **Simon Gindikin,** Tube domains and the Cauchy problem, 1992
110 **B. V. Shabat,** Introduction to complex analysis Part II. Functions of several variables, 1992
109 **Isao Miyadera,** Nonlinear semigroups, 1992
108 **Takeo Yokonuma,** Tensor spaces and exterior algebra, 1992
107 **B. M. Makarov, M. G. Goluzina, A. A. Lodkin, and A. N. Podkorytov,** Selected problems in real analysis, 1992
106 **G.-C. Wen,** Conformal mappings and boundary value problems, 1992
105 **D. R. Yafaev,** Mathematical scattering theory: General theory, 1992
104 **R. L. Dobrushin, R. Kotecký, and S. Shlosman,** Wulff construction: A global shape from local interaction, 1992
103 **A. K. Tsikh,** Multidimensional residues and their applications, 1992

(See the AMS catalog for earlier titles)

Nontraditional Methods in Mathematical Hydrodynamics

Translations of
MATHEMATICAL
MONOGRAPHS

Volume 144

Nontraditional Methods in Mathematical Hydrodynamics

O. V. Troshkin

American Mathematical Society
Providence, Rhode Island

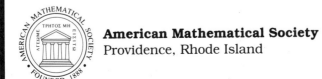

О. В. Трошкин

НЕТРАДИЦИОННЫЕ ЗАДАЧИ МАТЕМАТИЧЕСКОЙ ГИДРОДИНАМИКИ

Translated by Peter Zhevandrov from an original Russian manuscript
Translation edited by Simeon Ivanov

1991 *Mathematics Subject Classification.* Primary 76C05, 76D05, 76Fxx; Secondary 35Q30, 76Bxx, 76Cxx, 76Exx.

ABSTRACT. The main goal of the book is to present a number of features of mathematical models for incompressible fluids. Three basic systems of hydrodynamical equations are considered, namely, the system of stationary Euler equations for flows of ideal (nonviscous) fluid, stationary Navier-Stokes equations for flows of a viscous fluid, and Reynolds equations for the mean velocity field, pressure, and pair one-point velocity correlations of turbulent flows. The analysis concerns algebraic or geometric properties of vector fields generated by these equations, such as the general arrangement of streamlines, the character and distribution of singular points, etc. The information provided by this analysis is used for the investigation of the conditions for unique solvability of a number of problems for these quasilinear systems. The book contains a lot of examples of particular phenomena illustrating general ideas in the book. The book can be used by researchers and graduate students working in mathematical physics and hydrodynamics.

Library of Congress Cataloging-in-Publication Data

Troshkin, O. V.
[Hetraditsionnye zadachi matematicheskoĭ gidrodinamiki. English]
Nontraditional methods in mathematical hydrodynamics / O. V. Troshkin.
p. cm. — (Translations of mathematical monographs, ISSN 0065-9282; v. 144)
Includes bibliographical references p. –).
ISBN 0-8218-0285-2 (acid-free)
1. Hydrodynamics. I. Title. II. Series.
QA911.T7413 1995
532′.58—dc20 94-48581
CIP

This volume was typeset using $\mathcal{A}\mathcal{M}\mathcal{S}$-TEX,
the American Mathematical Society's TEX macro system.

10 9 8 7 6 5 4 3 2 1 00 99 98 97 96 95

Contents

Introduction

Mathematical hydrodynamics is a branch of mathematical physics that deals with the investigation of hydrodynamic equations. The principal equations of this kind are the three widely used systems of quasilinear partial differential equations of various orders, named after Euler, Navier and Stokes, and Reynolds, describing various regimes of flows exhibited by an incompressible fluid or a compressible gas. The investigation of these equations is complicated by their nonlinearity. Nonlinear models do not easily lend themselves to classification. The existing analytic tools are mainly oriented toward the investigation of classes of equations, while monographs are devoted, for the most part, to particular phenomena. The goal of this book is to compensate, to some extent, for the missing link between the first and the second by presenting a number of qualitative features of mathematical models for incompressible fluids.

Namely, in what follows we consider the system of stationary Euler equations for plane (plane-parallel) and axisymmetric flows of an ideal (nonviscous) fluid, stationary Navier-Stokes equations for plane flows of a viscous fluid, and Reynolds equations for the mean velocity field, pressure, and pair one-point velocity correlations of turbulent flows. Attempts to apply well-known methods (including functional ones) from the theory of boundary value problems to the equations mentioned above lead to a number of questions of "qualitative" character that are of interest from the point of view of physics as well as mathematics, but are beyond the scope of standard theorems and concern algebraic or geometric properties of vector fields determined by them, i.e., the general arrangement of streamlines, the character and distribution of singular points (zeros of the velocity field), conditions for their absence (or appearance), etc. The constructions in the text are devoted mainly to the analysis of the above-mentioned problems and to systematic application of the information provided by them to the investigation of the conditions for unique solvability of a number of problems for these quasilinear systems.

The Euler equations on the plane illustrate the character of the difficulties that arise. The well-known theorem of Yudovich **[109]** concerning the existence and uniqueness of the solution of the initial-boundary value problem for nonstationary flows of an ideal fluid in a bounded domain on the plane, considered on an arbitrary finite time interval, is proved under the following boundary conditions: the normal velocity component is given on the boundary and the vortex (the vorticity of the velocity field) is given on the influx regions (if any); the problem without influx regions (the boundary is impermeable) was also treated by Kato **[148]**; an earlier result in this direction is due to Wolibner **[190]** who used Lagrangian coordinates. The solution of the corresponding stationary problem under the above boundary conditions may be nonunique. This fact is connected with the existence of topologically different flows. Thus, a stationary solution of the passing flow type (all the streamlines are smooth arcs starting on one side of a curvilinear quadrangle and ending on the opposite side) is uniquely defined by the imposed restrictions [**4**]. In another example [**36**], there are infinitely many arbitrarily smooth solutions of the stationary problem without infux regions; the solutions have the structure of one or several circular vortices (domains filled out by closed streamlines) frozen in the fluid at rest (the velocity components vanish in the exterior of the indicated domains together with the partial derivatives of all orders). On the other hand, an analytic solution (the velocity components are analytic functions in the domain of definition) of the general stationary problem (with influx regions) is unique independently of the topology of the flow [**96, 98, 99**]. This problem involves a dimensionless parameter analogous to the Reynolds number. This parameter is the ratio of the product of two characteristic quantities: the vorticity on the influx regions and the dimension of the flow domain, to the normal velocity component on these regions. Its increase leads to the appearance and growth of one or several vortex zones in the domain of an analytic flow. When these zones develop to such an extent that the angular discharge of the fluid intermixed by them exceeds the one measured on the influx regions (so that on the flow domain as a whole the maximum principle for the stream functions is violated), the analytic solution bifurcates into infinitely many arbitrarily smooth nonanalytic solutions which are small deformations of the analytic one. Nevertheless, nonanalytic flows without closed streamlines are uniquely determined by the given boundary conditions even under minimal smoothness conditions [**96, 98, 99**].

The next class of solutions of Euler equations, from the point of view of simplicity, are spatially axisymmetric flows of an ideal fluid. The general difficulty of the three-dimensional case is in the absence of theorems on unique solvability of nonstationary Euler equations on an arbitrary time interval. Here, the possibility of a solution becoming singular for finite values of time is not yet refuted [**120, 137, 138, 181**]. The same difficulty remains when passing to the Navier-Stokes equations [**56, 155, 156**], where it is also not quite clear whether the appropriately defined "strong" and "weak" solutions of the nonstationary problem are equivalent, i.e., the "strong" solution is unique but is defined on a small time interval, while the "weak" solution exists for arbitrarily large values of time but its uniqueness is not proved [**56**]. In Ladyzhenskaya's example [**56**], the absence of the required smoothness of the nonstationary axisymmetric flow of a viscous incompressible fluid in an expanding space between coaxial cylinders with a given normal velocity component and vorticity on the boundary (the slip of the particles along the impermeable surface is admissible under the condition of absence of tangential stresses which is characteristic for interfaces of liquid media) leads to nonuniqueness of the flow for sufficiently large values of time in the class of "weak" E. Hopf's solutions [**146**]. The analysis of smoothness of solutions of the nonstationary problem is given in [**27, 56, 89, 90, 93, 143, 144**]. It is possible that the singularities arising with time are localized in the same way as for the solution of the mixed problem for a weakly nonlinear parabolic equation [**185**]. In this case it makes sense to speak about "partial regularity" of a solution; this notion was analyzed in [**124, 174–178**]. The question of formation of discontinuities in a three-dimensional flow due to convergent localized vortex rings was recently considered in [**170**]; however, the authors' conclusion that the discontinuity is formed in finite time needs additional mathematical justification.

In contrast to nonstationary flows, the stationary problem for an axisymmetric velocity field of an ideal fluid in a bounded domain admits an analysis similar to that for plane flows. The results on the uniqueness of a smooth stationary flow are given below in Chapter 3 and are based on a preliminary analysis of the topological structure of an analytic solution discussed in Chapter 2. The possibility of a topological approach to problems of partial differential equations is discussed in the Introduction to monograph **[69]**. In **[34]**, by means of examples of plane flows, it is shown that, using Morse's constructions **[69]** combined with existing analytical tools (the maximum principle, the lemma on the normal derivative, etc. **[59]**), one can make rigorous conclusions on the flow portrait without using the direct solution of the boundary problem posed. A part of the examples from this paper is given in §4, where the portraits of plane potential flows are analyzed together with analogous axisymmetric flows. Examples of nonpotential flows are also analyzed (§§5, 6).

The advantage of a two-dimensional flow of an incompressible fluid consists in the existence of the stream function whose level lines coincide with the streamlines, and the critical points, with singular points of the flow (of the velocity field). Only centers and saddles (of arbitrary, but finite multiplicity) can be isolated singular points. The separatrices of the saddles divide the flow domain into chambers. The analysis of flow portraits (§§4–6) is preceded by a list of topological restrictions imposed on the flow structure by the structure of the boundary conditions and a classification of chambers (§3). Possible applications of this method are limited to two-dimensional fields. The next step in constructions of this sort is the passage to the real three-dimensional case, where there are two stream functions and the streamline lies on the intersection of their level sets. Here, one encounters other difficulties. Moreover, the streamlines cannot always be described by means of two stream functions due to topological reasons. For example, streamlines can everywhere densely fill out two-dimensional surfaces **[116]**. According to computer experiments **[133, 142]**, "nonintegrable" cases of ABC-flows (proportionally of the curl of the velocity to its vector) with velocity components

$$u = A\sin z + C\cos y, \qquad v = B\sin x + A\cos z, \qquad w = C\sin y + B\cos x,$$
$$A, B, C = \text{const},$$

(see **[116, 128, 133]**) can serve as examples of specifically three-dimensional Euler fields. In these cases, single trajectories densely fill out spatial domains.

On the whole, the following approaches to the Euler equations have been developed recently:

1. *Group theoretic approach*, which use the analogy between the hydrodynamic equations and the equations of the top, i.e., a rigid body with one fixed point **[11, 26, 106, 115, 117, 137, 164]**. This analogy was noted in **[164]** and developed in **[117]**, where the infinite-dimensional group of volume-preserving diffeomorphisms of the flow domain corresponds to the group of rotations of the solid. As a result, a multidimensional version has been obtained of the Euler-Poinsot theorem **[11, 20]** on the stability of the stationary rotation of a top around the maximal or minimal axis of its ellipsoid of inertia and an analog of the Rayleigh stability condition was found **[11, 117]**. The paper **[117]** started a series of papers with the keywords "Arnold's stability". Among recent publications devoted to this topic we note a theorem on zonal symmetry of stable, according to Arnold, flows **[114]** and its subsequent revisions and generalizations **[125, 127]**. The reformulation of the mentioned analogy in terms of Lie algebras and its transfer to motions with friction lead to nonlocal (i.e., associated to arbitrary Reynolds numbers) theorems on uniqueness and stability of two-dimensional stationary flows of a viscous incompressible fluid in a bounded domain **[97]**; they are considered in Chapter 4.

2. *Topological approach,* providing analysis of the topology of stationary flows **[9–11, 34, 70, 98, 116, 117, 133, 139, 140, 142, 160–162]**. For example, three-dimensional flow domains with a nonconstant Bernoulli integral split into cells consisting of coaxial tori (the Arnold cell theorem) **[10, 11, 116, 117]**. Note that for an ABC-flow, where some of the streamlines are likely to densely fill out the volume, this integral is constant. Note also that an analogous example of a spiral flow, where the curl of the velocity is proportional to the velocity itself with coefficient $3^{1/2}$ (in the example of the ABC-flow given above, it is equal to 1) was considered by Naumova and Shmyglevskiĭ **[70]**. In **[9]** it is assumed that an ABC-flow possesses the property of exponential stretching of fluid particles similar to the one characterizing a dynamical U-system of Anosov **[5, 12]**; an example of an Euler flow on a special manifold exhibiting this property is given. In connection with nontrivial topology, we note exact solutions of Euler equations for compressible media from the paper **[3]** by Aleksandrov and Shmyglevskiĭ; they provide interesting flow portraits resembling the motion of clouds.

3. *Invariant-geometric approach* that allows one to describe integrals of motion by means of differential-invariant formulation of the Euler equations **[10, 11, 33, 36, 37, 115, 117, 179, 180]**. Note here the multidimensional generalizations of H. Hopf's asymptotic invariant (which are also called *helicity* in physics) for a three-dimensional field **[10]** in **[179, 180]**. Along with the well-known representation of the Euler equations **[117]**, we consider less known forms. In **[33]**, the problem of finding a combinatorial analog of the two-dimensional Euler equations on a rectangular grid identified with a two-dimensional complex (in the sense of homology theory **[79]**) was posed, and a model flow consisting of two streamlines illustrating the mechanism of violation of uniqueness of a stationary solution due to the appearance of singular points was considered. A presentation of the corresponding material is given in the monograph **[37]**. In **[36]**, a new invariant form of the Euler equations based on the use of a multidimensional analog of the integrating factor of a vector field is given (Euler equations on the plane are frequently expressed in terms of this quantity **[37, 55, 95]**). In the Appendix this form is deduced from the stationary action principle.

4. *Variational approach* is oriented on the investigation of the stability of stationary flows by means of the variational form of stationary equations **[6–8, 10, 106, 117, 118, 160–162]**. In the three-dimensional case this leads to the following unusual variational problem **[10]**: one has to minimize the energy (the integral of the scalar square of a field with zero divergence) transforming the field by means of volume-preserving diffeomorphisms of the flow domain. The two-dimensional analog of this problem is the following. One has to minimize the Dirichlet integral of a function in a two-dimensional domain (e.g., in a disc) over all the functions equivalent to a given one (i.e., functions obtained from a given one by an area-preserving transformation of the domain onto itself). If the initial function on the disc has zero boundary values and one critical point, then the extremum is attained on the "symmetrization" of this function (the symmetrized function depends solely on the distance from the center of the circle) **[126, Theorem 3.1]**. If the topological type of the initial function is more complicated (e.g., it has two maxima and a saddle), then the extremals are unknown. Physicists claim that in this case the extremal must have singularities **[160–162]**.

The equations of motion of an ideal fluid were obtained by Euler by means of a straightforward transfer of the Newton's dynamical law to the volumes of a continuous medium. This model adequately describes, e.g., comparatively slow water flows under normal physical conditions outside a thin boundary layer adjoining a solid surface.

The domain of the boundary layer cannot be described by this model. The presence of this domain gives an indirect evidence that there exists internal friction of molecular origin appearing between the contacting layers of a deformable medium. The internal friction is taken into account by the main hydrodynamic model, the Navier-Stokes equations. Formally, the latter can be considered as singular perturbations of the Euler equations (the presence of a small coefficient at higher derivatives) but exhibit nonlinear effects connected with the loss of stability of some basic stationary flow and with the appearance of a new velocity field when the positive parameter at the higher derivatives is decreased and passes through a certain (critical) value. The dimensionless form of these equations shows that the Reynolds number (the ratio of the product of characteristic values of length and velocity to the coefficient of the kinematic (or molecular) viscosity) plays the role of the varying parameter.

The Couette flow between the rotating coaxial cylinders and the Poiseuille flow in a plane channel or a pipe with circular cross-section maintained by a constant gradient of pressure serve as classical examples of basic flows **[47]**. Their common feature is the following straightforwardly verified property: the velocity field of a basic flow satisfies both the Euler and Stokes equations (the Euler equations are obtained by omitting the highest derivatives in the Navier-Stokes equations, the Stokes equations, by ignoring convective terms). Secondary hydrodynamic regimes bifurcating from the basic ones **[38]** are characterized by the large scale of the arising multivortex motion; for this reason they are often called "preturbulent". For example, the Taylor vortices **[129]** are secondary with respect to the Couette flow . The list of basic and secondary flows can be extended. The questions arising here concerning the bifurcation are closely connected with the classical results **[21]** but are associated with the analysis of the Orr-Sommerfeld equation (or equations of a similar type) **[38, 61, 94, 110–113, 171]** and are often assumed to involve Krasnoselskiĭ's theorem **[48]** (or analogous theorems) on bifurcation **[64, 73, 94, 110, 171]**. The existence of a solution of the Taylor vortices type which bifurcates from the Couette flow was established by Ivanilov and Yakovlev **[43]**, Yudovich **[111, 113]**, and Velte **[184]** (note also the papers by Berger and Veinberger in the collection **[94]** as well as Chapter 7 in the monograph by Shechter **[64]** devoted to this problem). The instability of the Poiseuille flow is shown in Krylov's paper **[51]**.

When investigating the bifurcation and the change of stability of solutions of the Navier-Stokes equations themselves, the standard boundary no-slip conditions of the fluid on a solid surface often become too restrictive and must be replaced by other conditions, e.g., the conditions of impermeability and absence of vorticity sources on the boundary, periodicity with respect to all spatial variables (flows on a closed manifold), or with respect to a part of them (flows on a manifold with a boundary), etc. In this case the requirement that the problem under investigation be well posed with respect to the linear Stokes equations is a general criterion of "correctness" of the imposed restrictions. Following the well-known constructions **[56, 154]** (i.e., obtaining and using some estimates for the norms and operations generated by the equations), one can show that for the indicated alternative conditions (under the requirement that the above-mentioned criterion holds), the corresponding stationary problem also conserves the main features of the classical one (with the no-slip conditions), i.e., it possesses at least one generalized solution for any value of the Reynolds number (the global existence theorem), which is unique for its sufficiently small values (the local uniqueness theorem). The required smoothness of the generalized solution (its equivalence to the classical one) is provided by the uniform ellipticity of the stationary system considered **[2, 56, 121]**.

Along the same lines, a functional equation of a special form with a quadratic nonlinearity is investigated below (Chapter 4). The solutions of the latter are interpreted as stationary rotations of either an abstract top with friction, or a "dissipative" top considered in a linear space with inner product (a pre-Hilbert space) and commutator (simultaneously a Lie algebra) consistent with the inner product (the invariance of the corresponding mixed product with respect to the cyclic permutation of cofactors, as for a Lie algebra of a compact Lie group [**19, 80**]). In the simplest case of an ordinary (three-dimensional) top, the angular velocity of rotation is the solution of the equation and the commutator is defined by the vector product of vectors in three-dimensional Euclidean space. Energy dissipation is generated by the momentum of friction forces present in the equation and proportional to the angular velocity or the moment of inertia. The tops with arbitrary linear uniform dependence of the moment of friction forces on the angular velocity are also admissible. In addition, strict positive definiteness of the matrix of this linear dependence is required; this ensures the required energy dissipation. In what follows, these matrices are called *dissipation matrices*. They are divided into two classes, i.e., of matrices commuting with the matrix of the moments of inertia of the top and of matrices not commuting with it. The tops of the first class are called *commutative* and those of the second class, *noncommutative*. The friction coefficient, i.e., the positive coefficient at the dissipation matrix, serves as a variable parameter. For any given moment of external forces, a global (i.e., corresponding to an arbitrary friction coefficient) theorem on the existence of a stationary rotation of a dissipative top, as well as a local (for sufficiently large values of the indicated coefficient) theorem on its uniqueness are proved. For a commutative top, it is established that the Euler-Poinsot theorem holds; moreover, the stationary rotation under consideration proves to be asymptotically stable, simultaneously the corresponding nonstationary equation is written out. Note also a version (given in [**26**]) of the corresponding alternative statement (on a nonstable rotation with friction around the intermediate axis of the moment ellipsoid).

The results concerning the uniqueness and stability are transferred to dissipative tops of an arbitrary finite dimension, which are derived as a result of finite-dimensional approximations (Galerkin expansions) of the solutions of the Navier-Stokes equations [**26, 85**]. For computational purposes one can make do with this straightforward generalization. In this case, the theorems obtained appear to be an adapted (taking into account the friction forces) translation of the paper [**117**] from the language of Lie groups into that of Lie algebras. Taking the friction forces into account is essential when passing to the infinite-dimensional case (which is of principal importance). The constructive approach used in [**117**] (based on the mentioned analogy between the groups of rotation and diffeomorphisms) becomes cumbersome in this case. In what follows, the theory of the dissipative top is presented axiomatically and the cases of finite and infinite dimensions are considered simultaneously.

The generalized form of the equation of the top used jointly with the standard notation allows us to include into consideration three-dimensional flows of a viscous incompressible fluid as well. However, the applications considered are again limited to two-dimensional fields. In this case, the Laplace operator with the minus sign corresponds to the matrix of moments of inertia, the biharmonic operator, to the dissipation matrix, the inverse of the Reynolds number, to the friction coefficient, the stream function, to the angular velocity, the intermixing force (the vorticity of the field of the mass forces), to the moment of external forces, etc. The no-slip conditions generate a noncommutative top; conditions on the interface of liquid media or periodicity conditions generate a commutative top. The case of a noncommutative top (with respect to globalization of the local uniqueness theorem) is not analyzed in detail. For a commutative top, we indicate a right-hand side of the equation such that the stationary solution is unique and stable for any friction coefficient.

The infinite-dimensional top generated by the well-known Kolmogorov problem or the problem of stability of the basic flow on the two-dimensional torus in the presence of an intermixing force provides an example of a commutative top **[13]**. The requirement of minimality of the principal rotation (in the sense of the Euler-Poinsot theorem) is equivalent here to the condition for stability in linear approximation due to Meshalkin and Sinaĭ **[66]**, the latter being in fact a criterion for global uniqueness and stability of the corresponding stationary flow (modeled experimentally **[26]**). This criterion can be considered as sufficiently complete due to the results of Yudovich **[110]** and the corresponding alternative statement on instability and bifurcation is analogous to the above-mentioned statement for a three-dimensional top. The statement on the global uniqueness and stability of the Kolmogorov flow under the Meshalkin-Sinaĭ condition will result, in what follows, from the theorems concerning the commutative top. Another interesting consequence of the latter is the global stability of two-dimensional stationary "vortex lenses", i.e., bounded vortex zones in a two-dimensional viscous incompressible fluid considered in §14 in connection with the use of the impermeability conditions and the absence of stresses on the boundary.

The problems of stability of basic flows belong to the intermediate region between the image of an incompressible fluid flow as of a gradually developing process, on one hand, and numerous facts testifying to the contrary, on the other. The vector velocity field and the scalar specific (i.e., divided by the constant density) pressure field are still considered to be the basic physical fields. At relatively small Reynolds numbers, these fields are sufficient for a complete description of the motion (including secondary flows). At large Reynolds numbers the predominant feature of the motion is a random behavior of the observed fields. The latter compels one to average (with respect to time or statistical ensemble **[47, 68]**) the measured values and to extend simultaneously the set of the averaged variables. In the classical approach of Reynolds **[47, 172]** this extension is carried out by means of introducing, along with the mean velocity and pressure, the mean values of the pair products of the components of the pulsational part of the velocity (the difference between the instantaneous and mean velocities) taken simultaneously at the same point in space, which are called *pair one-point* velocity correlations. The latter generate additional stresses in a turbulent medium (*Reynolds stresses*) and form the Reynolds tensor. The basic equations are still the Navier-Stokes equations, but they are considered on a richer manifold which includes "random functions" **[45]**. The averaging of these equations yields new equations of motion formally coinciding with the initial ones but containing additional terms defined by the divergence of the Reynolds tensor. The subtraction of the relations obtained from the initial ones gives equations for the pulsation components of velocity. Their use allows one to obtain evolution equations for the Reynolds stresses. However, the system obtained containing ten equations for the basic values (three mean velocity components, pressure, and six independent components of the symmetric Reynolds tensor) includes new statistical moments and, thus, turns out to be nonclosed. The new moments are expressed via the basic ones by means of approximate semiempirical dependencies modeling the "turbulent" diffusion processes of the Reynolds stresses and their relaxation to an equilibrium; a certain part of the investigations concerning the corresponding second-order closure schemes **[102, 186]** is devoted to these manipulations.

A different problem is solved in §15. From the formal point of view, we analyze here the structure of the matrix differential operator which is defined by the exact (instead of approximate) terms of the Reynolds system. They are generated by the quasilinear terms with partial derivatives of the first order present in the Reynolds system. These terms, called *generating terms,* enter all known second-order closure schemes of the above-mentioned type and are responsible for the energy exchange between macroscopic parameters of the flow (mean velocity and pressure) and the microstructure of turbulent pulsations (the Reynolds stresses). The reason for their investigation is a curious phenomenon of the hyperbolicity of the matrix differential operator generated by these terms. It is of interest that the corresponding one-dimensional wave equation occurs in Thomson's investigations **[63, 182]**. Turbulence models based on the closure relation of Nevzglyadov-Dryden **[71, 135]** degenerate into simplest hyperbolic systems if diffusion terms are omitted **[41, 42, 122]**. Beginning with the Rotta model **[173]**, other second-order closure schemes **[102, 134, 152, 158]** used for the spatial flows also exhibit, as a result of this omission, the characteristic cone (analogous to the Mach cone in acoustics) **[100]**. The latter is conserved when passing to the third-order closure schemes **[53]**. The questions arising here were analyzed in **[100, 101, 183]** under simplifying assumptions concerning the "nonclosed" terms. In what follows, the Rotta approximation is used for "relaxation" terms **[173]**. The terms of molecular and turbulent diffusion are recalled when necessary. Still the main goal is to study the property of an incompressible turbulent medium (revealed by the presence of the mentioned cone) to transfer the coupled oscillations of mean velocity and Reynolds stresses. The equation of the characteristic cone is used to calculate the slope angles of the outer and inner boundary lines of the mixing layer of the initial part of a plane submerged turbulent jet; to this end these lines are identified with the lines of weak discontinuities of the trapezoidal profile of the longitudinal velocity component propagating along them and approximating the profile observed at the efflux cross-section. The angles in question are associated with the mean velocity profile and the correlations of statistically stable one-dimensional turbulent flow in a channel. Spatial waves of small amplitude in a homogeneous and isotropic turbulent medium generated by solutions of the linearized Reynolds system closed on the second-order statistical moments are also considered. In particular, it is proved that these waves propagate in the form of intermittent transverse oscillations of vectors of small perturbations of vorticity and the turbulent energy flux; their structure is identical to that of electromagnetic waves. In this case, the propagation velocity turns out to be proportional to the intensity of small-scale pulsations. The equations for small perturbations are reduced after simple transformations to Maxwell equations. Moreover, the damping influence of molecular and turbulent diffusions is taken into account and the corresponding dispersion relation is considered **[193]**. In the conclusion of the section, some indirect evidence of the assumed existence of the indicated transverse waves is mentioned. On the whole, the results of §15 can be considered as providing a possible alternative approach to the understanding of large-scale turbulence phenomena based on the hypothesis (accepted by a great number of specialists) of their weak dependence on the Reynolds number.

The wave properties of a turbulent medium under investigation indeed weakly depend on molecular diffusion (viscosity) and, hence, on the Reynolds number. The assumed regions of their existence should be the turbulent kernel or the intermixing layer. In the boundary layer on a solid surface the "viscous" mechanism is prevalent. Moreover, the molecular viscosity also defines the passage to the turbulent regime (the increase of the Reynolds number). The description of this passage is one of the most interesting hydrodynamic problems; its investigation involves a great variety of approaches [**38, 58, 61, 65, 78, 85, 86, 105, 145, 147**]. In [**41**] the problem of describing this passage in terms of a certain bifurcation of the Poiseuille flow was considered. The corresponding weakly nonlinear equation straightforwardly follows from the Reynolds equations and the above-mentioned Nevzglyadov-Dryden relation. This equation admits a functional analogous to the Ginzburg-Landau functional in the theory of phase transitions. In this case, the appropriate "order parameter" is the correlation of longitudinal and transverse pulsations of the velocity, the varying scalar parameter is the Reynolds number. It turns out that the bifurcating regime is a statistically stable one-dimensional turbulent flow with a characteristic profile of the longitudinal velocity component. The plot of the resistance curve, the velocity profiles and correlations resulting from this model are obtained in subsequent calculations (e.g., in [**42**]) and presented in the concluding section of Chapter 5, where a detailed analysis of this model is also performed.

Formulas in a section are numbered throughout; the numbers of formulas from different sections are supplied by the number of the corresponding section separated by a period. The same scheme is also used to mark subsections. For example, §1.2 means the second subsection of the first section.

CHAPTER 1

Stationary Flows of an Ideal Fluid on the Plane

§1. Pressure field in a neighborhood of a singular point

1.1. Nondegenerate singular points. Let us start with the consideration of a hydrodynamic example. Having stirred some tea, with the tea-leaves floating at the bottom of the cup, we notice, after we have stopped doing this, that instead of the expected dispersion of the tea-leaves towards the walls they begin to cluster near the center of rotation. The following elementary analysis of this phenomenon differs from the well-known attempts to explain it by a supposed circulation of the liquid in the plane passing through the axis of rotation [**107**] or by the influence of viscosity forces [**55**]. The arguments used here are based on a simple idea that, in a neighborhood of the axis of rotation, the flow is plane-parallel and stationary (this can be easily verified experimentally), while the liquid is incompressible and locally obeys Newton's law of dynamics (the influence of viscosity is negligible). Hence, we have the required pressure distribution in a neighborhood of the axis of rotation, i.e., the pressure increases with the distance from the center. The tea-leaves are pushed out by the liquid from regions with higher pressure to those with lower pressure and, thus, cluster at the center.

By a *singular point* of a smooth vector field $\mathbf{u} = (u, v)$ with the components $u = u(x, y)$ and $v = v(x, y)$ (continuously differentiable real functions defined in a given domain of the Euclidean plane, equipped with a system of Cartesian orthogonal coordinates x, y, where u is the velocity component along the horizontal axis $y = 0$ and v is the component along the vertical axis $x = 0$) we mean a zero $P_0 = (x_0, y_0)$ of the function $u^2 + v^2$,

$$u(P_0) = v(P_0) = 0.$$

The singular point P_0 is *nondegenerate* if the determinant

$$\Delta_0 \equiv u_x^0 v_y^0 - u_y^0 v_x^0, \qquad u_x^0 \equiv \frac{\partial u}{\partial x}(P_0), \qquad v_x^0 \equiv \frac{\partial v}{\partial x}(P_0),$$

does not vanish. A nondegenerate singular point P_0 is necessarily isolated (locally unique); moreover, it is typical in the sense that it is preserved (does not vanish, does not decompose, and remains nondegenerate, only slightly changing its position) under a small perturbation of the field $\mathbf{u}$ in the class of smooth vector fields [**12**]. Furthermore, its *character*, i.e., the flow portrait in a small neighborhood of P_0 defined up to a topological equivalence, is preserved. By a *flow portrait* we mean the set of integral curves of the ordinary differential equation

$$v(x, y)\, dx = u(x, y)\, dy,$$

coinciding in this case with the streamlines. *Topologically equivalent* are, by definition, flow portraits obtained from one another by a *homeomorphic* (i.e., one-to-one and continuous together with its inverse) transformation of the corresponding flow domains. These properties of a nondegenerate singular point can be briefly expressed by saying that it is structurally stable. As a rule, structurally stable flow portraits correspond to those obtained experimentally. The center of rotation mentioned above can be naturally identified with one of the nondegenerate singular points.

The incompressibility condition means that the field $\mathbf{u} = (u, v)$ is *solenoidal*, i.e., satisfies the identity

$$\frac{\partial u}{\partial x} + \frac{\partial v}{\partial y} = 0 \tag{1}$$

in the domain of definition. Our next goal is to prove the following proposition:

A nondegenerate singular point of a solenoidal vector field can only be a center or a simple saddle, i.e., in a neighborhood of it the flow portrait is topologically equivalent to that defined by the set of level lines of the function $x^2 + y^2$ *(for a center) or* $x^2 - y^2$ *(for a saddle) in a neighborhood of the origin* $x = y = 0$ *(corresponding to the point under consideration).*

1.2. The stream function. Indeed, in a simply connected neighborhood of the point P_0 (e.g., in a disc with center at P_0), there exists *the stream function* associated with the field $\mathbf{u} = (u, v)$, i.e., the smooth real function $\psi = \psi(x, y)$ defined up to an arbitrary additive constant and connected with u, v by the following relations:

$$u(x, y) = \frac{\partial \psi}{\partial y}, \qquad v(x, y) = -\frac{\partial \psi}{\partial x}. \tag{2}$$

The function $\psi(x, y)$ can be calculated by means of the contour integral

$$\psi(P) = \psi(P_0) + \int_\gamma -v(x', y')\, dx' + u(x', y')\, dy' \tag{3}$$

taken along an arbitrary smooth curve γ connecting, in this neighborhood, the fixed point $P_0 = (x_0, y_0)$ with the varying point $P = (x, y)$. From the above identities it straightforwardly follows that the streamlines coincide with the *level lines* (or with the lines of the level set, i.e., $\psi(x, y) = \text{const}$), and the singular points coincide with the *critical points* (zeros of the gradient) of the function ψ. Moreover, the following equality is valid:

$$\Delta_0 = -\psi_{xy}^2(P_0) + \psi_{xx}(P_0)\psi_{yy}(P_0),$$

hence, for $\Delta_0 > 0$ (for $\Delta_0 < 0$) the point P_0 is a center (a simple saddle). The existence of a homeomorphic transformation of the family of level lines of the function $\psi(x, y)$ into the family of level lines of the function $x^2 + y^2$ (for $\Delta_0 > 0$) or $x^2 - y^2$ (for $\Delta_0 < 0$) follows, in this case, from *Morse's lemma* **[81, 165]**, which thus implies the italicized statement from the previous subsection.

1.3. Euler fields. Expanding the convective derivatives of the velocity components in a series in powers of $x - x_0$ and $y - y_0$ in a neighborhood of the point P_0 and retaining only the linear terms of the expansion, we obtain, taking into account the equalities $u(P_0) = v(P_0) = 0$,

$$u\frac{\partial u}{\partial y} + v\frac{\partial u}{\partial y} = (u_x^{0\,2} + u_y^0 v_x^0)(x - x_0) + u_y^0(u_x^0 + v_y^0)(y - y_0) + o(r),$$
$$u\frac{\partial v}{\partial x} + v\frac{\partial v}{\partial y} = v_x^0(u_x^0 + v_y^0)(x - x_0) + (u_y^0 v_x^0 + v_y^{0\,2})(y - y_0) + o(r),$$
$$r \equiv ((x - x_0)^2 + (y - y_0)^2)^{1/2} \qquad (o(r)/r \to 0 \quad \text{as} \quad r \to 0),$$

or, taking into account that **u** is solenoidal (i.e., $u_x^0 + v_y^0 = 0$),

$$u\frac{\partial u}{\partial x} + v\frac{\partial u}{\partial y} = -\frac{\partial}{\partial x}\left(\Delta_0\frac{r^2}{2}\right) + o(r),$$
$$u\frac{\partial v}{\partial x} + v\frac{\partial v}{\partial y} = -\frac{\partial}{\partial y}\left(\Delta_0\frac{r^2}{2}\right) + o(r).$$

The fact that the flow locally obeys Newton's law means that there exists a smooth real function $p = p(x, y)$ such that the velocity components $u = u(x, y)$, $v = v(x, y)$ satisfy the *Euler dynamical equations* whose classical form is as follows:

$$u\frac{\partial u}{\partial x} + v\frac{\partial u}{\partial y} + \frac{\partial p}{\partial x} = 0, \qquad u\frac{\partial v}{\partial x} + v\frac{\partial v}{\partial y} + \frac{\partial p}{\partial y} = 0.$$

The function $p(x, y)$ corresponds to the *specific* (i.e., pertaining to constant density) *pressure* of the fluid at the point (x, y). Solenoidal fields $\mathbf{u} = (u, v)$ satisfying, for some pressure $p(x, y)$, the Euler equations will be called in the sequel *Euler fields* (or *Euler flows*).

As a consequence, we have

$$\frac{\partial p}{\partial x} = \frac{\partial}{\partial x}\left(\Delta_0\frac{r^2}{2}\right) + o(r)$$

and

$$\frac{\partial p}{\partial y} = \frac{\partial}{\partial y}\left(\Delta_0\frac{r^2}{2}\right) + o(r),$$

or, after integrating over x and y, respectively,

$$p(x, y) = p(x_0, y_0) + \frac{\Delta_0}{2}(r^2) + o(r^2).$$

Thus we have established the following required statement:

The pressure at a nondegenerate center (saddle) of an Euler field is greater (smaller) than that at any other sufficiently close point of the flow domain.

Taken with the opposite sign, the gradient of the function p,

$$-\operatorname{grad} p = (-\partial p/\partial x, -\partial p/\partial y)$$

corresponds to the force field pushing the suspended particles into the region of lower pressure. This vector field, considered in a neighborhood of a nondegenerate critical point of an Euler flow, is shown in Figure 1.

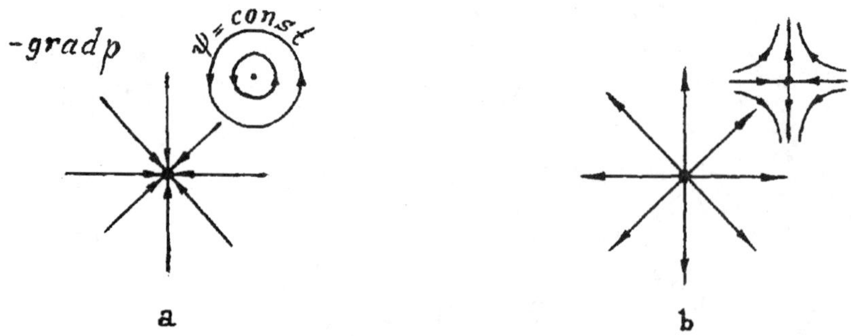

FIGURE 1. The pressure field of an Euler flow in a neighborhood of a nondegenerate center (a) and saddle (b).

§2. Other properties of plane flows

2.1. Equivalent forms of Euler's equations. The requirements that the field $\mathbf{u}$ be solenoidal and that the field $u\partial\mathbf{u}/\partial x + v\partial\mathbf{u}/\partial y$ be gradient lead to the *complete system of Euler equations*:

$$(1)\qquad u\frac{\partial u}{\partial x} + v\frac{\partial u}{\partial y} + \frac{\partial p}{\partial x} = 0, \qquad u\frac{\partial v}{\partial x} + v\frac{\partial v}{\partial y} + \frac{\partial p}{\partial y} = 0, \qquad \frac{\partial u}{\partial x} + \frac{\partial v}{\partial y} = 0.$$

Straightforward investigation of this simplest nontrivial system for an incompressible fluid runs into considerable difficulties. It is often more convenient to use its well-known equivalent forms. For example, introducing the *vorticity* (or the *curl*)

$$\omega = \frac{\partial v}{\partial x} - \frac{\partial u}{\partial y}$$

of the field $\mathbf{u} = (u, v)$ and the *Bernoulli integral*

$$(2)\qquad \gamma = p + \frac{u^2 + v^2}{2},$$

one can rewrite the first two equations of system (1) in the so-called *Gromeka-Lamb form*:

$$(3)\qquad -\omega v + \frac{\partial \gamma}{\partial x} = 0, \qquad \omega u + \frac{\partial \gamma}{\partial y} = 0.$$

The equivalence of the equations obtained to the initial ones is obvious: for a given function $\gamma(x, y)$ in (3), the pressure p is determined from (2).

Formula (3) implies that the field $\omega\mathbf{u} = (\omega u, \omega v)$ is solenoidal; and since $\mathbf{u}$ is also solenoidal, this yields another form of Euler's equations,

$$(4)\qquad \frac{\partial u}{\partial x} + \frac{\partial v}{\partial y} = 0, \qquad \frac{\partial v}{\partial x} - \frac{\partial u}{\partial y} = \omega, \qquad u\frac{\partial \omega}{\partial x} + v\frac{\partial \omega}{\partial y} = 0,$$

bearing the name of *Helmholtz*, due to the third equation, and also equivalent to (1) at least in a simply connected flow domain.

Indeed, by the first and the third relations of system (4), the field $\omega\mathbf{u}$ is solenoidal. Hence, in a simply connected domain this field possesses a single-valued flow function $\gamma(x, y)$ connected with it by equalities (3). Then, calculating p by means of (2), we obtain (1).

The validity of the following more general statement is also evident.

A flow $\mathbf{u} = (u, v)$ *satisfying equations* (4) *is Euler in a domain where the field* $\omega\mathbf{u}$ *has a stream function.*

In a domain where the stream function $\psi(x, y)$ exists, the first equation of system (4) becomes an identity and the other two acquire the following form:

$$(5) \qquad -\Delta\psi = \omega, \qquad \frac{\partial\omega}{\partial x}\frac{\partial\psi}{\partial y} - \frac{\partial\omega}{\partial y}\frac{\partial\psi}{\partial x} = 0 \qquad \left(\Delta \equiv \frac{\partial^2}{\partial x^2} + \frac{\partial^2}{\partial y^2}\right).$$

With respect to the number of the unknowns, system (5) is the simplest among the forms of Euler's equations given above. It is associated with the possibility of investigating Euler flows by stipulating the stream function ψ or the vorticity ω. If ψ is given, then the flow is completely defined; if ω is given, then the problem reduces to a linear one ($-\Delta\psi = \omega$). However, not every function $\psi(\omega)$ turns out to be admissible, i.e., defining an Euler flow. Our next goal is to investigate the necessary restrictions for ψ, ω.

2.2. Reducible fields. The second relation in (5) makes it natural to introduce a special class of solenoidal fields **u** with stream function ψ in which one of the following relations is satisfied:

$$(6) \qquad \omega(x, y) = \mu[\psi(x, y)] \qquad \text{or} \qquad \psi(x, y) = \nu[\omega(x, y)].$$

Here μ and ν are real functions defined on the ranges of the functions ψ and ω, respectively. The introduced flows will be called *reducible* and a field **u** satisfying the first (the second) relation (6) will be called a *μ-field* (*ν-field*). The stream function and the vorticity of a reducible field are connected by obvious equalities

$$(7) \qquad -\Delta\psi = \mu(\psi),$$

$$(8) \qquad -\Delta\nu(\omega) = \omega$$

for a *μ-field* and a *ν-field*, respectively. Moreover, a reducible field is Euler.

Indeed, in the case of a μ-field, it is sufficient to substitute the first equality from (6) in (3) and, by means of (1.2), to verify the existence of the corresponding function γ. In the case of a ν-field, using (1.2) and substituting

$$u = \nu'\frac{\partial\omega}{\partial y}, \qquad v = -\nu'\frac{\partial\omega}{\partial y}$$

in (3), we again verify the existence of γ.

The statement that a reducible flow is Euler is not very informative. The existence of a dependence between the functions ψ and ω expressed by the second relation of system (5), shows that its converse is quite evident, i.e., that an Euler flow is reducible. However, it should be stressed that this inversion (based on the implicit function theorem) is of a strictly local character and involves the *regularity* of the function ψ or ω (i.e., the absence of zeros of the gradient of ψ or ω, respectively). In this section the main problem under consideration is, in fact, finding conditions for global (i.e., in the flow domain as a whole) inversion of the statement that a reducible flow is Euler. The corresponding main statements (Theorems 1 and 2) are given in §§2.4, 2.5. In this and subsequent subsections some examples are considered.

It is well known that relations (7) can be used to investigate plane Euler flows. We note some ways of using the less studied relation (8).

PROPOSITION 1. *In the domain* $0 < x, y < 1$ *there does not exist a smooth* ν*-flow with the vortex* $\omega(x, y) = xy$.

Indeed, assuming the converse, we obtain from (8) the equalities

$$-\Delta\nu(xy) = -(x^2 + y^2)\nu''(xy) = xy,$$

and (since in the domain under consideration $x, y > 0$)

$$-\frac{\nu''(xy)}{xy} = \frac{1}{x^2 + y^2},$$

which is absurd. It remains to prove the required smoothness of the function ν.

The smoothness of $\mathbf{u}$ assumed in the formulation implies that ψ is twice continuously differentiable (equalities (1.2)). By setting $x = y$ in $\psi(x, y) = \nu[\omega(x, y)]$, we obtain the explicit form of ν,

$$\psi(x, x) = \nu(x^2) \qquad \text{or} \qquad \nu(t) = \psi(t^{1/2}, t^{1/2}), \qquad 0 < t < 1.$$

The ν thus obtained is twice continuously differentiable, hence the proposition is proved.

In some cases, by fixing $\omega(x, y)$ one can use (8) in order to find the general form of the functions ν and μ entering (6) and thus describe all the corresponding stream functions ψ without integrating the Poisson equation from (5).

EXAMPLE 1. Let the vortex of a ν-field be given by the equality $\omega = x^2 + y^2$. Then (8) acquires the form

$$\omega\nu'' + \nu' = -\omega/4.$$

The general solution of this equation (for $\omega > 0$) has the form

$$\nu(\omega) = -\frac{\omega^2}{16} + C_1 \log\omega + C_2, \qquad C_1, C_2 = \text{const}.$$

The next example shows the existence of irreducible Euler fields.

EXAMPLE 2. The Euler flow corresponding to the functions

$$\psi = x - x^4, \qquad \omega = 12x^2$$

(which satisfy (5)) is not a μ-field in a neighborhood of the line $x = 4^{-1/3}$ and not a ν-field in a neighborhood of the line $x = 0$.

Indeed, in this example the function $\psi(x)$ (the function $\omega(x)$) has its absolute extremum at the point $x = 4^{-1/3}$ (at the point $x = 0$) and, thus, has the same value at two different points in an arbitrary sufficiently small neighborhood of this point. At the same time the values of the function $\omega(x)$ (the function $\psi(x)$), which is strictly monotone in this neighborhood, are different at these points.

2.3. Frozen-in vortices. The Euler fields are not exhausted even by the flows with more general (compared with (6)) dependencies of the form

$$\psi(x, y) = \alpha[f(x, y)], \qquad \omega(x, y) = \beta[f(x, y)], \tag{9}$$

where f and α, β are smooth functions defined on the flow domain and the range of the function f, respectively, that satisfy the identity

$$-\Delta\alpha[f(x, y)] = \beta[f(x, y)], \tag{10}$$

which guarantees the validity of (5). We shall pass to the proof of the formulated statement after considering two examples.

EXAMPLE 3. Let

$$r_P(x, y) = ((x - x_P)^2 + (y - y_P)^2)^{1/2}$$

be the Euclidean distance from a fixed point $P = (x_P, y_P)$ to the varying point (x, y) and let

$$\lambda(t) = \begin{cases} 0, & \text{if } t \leqslant 0, \\ e^{-1/t}, & \text{if } t > 0. \end{cases}$$

For any $\varepsilon > 0$ the stream function ψ defined by (9) with

$$f = r_P(x, y), \qquad \alpha(t) = \lambda(\varepsilon^2 - t^2)$$

is infinitely differentiable on the plane and is an example of a circular vortex frozen in the fluid at rest and mentioned in the Introduction. Note that the flow defined by ψ is Euler since $(r = r_P(x, y))$

$$\omega = -\Delta\alpha(r) = -\frac{1}{r}\frac{d}{dr}\left(r\frac{d}{dr}\alpha(r)\right) \equiv \beta(r).$$

EXAMPLE 4. In this example ψ has the form

$$(11) \qquad \psi(x, y) = \sum_{k=1}^{n} C_k \lambda(\varepsilon_k^2 - r_{P_k}^2(x, y)), \qquad n \geqslant 2,$$

and defines a system of n circular vortices with centers at distinct points $P_k = (x_k, y_k)$, $k = 1, \dots, n$, and radii $\varepsilon_1, \dots, \varepsilon_n = \text{const} > 0$ not exceeding the smallest distance between all possible pairs of these points; $C_1, \dots, C_n \neq 0$ are some constants. Two frozen-in vortices are shown in Figure 2.

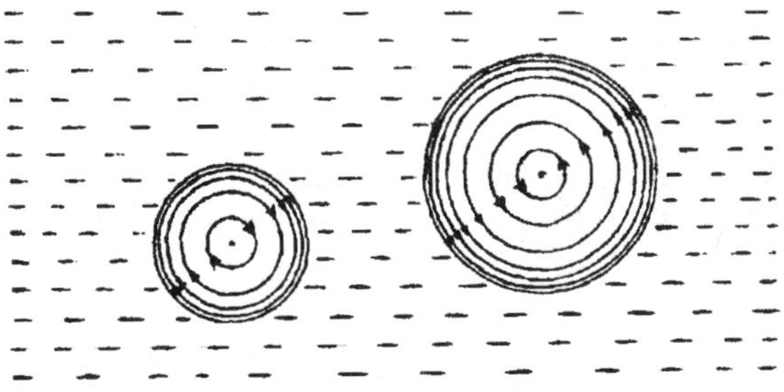

FIGURE 2. Frozen-in vortices.

Let us show that the flow defined by ψ from (11) is Euler. Indeed, in each disc $r_{P_k} = r_{P_k}(x, y) < \varepsilon_k$ we have

$$\omega \equiv -\Delta\psi = -C_k \Delta\lambda(\varepsilon_k^2 - r_{P_k}^2),$$

which implies the validity of equalities (9) and (10) in this disc (Example 3) and, hence, the validity of equality (5). Outside these discs, we have $\psi \equiv \omega \equiv 0$ (i.e., ψ, ω also satisfy (5)).

Nevertheless, *globally in an arbitrary domain containing the system of vortices from* (11) *for equal*

$$\varepsilon_1, \dots, \varepsilon_n$$

and pairwise different

$$C_1, \dots, C_n,$$

the Euler field defined by ψ *is not reducible to the dependencies* (9) *for any* λ, β, *and* f.

Indeed, the function ψ from (11) can be represented in the disc $r_{P_1}(x, y) < \varepsilon_1$ by relation (9) with $\alpha = \alpha_1(t) = C_1\lambda(\varepsilon_1^2 - t^2)$, and in the disc $r_{P_2}(x, y) < \varepsilon_2$ by the same dependence, but with $\alpha = \alpha_2(t) = C_2\lambda(\varepsilon_2^2 - t^2)$; for equal ε_1, ε_2 and different C_1, C_2 this obviously results in violation of this global relation in the flow domain. Obviously, the same is true for the function $\omega = -\Delta\psi$ with respect to the second relation in (9).

Thus, *the Euler flows admit the relations* (6) *and* (9) *but are not exhausted by them.*

2.4. Simple domains. The question of reducing an Euler field to the simplest of the above-mentioned relations, i.e., the identity $\omega = \mu(\psi)$, is also of considerable interest. In this case, the initial system (1) is reduced to the comparatively well-studied weakly nonlinear elliptic equation (7). To carry out the necessary analysis, we pass to the formulation of the first of the two above-mentioned theorems converse to the statement that a reducible flow is Euler. The following notion will play the major part here. A bounded domain on the plane will be called *simple* with respect to a smooth function $\psi(x, y) \neq \text{const}$ if it contains no critical points of ψ and for any constant C the set of points of the level $\psi(x, y) = C$ located in the closure of the indicated domain (the closure is obtained by adjoining the boundary) is connected (or empty). Note that, generally speaking, a simple domain is not assumed to be simply connected and the presence of critical points of ψ on its boundary is not excluded.

THEOREM 1. *A smooth Euler field with a stream function* ψ (*with a smooth curl* ω) *defined in a domain simple with respect to* ψ (*with respect to* ω) *is a* μ*-field* (ν*-field*) *with a uniquely defined function* μ (*function* ν, *respectively*).

To prove the theorem, we select one point (x_C, y_C) on each level $\psi = C$ in the domain G which is simple with respect to the function ψ. To each value C of ψ in G there corresponds a unique value $\omega(x_C, y_C)$ of the function ω at the selected point. Thus, on the set of all indicated values C a function μ with values $\mu(C) = \omega(x_C, y_C)$ is defined. An arbitrary point (x, y) of the domain G lies on some level $\psi = C$. By the implicit function theorem, all the points of this level are covered by a finite or countable system of arcs (the absence of critical points of ψ in G is used). Simultaneously, the set $\psi = C$ is connected in the domain G (simple for ψ) and, hence, there is a smooth curve l_C passing through the points (x, y) and (x_C, y_C). If the function ω is connected with ψ in G by the second relation from (5), then ω is constant on l_C. Hence,

$$\omega(x, y) = \omega(x_C, y_C) = \mu(C) = \mu[\psi(x_C, y_C)] = \mu[\psi(x, y)].$$

The uniqueness of the function μ is obvious: if for all x, y from G and some real $\tilde{\mu}$ we have $\omega(x, y) = \tilde{\mu}[\psi(x, y)]$, then, for any value C of ψ in G, the value

$$\tilde{\mu}(C) = \tilde{\mu}[\psi(x_C, y_C)] = \omega(x_C, y_C) = \mu(C).$$

The case of a ν-field is reduced to the one considered above by substituting ω for ψ and ψ for ω, respectively. Theorem 1 is proved.

Note that the first relation of system (5) was not used in the proof Theorem 1. The statement of the theorem concerns arbitrary pairs of smooth *dependent* functions (connected by the second relation of this system).

EXAMPLE 5. The functions

$$f = xy \qquad \text{and} \qquad g = \begin{cases} x^2y^2, & \text{if } \ y \geqslant 0, \\ -x^2y^2, & \text{if } \ y \leqslant 0, \end{cases}$$

considered in the disc $x^2 + y^2 < 1$ are evidently dependent and the domain indicated is not simple for f. Two values of the function g on the level $f = \varepsilon$ correspond to each sufficiently small value $\varepsilon > 0$ of the function f: $g = \varepsilon^2$ for $x > 0$, $y > 0$, and $g = -\varepsilon^2$ for $x < 0$, $y < 0$. As a simple domain, one can take, e.g., the semidisc $y > 0$. Then $g = f^2$. This consideration admits quite a number of different versions.

PROPOSITION 2. *In the domain* $0 < x, y < 1$ *there is no Euler flow with the vortex* $\omega(x, y) = xy$.

Indeed, since the domain is simply connected, there exists the stream function ψ in it. Simultaneously, it is simple for the function $\omega(x, y)$ under consideration. According to Theorem 1, for an Euler field with a given vortex ω, in this domain there must exist a function v satisfying the relation $\psi(x, y) = v[\omega(x, y)]$. But it follows from Proposition 1 that this is impossible.

Now let V be a bounded simply connected domain on the plane and let a solenoidal field $\mathbf{u}$ be continuous in the closure of V. Let the simple domain G of the stream function ψ of this field be obtained by excluding the critical points of ψ from the domain V. Denote by $\{a\}$ and $\{b\}$ the sets of points at which the function ψ (continuous in the closure of V) has its smallest and greatest values a and b, respectively. It is clear that $a < b$ (ψ is regular and, hence, is not constant in G), and $\{a\}$, $\{b\}$ lie on the boundary of the domain G (due to regularity, the function ψ provides an *open mapping* of the domain G, so that the image of any neighborhood of a point is a neighborhood of the image of this point, and, thus cannot attain its smallest and greatest values in this domain). It is natural to distinguish the following three main cases:

1) $\{a\}, \{b\}$ are simple arcs,
2) $\{a\}$ is a point, $\{b\}$ is an arc,
3) $\{a\}, \{b\}$ are points.

Consider the corresponding examples.

EXAMPLE 6. The sets $\{a\}$, $\{b\}$ are open simple arcs forming the sides AB, CD of a curvilinear quadrangle $ABCD$. The domain $V = G$ contains no critical points. The levels $\psi = c$, $a < c < b$, are simple arcs with boundary points on the sides AD, BC (simple passing flow, Figure 3a).

EXAMPLE 7. The set $\{a\}$ coincides with the boundary of the domain V; the set $\{b\} = P_b$ is an inner point of this domain. The point P_b is a local maximum of ψ. The domain G is an annulus whose inner boundary is contracted to the point P_b; the levels $\psi = c$, $a < c < b$, are the corresponding closed curves (Figure 3b).

These examples correspond to cases 1) and 2) mentioned above. Case 3) will be illustrated by the two following examples, where $\{a\} = P_a$, $\{b\} = P_b$ are the points of maximum and minimum of the function ψ, and ψ has a constant value d, $a < d < b$, on the boundary of V.

EXAMPLE 8. The simple domain G is a quadrangle, additionally divided into two parts by an arc of the (boundary) level $\psi = d$ meeting the boundary at two different points P', P''. In the part of G containing P_b the configuration of the level lines is the same as in Example 6; in the other part it is analogous but the role of P_b is performed by the point P_a (local minimum). The points P', P'' are critical for ψ (they lie on the boundary of G, Figure 3c).

EXAMPLE 9. The simple domain G is a quadrangle divided into two parts by a closed arc of the level $\psi = d$ lying in G and having with the boundary a unique common point P'. In each part (one of them is "inside" the other) the configuration of the level lines is the same as in Example 7. The point P' is critical for ψ. Up to a topological equivalence, the configuration of the level lines in this example is given in Figure 3d.

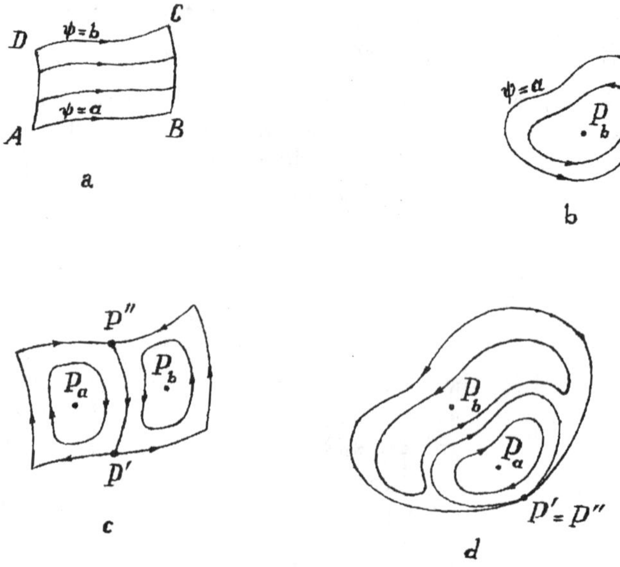

FIGURE 3. Simple domains.

2.5. Holomorphic flows. Usually, the vector fields $\mathbf{u} = (u, v)$ whose components $u(x, y)$ and $v(x, y)$ are analytic functions in the flow domain or can be represented in a neighborhood of each point of the flow domain by power series in the increments of the arguments x and y are called *holomorphic* (*or analytic*) flows. Quite often, the possibility of representation of this type itself is not as essential as its well-known corollary, i.e., the coincidence of two analytic functions in the common domain of definition if they coincide on some subdomain of this domain. This statement implies the second of the above-mentioned theorems on the reducibility of a Euler field. Before passing to its formulation, we note the following statement.

PROPOSITION 3. *Under the additional assumption that the Euler field from Theorem* 1 *is holomorphic, the function* μ (*the function* ν) *is analytic on the intervals of values of the function* ψ (*the function* ω, *respectively*).

Indeed, in the proof of Theorem 1, the values of the function μ were given by the equality $\mu(C) = \omega(x_C, y_C)$ and the point (x_C, y_C) was defined by an arbitrary solution of the equation $\psi(x_C, y_C) = C$ within the boundaries of the domain indicated. Now let us fix an arbitrary number C_0 from the interval of values of ψ as well as some point (x_0, y_0) of the level $\psi = C_0$ in this domain (a solution of the equation $\psi(x_0, y_0) = C_0$), and for C close to C_0, $|C - C_0| < \varepsilon$, let us choose in a neighborhood of (x_0, y_0) the solutions (x_C, y_C) as points of a regular analytic curve l parametrized by C, passing through (x_0, y_0), and intersecting the arcs of the levels $\psi = C$ at nonzero angles. Under this choice of solutions (x_C, y_C), the dependence of μ on C is analytic, since, in this case, they form the analytic curve

$$l = \{(x_C, y_C) : |C - C_0| < \varepsilon\};$$

hence, $\mu(C) = \omega(x_C, y_C)$ is analytic at C_0 as a superposition of analytic functions (the analyticity of the function $\omega(x, y)$ is a consequence of the analyticity of the derivatives of the components of the field $\mathbf{u}$). The choice of l is quite simple when $\psi_y(x_0, y_0) \neq 0$. Then l is an interval of the vertical line $x_C = x_0, |C - C_0| < \varepsilon$. Indeed, substituting $x = x_0$ in the equation $\psi(x, y) = C$ and using both the implicit function theorem and the theorem on the analyticity of the function inverse to an analytic one, we come to the conclusion that the solution of the equation $\psi(x_0, y) = C$ for

$$|C - C_0| < \varepsilon, \qquad |y - y_0| < \delta,$$

and small $\varepsilon, \delta > 0$ is an analytic function $y = y(C) \equiv y_C$, $|C - C_0| < \varepsilon$, such that $y(C_0) = y_0$ (the analyticity of the stream function ψ is taken into account). The general case

$$\psi_x^2(x_0, y_0) + \psi_y^2(x_0, y_0) \neq 0$$

is reduced to the one considered above by a suitable rotation of the coordinate axes. The case of a ν-field is analyzed along the same lines.

The validity of Proposition 3 is established.

THEOREM 2. *A holomorphic Euler field with a stream function ψ (with vortex ω) defined in a domain with a subdomain that is simple with respect to ψ (ω, respectively) is a μ-field (ν-field) if the function μ (function ν) obtained in the simple domain admits an analytic continuation to the whole real axis.*

Indeed, for a μ-field the functions $\omega(x, y)$ and $\mu[\psi(x, y)]$ are analytic, defined everywhere in the flow domain, and coincide in some subdomain (simple for ψ). Consequently (by the uniqueness of the analytic continuation), they also coincide in the exterior of this subdomain. The arguments for a ν-field are analogous.

Theorem 2 is proved.

Note that Theorem 2 does not hold for nonanalytic flows from Examples 3 and 4. On the other hand, in Example 2, the functions $\psi = x - x^4$ and $\omega = 12x^2$ are analytic, but the corresponding $\nu(t) = (t/12)^{1/2} - (t/12)^2$, $t > 0$, cannot be extended analytically to the whole set $-\infty < t < \infty$; this does not allow us to extend the dependence $\psi = \nu(\omega)$ from the domain $x > 0$ (simple for ω) to the whole flow plane. Some other examples using Theorem 2 will be considered in Chapter 3.

CHAPTER 2

Topology of Two-Dimensional Flows

§3. An extended version of some Morse constructions

3.1. Admissible stream functions. In this section we analyze the topological structure of the family of streamlines of a two-dimensional stationary flow of an incompressible fluid in the part V of a closed surface (two-dimensional compact connected smooth closed oriented manifold) of genus $g = 0, 1, \ldots$ ($g = 0$ corresponds to the plane, $g = 1$ to the torus, $g = 2$ to the sphere with two handles, as is shown in Figure 4, etc. **[14]**), which is the complement of $\nu = 1, 2, \ldots$ *discs* (closed domains homeomorphic to the ordinary disc) or coincides with the surface (then $\nu = 0$).

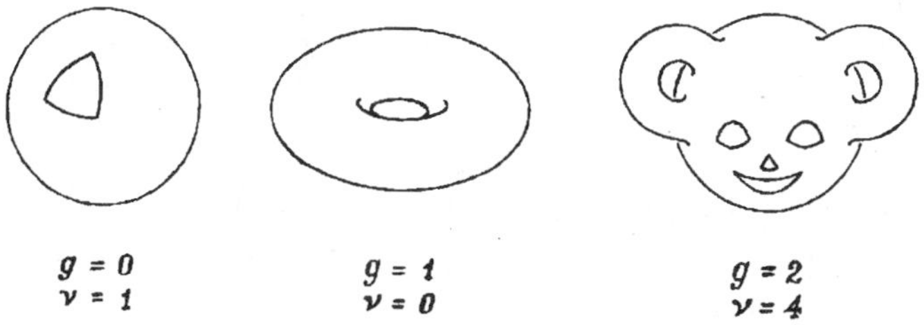

FIGURE 4. Types of two-dimensional domains.

The streamlines are identified with the level lines of the stream function, i.e., of the continuous function ψ in the closure of the domain V. The function is supposed to be *admissible* in the following sense: if the *regular points* of ψ are those interior points of the domain V where ψ is continuously differentiable and which are not critical for ψ, then the corresponding irregular, or *singular*, points lying in V are assumed to be isolated (i.e., they have neighborhoods not containing other singular points of ψ). By a *level* C of the function ψ we mean the complete inverse image of its value C (the set of all solutions P of the equation $\psi(P) = C$), by a *level line* a connected level component that does not coincide with a point (this corresponds to the definition of a line as a *continuum* (a connected compact set) of (topological) dimension 1 **[14, 75]**, i.e., the dimension of the levels considered does not exceed 1 (singular points are isolated), the levels are compact (since they belong to a compact surface), and the continua of dimension 0 are points). A line of level C *separates* the signs of the difference $\psi - C$ if each interior point of this line has a neighborhood divided by this line in two parts, in one of which $\psi - C$ is positive and in the other negative. The homeomorphic image of an interval (with noncoinciding images of the endpoints) is called a (*simple*) *arc*, and the boundary of a disc is called a *cycle*. The interior of a disc will often be called a *Jordan domain* and the cycle, a *Jordan closed curve*. The level arcs separating signs are assumed to be *oriented* in the domain V according to the requirement that the points of the domain V lying close to this arc and located *above* ($\psi > C$) or *below* ($\psi < C$) the level C be enclosed by the arc in *clockwise* and *counterclockwise* directions, respectively (the ordinary direction of the flow).

Let P be an arbitrary fixed point of the domain V. With respect to the set of points of its level $C = \psi(P)$, this point P is *isolated* (in a neighborhood of P there are no points of the level $\psi = C$ different from P) or is a *limit* point (in any neighborhood of P there are points of the level $\psi = C$ different from P). The next theorem shows that singular points of an admissible function ψ are analogous to nondegenerate singular points (§1.1), i.e., are exhausted by vortex centers and saddles (of an arbitrary, but finite multiplicity).

THEOREM 1. *If P is an interior point of the domain of definition of an admissible function ψ lying on some level $\psi = C$, then*

(i) *either P is a limit point of the level $\psi = C$ and then a neighborhood N_P of it admits an arc-orientation preserving homeomorphic mapping onto the (ordinary) disc, under which P is taken to the center of the disc, and the other points of the level $\psi = C$ lying in N_P are taken to $2n$ radii ($n = 1, 2, \cdots < \infty$) bounding sectors of angles π/n and separating the signs of the difference $\psi - C$, and then either*

(ii) *for P there exists a sufficiently small $\varepsilon_0 > 0$ such that for any $0 < \varepsilon < \varepsilon_0$ the set of points of the level $\psi = C + \varepsilon$ (of the level $\psi = C - \varepsilon$) belonging to the preimage of a sector bounded by neighboring radii, where $\psi > C$ ($\psi < C$, respectively), is a simple arc dividing this preimage into two Jordan domains with different signs of the difference $\psi - (C + \varepsilon)$ (difference $\psi - (C - \varepsilon)$, Figure 5a), or*

(iii) *P is a local extremum of ψ, and then for some $\varepsilon_0 > 0$ and any C', $0 < |C' - C| < \varepsilon_0$, every nonempty set of the level $\psi = C'$ in a neighborhood of P is a cycle separating the signs of the difference $\psi - C'$ and bounding a Jordan domain containing the point P (Figure 5b).*

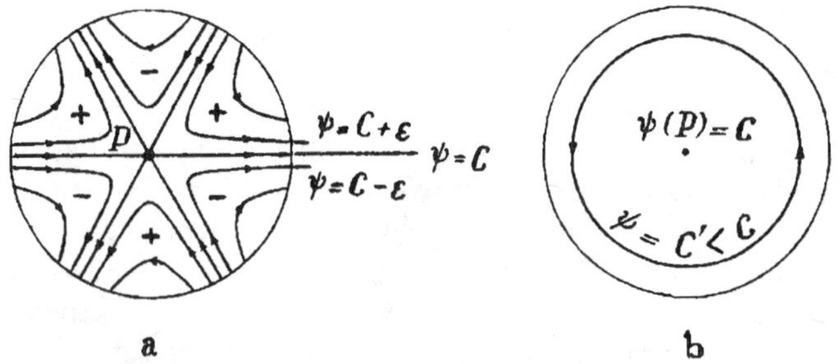

FIGURE 5. A saddle with six sectors (a) and a center generated by a local maximum (b).

The proof of Theorem 1 is given in subsection 4. A corollary of the theorem is the local topological equivalence of ψ to one of the following harmonic functions: the function $\varphi(x, y) = -\ln(x^2 + y^2)$ if P is a local maximum of ψ (P corresponds to the origin $x = y = 0$); the function $\varphi(x, y) = \ln(x^2 + y^2)$ if P is a local minimum; the real part $\varphi(x, y) = \operatorname{Re}(x + iy)^n$ of the integer positive power n of the complex variable $x + iy$ ($i^2 = -1$) if P is a limit point of its level $\psi = C$. *Topological equivalence* means that (in addition to §1) there exist an orientation preserving homeomorphism

h of a neighborhood N_P of the point P to the plane (mapping P to the origin) and a strictly monotonically increasing function η defined on the real line and sending the point at infinity of the line to $C = \psi(P)$ such that ψ can be represented in N_P by the sequence of mappings h, φ, and η, i.e., for any point P' from N_P, the equality $\psi(P') = \eta\{\varphi\left[h(P')\right]\}$ holds. Now Theorem 1 can be reformulated in the following way.

THEOREM 1$'$. *Any admissible function ψ is locally topologically equivalent to a harmonic function with, possibly, singular points of logarithmic type corresponding to the points of local extremum of ψ.*

The existence of such a homeomorphism h directly follows from statements (i) and (ii) of Theorem 1, and was established by Morse in an analogous situation **[69]**. For the case of a limit point of the level, the identical transformation $\eta(t) \equiv t$ can be taken as η **[69]**. The functions that are locally topologically equivalent to harmonic functions and admit logarithmic singularities, for which η is the identical transformation (they are obtained from harmonic functions only by means of an "inner", preserving the orientation, homeomorphic transformation of variables), were called by Morse *pseudoharmonic* (the basis of the corresponding topological theory of harmonic functions is due to Morse **[69]** and Stoïlov **[92]**). We complement Morse's considerations with the case where $\eta(t) \not\equiv t$ (e.g., when $\psi(x, y) = x^2 + y^2$, $P = (0, 0)$, and h is the identical transformation; the only η satisfying the required identity

$$\psi(x, y) = \eta\left[\ln(x^2 + y^2)\right]$$

is, obviously, $\eta(t) = e^t$ sending $-\infty$ to $\psi(0, 0) = 0$). Taking the latter into account, we call the functions satisfying the statement of Theorem 1 *pseudoharmonic* as well. Note that the requirement that η increases is necessary in order to preserve the orientation of the level lines. For example, for $P = (0, 0)$ and h the identity, in the equality $\psi(x, y) = \eta\left[-\ln(x^2 + y^2)\right]$ the orientation does not change for the increasing $\eta(t) = -e^{-t}$ and becomes the opposite for $\eta(t) = -e^t$.

Conversely, Theorem 1$'$ implies the statement of Theorem 1 (this is obvious). Simultaneously, it gives a rich class of examples of pseudoharmonic functions formed by continuous real functions $\psi(x, y)$ with isolated singular points. The essential requirement of continuity excludes from consideration other singular points (different from those considered above), e.g., those corresponding to a zero of the harmonic function

$$\operatorname{Re} e^{1/z} = \exp\left[x/(x^2 + y^2)\right] \cos\left[y/(x^2 + y^2)\right].$$

It would be of interest to analyze the possibilities of weakening this requirement. It is obvious here that even a complete elimination of this requirement will not allow us to cover all the isolated singularities of two-dimensional vector fields described by the Poincaré-Bendixson theory **[72]** (e.g., singular points with dense incoming or outcoming trajectories). The converse may also take place: real functions admitting discontinuities with isolated singular points of gradient fields may turn out to be locally topologically equivalent to the same harmonic functions admitting singularities of general type. From the hydrodynamic point of view, the first fact can be easily understood and the second is quite possible: the flow configurations of an incompressible fluid cannot cover all possible flow configurations of a compressible gas, but they can be covered by potential flow configurations with point vortices (corresponding to logarithmic singularities) or other types of singular points admissible for them. Now we pass to the investigation of the general configuration of a two-dimensional flow of an incompressible fluid modeled by potential flows with point vortices.

The subsequent constructions do not use the smoothness of the admissible function. The continuous real function ψ considered in the closure of the domain V (defined as above) is assumed to be pseudoharmonic (satisfying the statements of Theorems 1′ or 1). Only the information given by the statement of Theorem 1 on the local structure of the function under investigation is used. Following Morse's terminology **[69]**, the neighborhood N_P of the limit point P of the level C ($\psi(P) = C$) from item (i) of this theorem will be called *canonical* for P and the preimage of the radius together with P, the *radius of* N_P. The union of a pair of diametrically opposite *radii* of the canonical neighborhood N_P (preimages of diametrically opposite radii of the ordinary disc) will be called its *width* (or the *width of the level* C), and the domain between the two neighboring radii will be called a *sector* of N_P. If the number n of sectors of the canonical neighborhood N_P located below (or above) the level C is greater than 1, then P will be called a *saddle point* (or a *saddle*) of multiplicity $n - 1$. As usual, the point P from part (iii) of Theorem 1 will be called a *center*.

Assuming that there can by only a *finite number of centers* $q = 0, 1, \dots$, we shall determine the behavior of ψ in a neighborhood of the boundary of a domain V that does not coincide with a closed surface. For this purpose we introduce into consideration the maximal connected component of the nonempty set of the level $\psi = C$ on the boundary cycle J of the domain V whose closure ξ_C will be called the (*boundary*) *continuum of the level* C. A continuum ξ_C that does not coincide with its cycle J or a point is, obviously, an arc with endpoints. These arcs will be called *stable parts*. Stable parts correspond to streamlined impermeable walls at the boundary. The following conditions are assumed to be satisfied.

BOUNDARY CONDITIONS τ. *The trace of the function* ψ *on the boundary cycle is continuous and either attains a relative extremum on a finite set of points of stable parts, or is an identical constant.*

The given restrictions are an extended version of boundary conditions used by Morse **[69]** (which, in the case of a nonconstant boundary trace, imply that the latter has no stable parts and is strictly monotone between points of local extremum).

As a corollary of conditions τ, the set of all ξ_C of a fixed level C is finite. Hence, for every ξ_C located on some cycle J and not coinciding with it, an (open) arc λ of the cycle J is defined, which covers ξ_C and does not contain points of the level C that do not belong to ξ_C is defined. By a *relative neighborhood* $\Omega = \Omega(\xi_C)$ of such a ξ_C we mean a Jordan domain adjoining λ (the arc λ with the endpoints is a common part of the cycle J and the boundary Ω); when ξ_C coincides with J, the role of Ω will be performed by a curvilinear annulus bounded by J and a cycle lying in V (Figure 6). A continuum ξ_C is *limit* (*isolated*) if there exists a sequence of points of the level C from the domain V accumulating at ξ_C (conversely, there is no such sequence).

THEOREM 2. *For any continuum* ξ_C *of the level* C *lying on the boundary cycle* J *of the domain* V, *there exists a relative neighborhood* Ω *and a number* $\varepsilon_0 > 0$ *such that*

(i) *if* ξ_C *is a limit continuum, then the set of points of the level* C *in the domain* Ω *is the union of a finite number* $n = 1, 2, \dots$ *of pairwise nonintersecting simple arcs with one endpoint on* ξ_C (*arc endpoints may coincide on* ξ_C) *separating the signs of the difference* $\psi - C$ *and dividing* Ω *into* $n + 1$ *Jordan domains* (*sectors*) F *if* ξ_C *does not coincide with* J, *or into* n *sectors* F *if* ξ_C *coincides with* J, n *being even in the latter case* (*Figure* 6a);

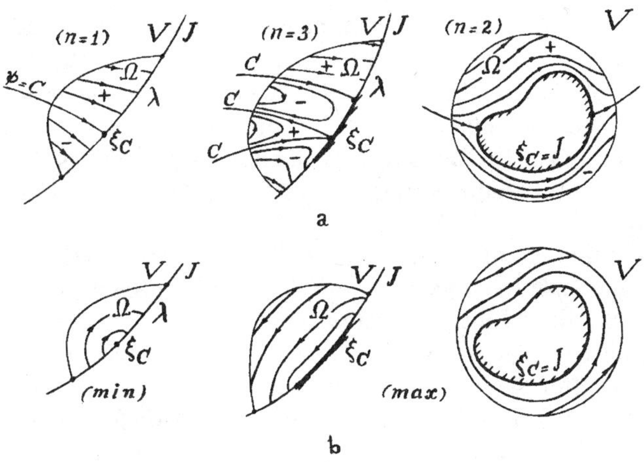

FIGURE 6. Boundary continua with relative neighborhoods.

(ii) *moreover, for any* $0 < \varepsilon < \varepsilon_0$, *the simple arc dividing F into two Jordan domains and separating the signs of the difference* $\psi - (C \mp \varepsilon)$ *is a nonempty set of points of the level* $\psi = C \mp \varepsilon$;

(iii) *if* ξ_C *is an isolated continuum where a relative minimum (maximum) of the function* ψ *is attained, then for any* $C < C' < C + \varepsilon_0$ (*for any* $C < C' < C - \varepsilon_0$, *respectively*) *the set of points of the level* $\psi = C'$ *in* Ω *is an arc if* ξ_C *does not coincide with* J, *or a cycle if* ξ_C *coincides with* J, *that separate the signs of the difference* $\psi - C'$ *and divide* Ω *into two domains, one of which is again a relative neighborhood of the continuum* ξ_C (*Figure* 6b).

The proof of Theorem 2 is given below in subsection 4. Now we shall obtain the basic integer relations associating the distribution of singular points of the function ψ with its behavior on the boundary of the domain V.

3.2. The basic integer relations. The relative neighborhood $\Omega = \Omega(\xi_C)$ of a limit continuum ξ_C from statements (i) and (ii) of Theorem 2 will be called *canonical*. By μ_C ($\overline{\mu}_C$) we denote the number of sectors of the neighborhood Ω located below (above) the level C. A limit continuum ξ_C not coinciding with its cycle J is called an *open* one and an open ξ_C, a *saddle* (*inverse-saddle*) *continuum of multiplicity* $\mu_C - 1$ (*multiplicity* $\overline{\mu}_C - 1$) if $\mu_C > 1$ for it ($\overline{\mu}_C > 1$, respectively). The corresponding *closed* limit continuum $\xi_C = J$ is also called a *saddle*, its *multiplicity* being the number $\mu_C = \overline{\mu}_C$.

Theorems 1 and 2 directly imply that for any C the corresponding set M_C^- (the set M_C^+) of points of the domain V and its boundary located *no higher* (*no lower*) than the level C is *triangulable*, i.e., it can be represented as the union of a finite number of points, arcs, or curvilinear triangles having only common points, or arcs, or not intersecting at all (they form the polyhedron of the two-dimensional complex **[14, 79]**). By subtracting from the sum of the total number of points and triangles of the indicated triangulation the number of arcs, we obtain its *Euler characteristic*, i.e., the number $\chi_C^{\mp} \equiv \chi(M_C^{\mp})$, which does not depend, as it can be easily verified, on the specific triangulation and, hence, is a topological invariant characterizing the set $M_C^{\mp}$ itself.

In particular, when C is less than the smallest value of ψ, the set M_C^- is empty and $\chi(M_C^-) = 0$. When C is greater than the greatest value of ψ, the set M_C^- coincides with the closure of V, and $\chi(M_C^-)$ with the *Euler characteristic of the domain* V (or its closure), i.e., with the number $\chi \equiv \chi(V)$ equal to 1 if V is a Jordan domain, 0 if V is an annulus, $2 - \nu$ if V is a plane domain bounded by ν cycles, $2 - 2g - \nu$ if V is a general domain on a two-dimensional closed (oriented) surface of genus g bounded by ν cycles **[14]**.

Varying C within the intermediate values of ψ and using the additivity of the Euler characteristic (the Euler characteristic of the union of nonintersecting triangulable sets is the sum of their Euler characteristics, which is obvious), we calculate the corresponding difference $\chi_{C+\varepsilon}^- - \chi_{C-\varepsilon}^-$ as the sum of *contributions*, i.e., of the analogous increments, defined for the Euler characteristics of the neighborhoods Ω of the inner points P of the domain V and boundary continua ξ_C from Theorems 1 and 2. The corresponding results are given in Figure 7, where $\chi_{C\mp\varepsilon}^-$ temporarily denotes the Euler characteristic of the intersection of the set $M_{C\mp\varepsilon}^-$ and the neighborhood Ω (the part of Ω where $\psi \leqslant C \mp \varepsilon$).

By dividing the domain V into a finite number of nonintersecting canonical neighborhoods from Theorems 1 and 2 and summing the local increments (contributions) of the characteristic χ_C^- for any C and any $0 < \varepsilon < \varepsilon_C$ (less than some fixed ε_C whose existence follows from the finiteness of the number of covering neighborhoods) we obtain

$$\chi_{C+\varepsilon}^- - \chi_{C-\varepsilon}^- = q_C + m_C - s_C - \sigma_C - p_C,$$

where q_C is the number of centers (local extrema of ψ in G) lying on the level $\psi = C$, m_C is the total number of points and open arcs (stable parts of the boundary) of V, where ψ attains the relative *minimum* C, and p_C, s_C, and σ_C are the sums of the multiplicities of the saddles of ψ lying in V, the open and closed *saddle* boundary continua located on the level $\psi = C$, respectively.

By substituting $-\psi$ for ψ, we obtain an analogous relation

$$\chi_{C-\varepsilon}^+ - \chi_{C+\varepsilon}^+ = q_C + \overline{m}_C + \overline{s}_C - \sigma_C - p_C,$$

where $\overline{m}_C$ is the total number of points and open arcs of the boundary of V where ψ attains a relative *maximum* C, $\overline{s}_C$ is the sum of the multiplicities of the open *inverse saddle* boundary continua of the level $\psi = C$. Here the nonnegative values $|\chi_{C\mp\varepsilon}^\mp|$, q_C, $m_C, \overline{m}_C$ are obviously bounded uniformly in all the admissible values of the parameter C, i.e., the first does not exceed $2 + 2g + \nu$, the second, the total number of centers in V which is assumed to be finite, and the other two, the total number of points and open arcs of the relative extremum of the boundary trace of the function ψ, which is assumed to be finite due to the boundary conditions τ.

Now let:

q be the *total number of centers* (*local extrema of* ψ) *in* G,

p be the *sum of multiplicities of saddles of* ψ *in* G,

s ($\overline{s}$) and σ be the *sums of the multiplicities of the open saddle* (*inverse saddle, respectively*) *and closed limit continua of the boundary of* V,

m ($\overline{m}$) be the *total number of points and stable parts of the boundary of* V *where the function* ψ *attains its minimum* (*maximum*).

The above-mentioned restrictions imply that the values $m, \overline{m} \geqslant 0$ are finite, and the obtained relations imply that the values $s, \overline{s}, \sigma \geqslant 0$ are also finite (this is equivalent to the boundedness of the nonnegative integers $s_C, \overline{s}_C, \sigma_C$ as the parameter C varies on

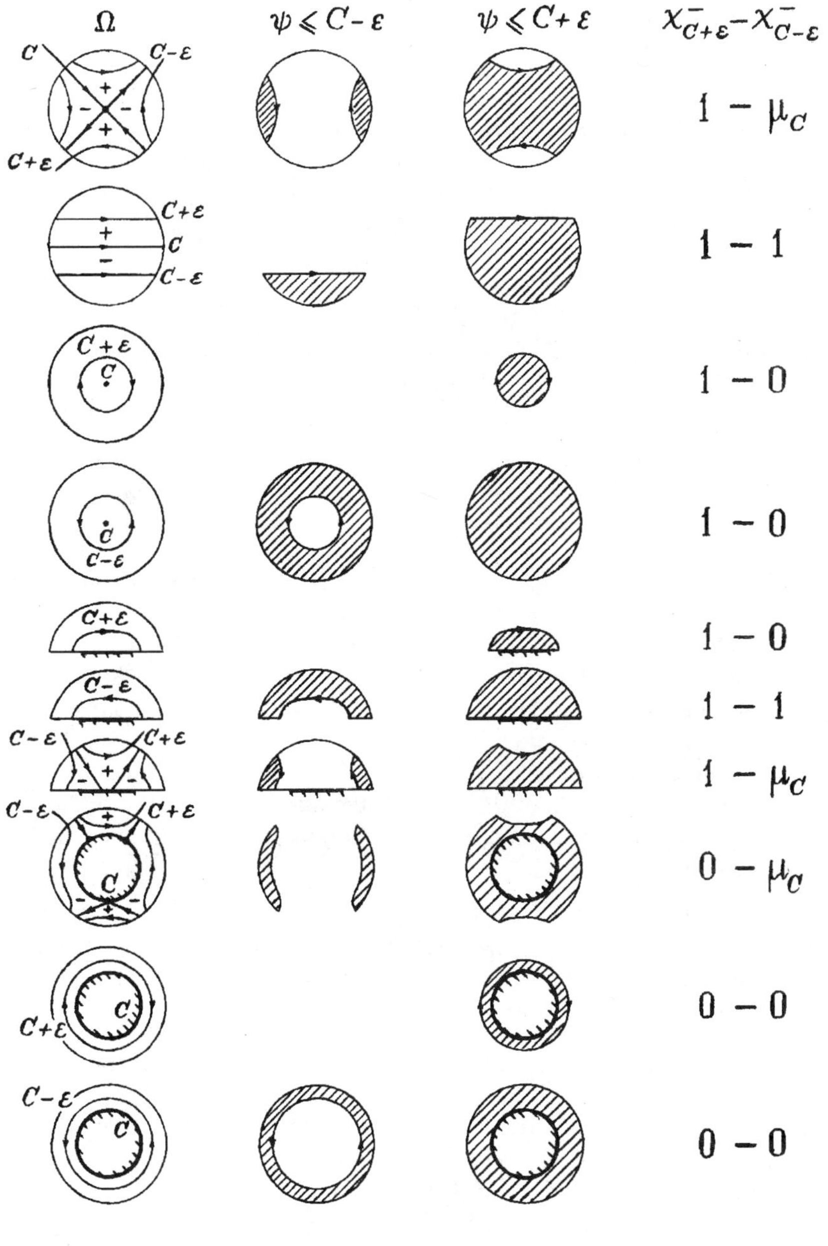

FIGURE 7. Local changes of Euler characteristic.

the interval of admissible values of the continuous function ψ defined on the (compact) closure of the domain V). These relations also imply that the set of levels $\psi = C$ such that the numbers entering the right-hand sides of these equalities are different from 0 is finite; on all the remaining levels $\psi = C$, the right-hand sides vanish.

Taking into account the above-mentioned observations, we enclose the interval $a \leqslant \psi \leqslant b$ of all the values of the function ψ (including the smallest and the greatest values a and b) in an arbitrary finite interval $a-\varepsilon < \psi < b+\varepsilon$, divide the latter by a sufficiently large number of points

$$a-\varepsilon<a<\cdots<C<C'<C''<\cdots<b<b+\varepsilon,$$

and sum, with respect to them, the left-hand sides of the mentioned equalities,

$$\chi_{b+\varepsilon}^{-}-\chi_{b}^{-}+\cdots+\chi_{C''}^{-}-\chi_{C'}^{-}+\chi_{C'}^{-}-\chi_{C}^{-}+\cdots+\chi_{a}^{-}-\chi_{a-\varepsilon}^{-},$$
$$\chi_{a-\varepsilon}^{+}-\chi_{a}^{+}+\cdots+\chi_{C}^{+}-\chi_{C'}^{+}+\chi_{C'}^{+}-\chi_{C''}^{+}+\cdots+\chi_{b}^{+}-\chi_{b+\varepsilon}^{+}.$$

The intermediate terms of the given sums cancel out and the remaining ones reduce to the following values: $\chi_{b+\varepsilon}^{-}$ and $\chi_{a-\varepsilon}^{+}$, to the Euler characteristic χ of the domain V, while $\chi_{a-\varepsilon}^{-}$ and $\chi_{b+\varepsilon}^{-}$, to 0 (the Euler characteristic of the empty set). The corresponding right-hand sides are given by the sums $q+m-s-\sigma-p$ and $q+\overline{m}-\overline{s}-\sigma-p$. Subtracting one of the obtained equalities from the other, we come to the required integer relations,

$$(1)\qquad q-p=s-m+\sigma+\chi,\qquad s-m=\overline{s}-\overline{m}\qquad (\chi=2-2g-\nu).$$

These relations hold even in the case when there exist singular points of the logarithmic type (in whose neighborhoods ψ is unbounded), i.e., even for flows with point vortices (in this case it is sufficient to cut out neighborhoods of centers bounded by cycles from statement (iii) of Theorem 1 and, by means of the given arguments, using the continuity of ψ, to obtain relations (1) for $q=0$, simultaneously substituting $\nu+q$ for ν which again gives (1) with an arbitrary q). All numbers entering these relations, with the exception of χ, take nonnegative (integer) values. For $\nu=0$ (in the case of a closed surface), the numbers s, $\overline{s}$, m, $\overline{m}$, σ are assumed to be zeros. In this case, relations (1) yield the statement of the Poincaré theorem "on the sum of indices" **[14]**. Another straightforward application of these relations is an elementary proof of the following well-known fact **[49]**:

A smooth function ψ defined on the closure of a two-dimensional domain V (in our case a part of a surface) and taking constant values on the cycles bounding V has at least one critical point in the domain V or on its boundary, if V does not coincide with an annulus.

Indeed, a smooth function ψ without critical points taking constant values on the boundary cycles of the domain V is necessarily admissible and satisfies the boundary conditions τ. Then relations (1) are valid for it and all the numbers entering them, except χ, are equal to 0. But then $\chi=2-2g-\nu=0$, which for $\nu\geqslant 1$ immediately implies the equalities $g=0$ and $\nu=2$. Hence, V is an annulus.

3.3. Graphs and chambers. Using relations (1), we classify the elements of a general flow portrait defined by the set of level lines of the function ψ. Its saddle point lying in V, or a boundary point of the domain V that is an endpoint for at least two arcs of this level lying in the closure of the domain V, one of which must belong to V (does not lie on the boundary of V), is called a *multiple point of the level $\psi=C$*. Multiple level points lying in the domain V are saddles. As a corollary of statement (i) of Theorem 2, there are only three types of multiple level points lying on the boundary of V:

(α) an isolated point of a local extremum of the trace of ψ, an endpoint for an even number of arcs of the level C belonging to V (Figure 8a),

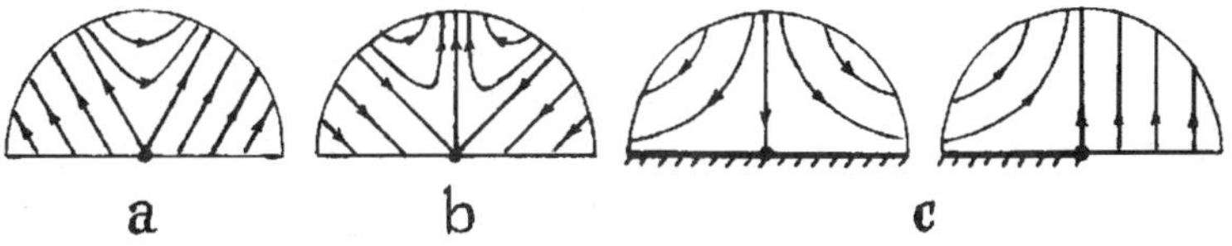

FIGURE 8. Types of multiple level points on the boundary.

(β) an isolated point of the set of points of the level $\psi = C$ lying on the boundary of V, not a point of a local extremum of the trace of ψ on the boundary of V, and is an endpoint for an odd number of three or more arcs of the level C lying in V (Figure 8b),

(γ) a point of the line of the level $\psi = C$ belonging to the boundary of V that is an endpoint for at least one arc of this level lying in V (Figure 8c).

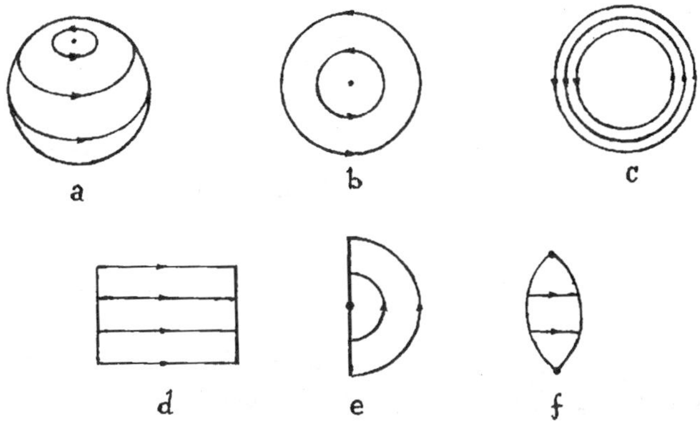

FIGURE 9. Chambers.

The above-mentioned boundedness of the numbers m and $\overline{m}$ (resulting from conditions τ) and the assumed finiteness of the number of centers q imply, as it was already mentioned, that the numbers p, s, $\overline{s}$ are bounded and, simultaneously, that the number of saddles and points of the type α or β is finite (since the latter obviously are either saddle or inverse-saddle). Assuming additionally the *finiteness of the set of stable parts*, we obtain the finiteness of sets of points of type γ and, hence, the finiteness of the set A_ψ of all multiple level points. A level $\psi = C$ will be called *critical* if it contains a point of the set A_ψ. The finiteness of the set A_ψ implies the finiteness of the set of critical levels. The points of the set A_ψ divide the maximal connected component l_C of the level $\psi = C$ into a finite set of curves not containing multiple level points and thus being arcs or *loops*, i.e., arcs with identified endpoints (cycles with a deleted point). The number of loops does not exceed $q + v$; due to statement (i) of Theorem 1, the function ψ satisfies the *strict maximum principle* in any subdomain of the domain V not containing centers (the function ψ considered in the closure of this subdomain, does not attain extremal values at interior points of this subdomain); in particular, ψ cannot take a constant value on the boundary of a disc that does not contain a center or a boundary cycle of the domain V. Now let B_ψ be the set of all the arcs and loops indicated. The sets A_ψ and B_ψ will be referred to as the *vertices* and *edges* of the *graph* $\tau_\psi = (A_\psi, B_\psi)$, which defines the *flow structure* described by the set of level lines of the function ψ. A maximal connected component G of the set obtained by the removal from V of vortices and edges of the graph τ_ψ will be called a *chamber*. The flow configuration as a whole is divided by the graph τ_ψ into chambers, i.e., the elements mentioned above.

Let us clarify the flow configuration inside a chamber. First, consider the simplest case when *the domain V coincides with a chamber* (the set A_ψ is empty). In this case, the numbers p, s, $\bar{s}$, σ are necessarily zeros. Using (1), we find that $q + 2g + v \leqslant 2$. Consequently, $g = 0$ (which is obvious for $v \geqslant 1$, while for $v = 0$ it follows from the necessary condition $q \geqslant 2$). Thus, all the chambers can be assumed to lie on a plane. Equalities (1) are written as follows:

$$q + m + v = 2, \qquad m = \overline{m}.$$

Examining the admissible combination of the numbers $v, q, m, \overline{m}, = 0, 1, 2$, we obtain (taking into account that $v = 0$ implies $m = \overline{m} = 0$) the following flow configurations:

q	v	m	Name of the chamber	Figure
2	0	0	axle	9a
1	1	0	vortex	9b
0	2	0	circulation	9c
0	1	1	passing flow	9d
0	1	1	semivortex	9e
0	1	1	sector	9f

The classification of chambers in the case $v = m = \overline{m} = 1$ is obtained by varying the boundary continua of a relative extremum of the function ψ. For a passing flow these are stable parts, for a sector they are points, and for a semivortex—a stable part and a point.

When the domain V does not coincide with a chamber, the boundaries of chambers G lie on the union of the set of the points of the graph τ_ψ and the boundary of V and admit points of self-intersection, necessarily coinciding with multiple level points (which is obvious). Following Morse **[69, §8]** and cutting the boundary G at the indicated points, we obtain a Jordan domain G_* covering G; the considerations concerning the above case can be applied to this domain and this results in the mentioned classification.

Chambers not containing centers ($q = 0$) will be called *laminar*. Laminar chambers necessarily adjoin the boundary of the flow domain V, preventing circulation, and, thus, are absent in the flow on a closed surface (the case $v = 0$). The possibility of covering the compact closure of the domain V by a finite number of neighborhoods from Theorems 1 and 2 implies that the number of chambers is finite. The following statement is the result of the considerations in this subsection:

For $v = 0$ and a finite number of centers, a chamber is an axle, a vortex, or a circulation, and for $v \neq 0$, the boundary conditions $\boldsymbol{\tau}$ fulfilled, and a finite number of both stable parts of the boundary of V and centers—it is a vortex, or one of the laminar chambers defined up to identification of boundaries at the multiple level points. The total number of chambers and multiple level points is finite.

In conclusion we note that the presence of an axle or a sector in the flow domain V implies that V coincides with these chambers. Note also that analogous considerations are possible for the case of a nonorientable two-dimensional manifold. Finally, we emphasize that for $\nu \neq 0$ relations (1) are similar to the argument principle **[54]**, but do not involve the regularity of ψ on the boundary of V. Simultaneously, these relations generalize the analogous Morse-Heins relations **[69, 166–168]** to the case when stable parts are present, also covering flows on the sphere **[29, 36]** and the torus **[11]**.

3.4. The structure of admissible functions. Now we pass to the proof of Theorems 1 and 2. We establish the validity of statement (i) of Theorem 1. First we prove the following statement.

LEMMA 1. *Let G be some subdomain of the domain V and let an admissible function ψ in the closure of V be given. The following statements are equivalent*:

(S1) *the function ψ performs an open mapping of the domain G (i.e., the image of an arbitrary neighborhood of any interior point of the domain G under this mapping is a neighborhood of the image of this point),*

(S2) *the values taken by the function ψ in any subdomain Ω of the domain G lie between the smallest and the greatest values taken by ψ on the boundary of Ω (the strict maximum principle),*

(S3) *ψ has no isolated level points (local extrema) in G.*

Indeed, the equivalence of the first and the second statements is evident and does not involve the fact that singular points are isolated. Further, by the implicit function theorem, the mapping ψ is open everywhere where it is regular. Moreover, the function ψ is admissible, hence, if (S3) is satisfied, then any point of the domain G is the limit of a sequence of regular points of its level. Consequently, (S3) implies (S1). Since this ψ, *a fortiori*, does not represent an open mapping in a neighborhood of a local extremum, the converse implication also holds. Lemma 1 is established.

Now let U be some connected neighborhood of an interior point P of the domain V that is the limit for the points of its level $C = \psi(P)$. Taking into account that ψ is admissible, we assume that the isolated neighborhood U' obtained by deleting P from U does not contain singular points of ψ (the point P itself may be singular). We denote by l_C the maximal connected component of the set of points of the level $\psi = C$ lying in U'.

PROPOSITION 1. *l_C is a simple arc with endpoints on the boundary of the neighborhood U'.*

Indeed, covering the set l_C by a locally finite system of arcs from the implicit function theorem, we obtain a regular curve (smooth, with no self-intersection points) with endpoints on the boundary U'. The coincidence of the endpoints of l_C would contradict the maximum principle (statement (S2) of Lemma 1), since, in this case, l_C would be bounding a domain (ψ cannot take a constant value on the boundary of a domain that does not contain centers).

PROPOSITION 2. *The arcs l_C have the following properties*:

(a) *they have no common points in U',*

(b) *the set of those having P as an endpoint is finite,*

(c) *among them there is an arc having P as an endpoint,*

(d) *P is isolated for the arcs l_C that do not have P as an endpoint,*

(e) *each l_C separates the signs of the difference $\psi - C$,*

(f) *the total number of arcs l_C ending at P is even.*

Indeed, properties (a) and (e) follow from the implicit function theorem, and property (f) from (e). The verification of the remaining properties is similar to that carried out below in the proof of Proposition 6 (the corresponding main scheme of argument is given in **[69,** Chapter 1, §5]).

Statement (i) of Theorem 1 directly follows from Propositions 1, 2. The validity of statement (ii) is established by arguments similar to those in the proof of statement (ii) of Theorem 2 below (or to those used in **[69,** §7, case I]). Theorem 1 is proved.

Let us prove statement (i) of Theorem 2. Since the total number of centers is finite, they are assumed to be absent in the neighborhood under consideration. Following **[69]**, we introduce "*maximal arcs of the level C*", analogous to the arcs l_C used above. In order to do this, we fix a domain Ω containing points of the level C and cover the set of these points in Ω by a locally finite system Σ_C of widths of the level C from statement (i) of Theorem 1,

$$\ldots, k_C, k'_C, \ldots, k''_C, \ldots.$$

By definition, a width k_C is equivalent to a width k'_C, or $k_C \sim k'_C$, if there are $s \geqslant 2$ widths $k^1_C, \ldots, k^s_C$ of the system Σ_C such that $k^1_C = k_C$, $k^s_C = k'_C$ and for any $j = 1, \ldots, s-1$ the intersection of k^j_C and k^{j+1}_C is a simple arc. Obviously, we have $k_C \sim k_C$; $k_C \sim k'_C$ implies $k'_C \sim k_C$; $k_C \sim k'_C$ and $k'_C \sim k''_C$ imply $k_C \sim k''_C$. Thus, the binary relation "$\sim$" introduced in Σ_C is an equivalence relation. The union of all equivalent widths of the covering Σ_C is called the *maximal arc of the level C in* Ω and is denoted by γ_C.

From this definition it follows that γ_C is a continuous curve in Ω whose endpoints (if any) necessarily belong to the boundary of Ω and possibly coincide. The curve γ_C can have multiple points or points of self-intersection, necessarily lying (by Theorem 1) inside Ω and coinciding with saddles of the function ψ.

PROPOSITION 3. *If the curve γ_C is not a simple arc, then the inclusion of the boundary points into γ_C leads to a new curve which contains a cycle (or coincides with a cycle).*

Indeed, if γ_C has a multiple point P_* lying within this curve, then, together with P_*, the maximal curve γ_C contains at least two equivalent widths of the level C lying in the canonical neighborhood of this point. The presence of two equivalent widths immediately leads to the existence of a cycle that connects them and consists of the widths of the level C belonging to γ_C. If γ_C does not have multiple points among its interior points, then it evidently forms a cycle if and only if its endpoints coincide. The validity of the proposition is proved.

Now let $\Omega = \Omega(\xi_C)$ be a relative neighborhood of a given limit continuum ξ_C of the level C lying on the boundary cycle J.

PROPOSITION 4. *γ_C is a simple curve.*

Indeed, otherwise γ_C has a cycle L_C (according to Proposition 3). If the continuum ξ_C does not coincide with J, then the domain Ω is simply connected and the cycle L_C bounds a domain. If ξ_C coincides with J, then Ω is an annulus. It can be easily seen in this case that either L_C itself is the boundary of a domain, or bounds a domain containing the cycle ξ_C. In either of these three cases ψ takes constant value C on the boundary of some subdomain of the domain Ω without local extrema of ψ, which contradicts the maximum principle. The proposition is proved.

We denote by Γ the intersection of the boundary of Ω with the domain V.

PROPOSITION 5. *The endpoints of the arc γ_c lie on the union of the sets ξ_C and Γ; if one of them belongs to ξ_C, then the second lies on Γ.*

Indeed, the endpoints of γ_C belong to the boundary of the neighborhood Ω. If ξ_C does not coincide with its cycle J, then (by the definition of Ω) the complement of the union of ξ_C and Γ to the whole boundary of Ω does not contain level points. If ξ_C coincides with J, then it is empty. Thus, in any case the endpoints of γ_C (since they lie on the level C) belong to the union of ξ_C and Γ. Finally, if both these points lie on ξ_C, then there exists a domain on whose boundary $\psi = C$, which, as we have already seen, is impossible. The proof is completed.

Statement (i) of Theorem 2 follows from Proposition 4 and the following result.

PROPOSITION 6. *The maximal arcs γ_C have the following properties in $\Omega = \Omega(\xi_C)$:*
(a) *the arcs ending in ξ_C do not have common limit points not belonging to ξ_C,*
(b) *the total number of arcs γ_C ending on ξ_C is finite,*
(c) *any sequence of points of the level C from Ω accumulating at ξ_C contains a subsequence converging to some point P of the continuum ξ_C,*
(d) *any point P of the continuum ξ_C, limit for the sequence of points of the level C from Ω, is an end of some arc γ_C,*
(e) *ξ_C is an isolated set for the arcs γ_C that do not end on ξ_C,*
(f) *the arcs γ_C ending on ξ_C separate the signs of the difference $\psi - C$,*
(g) *for $\xi_C = J$, the number of arcs γ_C ending on ξ_C is even.*

Indeed, if among the indicated arcs there is a pair with a common limit point not belonging to ξ_C, then together with this pair there exists a domain, partially or completely bounded by these arcs, on whose boundary $\psi = C$, which, as we have already mentioned, is impossible. Therefore, property (a) holds. Property (f) follows from Proposition 4 and property (a), which imply that canonical neighborhoods of interior points of the arcs γ_C ending on ξ_C contain precisely one width of level C. Property (g) follows from (f). Property (c) follows from the fact that the closure of the domain V (and, hence, that of the neighborhood Ω) is compact, and that the set ξ_C is closed and is isolated for the points of the level C lying on the cycle J containing ξ_C, $\xi_C \neq J$. To prove the remaining properties (b), (d), and (e), we fix in Ω a curve S such that S is a simple arc dividing Ω into two domains and ending at the endpoints of the arc λ (covering ξ_C on J and constituting a common part of the boundary of Ω and the cycle J) if $\xi_C \neq J$, and S is a cycle in Ω bounding, together with J, a new relative neighborhood of ξ_C if $\xi_C = J$. Let us establish that (b) is valid. Since for $\xi_C \neq J$ the endpoints of the arc λ do not lie on the level C (due to the choice of λ) and for $\xi_C = J$ the set S is closed, the set of points of the level C lying on S can always be covered by a finite number of widths of this level. By Proposition 5, any arc γ_C ending on ξ_C ends on Γ and, thus, intersects S. Hence, the total number of these arcs is finite (we additionally take into account the fact that at each saddle point of the level C on S only a finite number of maximal arcs can intersect according to statement (i) of Theorem 1). Let us prove (e). Let us assume (contrary to this statement) that there exists a sequence of points of level C lying on the arcs γ_C, not ending on ξ_C and accumulating at ξ_C. If the number of these arcs is finite, then, obviously, there exists among them one arc containing an infinite part of this sequence, necessarily ending on ξ_C (property (c)), which is in contradiction with our assumption. Thus, the number of arcs containing points of the sequence under consideration is infinite. Their endpoints lie on Γ (Proposition 5); a part of the interior points forms a sequence accumulating at ξ_C. Since each of these arcs can contain only a finite number of points of this sequence, an infinite part of the set of these arcs intersects the curve S. The latter, as it was already mentioned, is impossible, hence (e) is valid. Property (d) follows from (b) and (e); according to these two properties, the sequence mentioned in (d) has a subsequence lying on some arc γ_C. Obviously, such an arc necessarily ends at P. The validity of Proposition 6 and, simultaneously, of statement (i) of Theorem 2 is established.

Now we shall prove statement (ii) of this theorem. By substituting in Proposition 3 the domain Ω with the sector F from statement (i) of Theorem 2, and the level C with a level C' in F, we establish the validity of an analogous statement for the maximal arc $\gamma_{C'}$ of this level C' in F. By Proposition 3, $\gamma_{C'}$ is a simple arc. The canonical neighborhoods of the points of the arc $\gamma_{C'}$ necessarily contain one width of the level C', hence, $\gamma_{C'}$ separates the signs of the difference $\psi - C'$. In order to prove that $\gamma_{C'}$ is unique for C' sufficiently close to C, we use the considerations from **[69,** §7, Lemma 7.1]. We have two types of sectors: (1) a sector of ordinary type, bounded by two arcs of level C (lying in V, Figure 10a), and (2) the sector of "boundary" type with the boundary containing only one arc g_C of level C (Figure 10b). Omitting case (1) considered by Morse **[69,** Chapter 1, §7, case (i)], which can also be studied according to the scheme used for the analysis of the second case given below, we immediately turn to the case (2), in which, contrary to the analogous situation studied in **[69]**, a stable region can serve as the continuum $\xi_{C'}$ of level C'. Simultaneously, we shall show that in spite of the fact that such possibility exists, the assumption on the existence of two arcs $\gamma_{C'}$ and $\gamma'_{C'}$ of the same level C' lying in F will lead to a contradiction with the maximum principle.

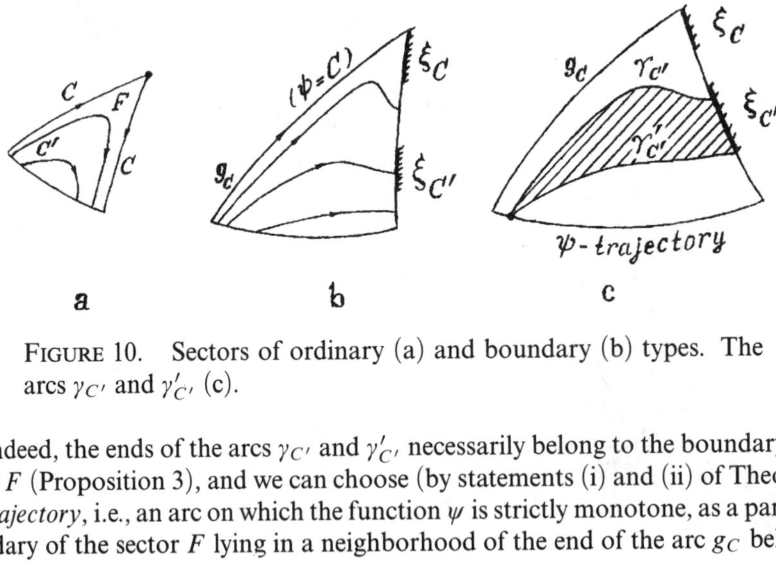

FIGURE 10. Sectors of ordinary (a) and boundary (b) types. The arcs $\gamma_{C'}$ and $\gamma'_{C'}$ (c).

Indeed, the ends of the arcs $\gamma_{C'}$ and $\gamma'_{C'}$ necessarily belong to the boundary of the sector F (Proposition 3), and we can choose (by statements (i) and (ii) of Theorem 1) a *ψ-trajectory*, i.e., an arc on which the function ψ is strictly monotone, as a part of the boundary of the sector F lying in a neighborhood of the end of the arc g_C belonging to V (Figure 10c). Due to these facts, the ends of the arcs $\gamma_{C'}$ and $\gamma'_{C'}$ coincide at one of the points of the ψ-trajectory. The other pair of these arcs necessarily belongs to the continuum $\xi_{C'}$ (it is taken into account that this continuum is unique for C' sufficiently close to C, which follows from the boundary conditions $\boldsymbol{\tau}$), or both ends of one of the arcs $\gamma_{C'}$, $\gamma'_{C'}$ belong to $\xi_{C'}$. For either one of these possibilities, there exists a subdomain of the sector F on whose boundary $\psi = C'$, which is impossible by the maximum principle.

The proof of statement (ii) is completed.

Let us show the validity of statement (iii) of Theorem 2. The proof of the part of this statement concerning the case when $\xi_C \neq J$ is analogous to that given above for a sector of the boundary type in the proof of statement (ii) (where we followed **[69,** §7]). In the remaining case $\xi_C = J$, we show the existence in Ω of a maximal arc γ'_C of level C' completely lying in Ω (including its limit points) for C' sufficiently close to C. Without loss of generality we assume that in this case the cycle Γ bounding the neighborhood Ω in the domain V does not contain points of the level C (here we use the assumption that the continuum ξ_C is isolated). Let us assume the converse, i.e., that a maximal arc γ'_C lying completely in Ω does not exist. Then there exists a sequence of numbers C_i, $i = 1, 2, \dots$, not equal to C and converging to C as $i \to \infty$ such that for each C_i in Ω there exists a maximal arc γ_{C_i} of the level C_i ending on Γ. Let P_* be a limit point of the ends of the arcs γ_{C_i} (the existence of P_* follows from the compactness of Γ). Using the continuity of the function ψ in the closure of Ω and, if necessary, passing to a subsequence of the sequence C_i, $i = 1, 2, \dots$, we find that $\psi(P_*) = C$. However, the latter is in contradiction with the assumption on the absence of points of the level C on Γ (because the continuum ξ_C is isolated) and, thus, completes the proof of Theorem 2.

§4. Portraits of potential flows

We consider hydrodynamic applications. Plane and spatial axisymmetric stationary flows of an ideal (nonviscous and incompressible) fluid considered in a bounded domain V of the upper halfplane $y \geqslant 0$ in the absence of nonpotential mass forces, is described by a vector field $\mathbf{u} = (u, v)$ whose components $u = u(x, y)$ and $v = v(x, y)$ satisfy a first-order system of quasilinear partial differential equations of the following form:

$$\text{(1)} \qquad \begin{gathered} \frac{\partial}{\partial x}\left(y^k u\right) + \frac{\partial}{\partial y}\left(y^k v\right) = 0, \qquad \frac{\partial v}{\partial x} - \frac{\partial u}{\partial y} = \omega, \\ u\frac{\partial}{\partial x}\left(\frac{\omega}{y^k}\right) + v\frac{\partial}{\partial y}\left(\frac{\omega}{y^k}\right) = 0. \end{gathered}$$

The parameter k is equal to 0 for a plane field and to 1 for an axisymmetric flow. In the latter case, the variable $y > 0$ is the distance from the axis of symmetry, the variables x, y are connected with the Cartesian coordinates x_1, x_2, x_3 by the equalities

$$x_1 = x, \qquad x_2 = y\cos\theta, \qquad x_3 = y\sin\theta \qquad (0 \leqslant \theta \leqslant 2\pi),$$

the spatial flow $\mathbf{u}' = (u, v\cos\theta, v\sin\theta)$, whose portrait (the family of the streamlines) is obtained by rotating the meridional plane $\theta = \text{const}$ with the portrait of the plane flow $\mathbf{u} = (u, v)$ around the axis of symmetry $y = 0$ (variation of the angle θ, Figure 11), corresponds to the plane field $\mathbf{u} = (u, v)$. The derivation of the equations for the case $k = 0$ is given in §2.1; for $k = 1$ it is given in Appendix (§17.6). Recall that the system of equations (1) unites the condition of the incompressibility of the flow, the definition of the vorticity ω, and the necessary compatibility condition for the initial Euler equations (the *Helmholtz relation*).

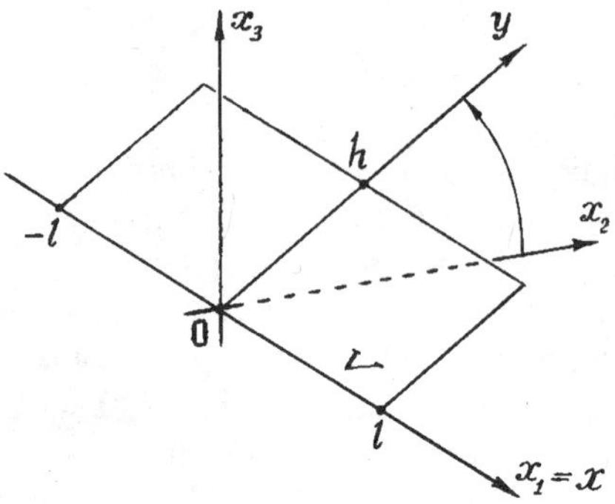

FIGURE 11. Meridional plane of an axisymmetric flow.

The problem considered below consists in studying the topological properties of vector fields mentioned in the Introduction and satisfying system (1). The necessary information can sometimes be obtained from the explicit formulas for the solution **[29, 47, 52, 54, 55, 57, 67, 77, 136, 139, 159]**. However, in most cases, to analyze these properties we have to use only the data of some boundary problem for system (1), which does not always admit a solution expressed by an analytic formula, and even when it does, most often it is cumbersome to obtain necessary information from it. A natural alternative way of studying the flow portrait is to obtain a numerical solution of the boundary problem under consideration. However, here we encounter some other difficulties. Moreover, for a specialist in hydrodynamics, it is essential to have a set of possible flow portraits which exhaust those admissible for the problem under consideration. Such a set can be obtained by using relations (1) from the previous section. The idea of the method was elaborated in **[34]**. Its detailed exposition is given below.

The simplest class of solutions of equations (1) is presented by *potential flows* or vector fields with zero vorticity $\omega \equiv 0$. The smooth components u, v of the velocity in this case satisfy the Carleman system of the following form:

$$\frac{\partial u}{\partial x} + \frac{\partial v}{\partial y} = -\frac{k}{y} v, \qquad \frac{\partial v}{\partial x} - \frac{\partial u}{\partial y} = 0,$$

and, hence, are analytic in the flow domain V **[24, 54]**. The stream function $\psi(x, y)$ of the potential flow $\mathbf{u} = (u, v)$,

$$u(x, y) = \frac{1}{y^k} \frac{\partial \psi}{\partial x}, \qquad v(x, y) = -\frac{1}{y^k} \frac{\partial \psi}{\partial x}, \tag{2}$$

satisfies in V the elliptic equation

$$L_k \psi \equiv -\frac{1}{y^k} \left(\frac{\partial}{\partial x} \left(\frac{1}{y^k} \frac{\partial \psi}{\partial x} \right) + \frac{\partial}{\partial y} \left(\frac{1}{y^k} \frac{\partial \psi}{\partial y} \right) \right) = 0 \tag{3}$$

(which follows from the second relation of system (1)). The following statement establishes the identity of the multiplicity of a saddle of a potential flow and the order of its degeneration.

PROPOSITION 1. *The stream function* $\psi \not\equiv \text{const}$ *of the potential flow* $\mathbf{u}$ *is admissible in the domain of definition* V *(in the sense of* §3.1*) and has no centers. If all the partial derivatives of the function* ψ *up to the order* $n \geqslant 1$ *inclusively vanish at a point* P_0 *of the domain* V, *then* P_0 *is a saddle of multiplicity* $n_* \geqslant n$. *If, simultaneously, at least one of the partial derivatives of* ψ *of order* $n+1$ *is different from zero at* P_0, *then* $n_* = n$.

Indeed, associating to the variable $P = (x, y)$ the complex variable

$$z = z(P) = x + iy, \qquad i^2 = -1,$$

and to the critical point P_0 the number $z_0 = z(P_0)$, we introduce into consideration the function

$$w(z) = u(P) - iv(P) = \frac{1}{y^k}\frac{\partial \psi}{\partial y} + \frac{i}{y^k}\frac{\partial \psi}{\partial x}. \tag{4}$$

As a consequence of the Vekua-Bers formula **[24, 54]** we have

$$w(z) = (z - z_0)^{n_*} h(z), \qquad |z - z_0| \leqslant \varepsilon, \tag{5}$$

where $\varepsilon > 0$ is a sufficiently small constant, $|z| \equiv (x^2 + y^2)^{1/2}$ is the absolute value of the complex number $z = x + iy$, $h(z)$ is a continuous complex function of the variable z, defined in the given disc and attaining at its center the nonzero value $h(z_0) \neq 0$, $n_* = 1, 2, \ldots$. According to (5), P_0 is a saddle of multiplicity n_*.

Let us show that $n_* \geqslant n$. Indeed, for $n_* \leqslant n - 1$ the identity (5) implies, for any $z \neq z_0$ sufficiently close to z_0, the following inequality:

$$\frac{|w(z)|}{|z - z_0|^{n-1}} = |z - z_0|^{n_* - (n-1)} |h(z)| \geqslant |h(z_0)| > 0,$$

which together with (4) provides the existence of a nonzero partial derivative of the function ψ of order n at P_0, and this contradicts our assumption.

Further, if some partial derivative of ψ of order $n+1$ is different from zero at P_0, then some partial derivative of order n of one of the components u or v of the velocity is not equal to zero at P_0 (identities (2)). Then there must exist a sequence of points z_s, $s = 1, 2 \ldots$, converging to z_0 as $s \to \infty$ such that

$$|z_s - z_0|^{n_* - n} |h(z_s)| = \frac{|w(z_s)|}{|z_s - z_0|^n} \to C_0 \neq 0 \qquad (s \to \infty).$$

The latter implies $n_* \leqslant n$ and completes the proof of Proposition 1.

In what follows, the flow domain is the rectangle

$$V : |x| < l, \quad 0 < y < h$$

$(l, h = \text{const} > 0)$, on whose boundary the normal component of the velocity $\mathbf{u} = (u, v)$ is given,

$$\begin{aligned} u(\mp l, y) &= U_{\mp}(y), \qquad 0 \leqslant y \leqslant h, \\ v(x, 0) &= v(x, h) = 0, \qquad |x| \leqslant l. \end{aligned} \tag{6}$$

Here $U_{\mp}(y)$ are sufficiently smooth (e.g., of class C^2) functions on the interval $0 \leqslant y \leqslant h$, subject to the requirement

$$\int_0^h y^k (U_- - U_+)\, dy = 0. \tag{7}$$

Under the assumption that the velocity components are continuous in the closure of the flow domain V, restriction (7) is obtained by integrating the first relation of system (1) over V. For $k = 0$ ($k = 1$) the boundary conditions (6) generate a fluid flow problem in a *channel* (in a *pipe of circular cross-section*, respectively) of length $2l$ and width (or radius) h, whose horizontal walls (the lateral surface) are assumed to be impermeable and on the vertical walls (on the ends) of which the distribution of the normal velocity component is fixed. The equality $v(x, 0) = 0$ in the case $k = 1$ is equivalent to the continuity of the spatial flow $\mathbf{u}'$ corresponding to $\mathbf{u}$ (and defined as above) on the axis of symmetry $y = 0$ and is in accordance with the assumption of absence of sources and sinks on this axis.

The existence of a flow $\mathbf{u}$, potential in V and continuous in the closure of this domain, is provided by the imposed restrictions in the case $k = 0$ **[54]**; for the case $k = 1$, under the additional requirement of smoothness on the interval $0 < y < h$ up to its ends of the derivatives $dU_{\mp}/dy$ divided by y, this is established in Chapter 3. Simple connectedness of the domain V under consideration implies the existence of a stream function ψ satisfying in V identities (2) and equation (3). Using (2), we calculate the boundary values of the function ψ,

$$\begin{gathered} \psi(\mp l, y) = \int_0^y s^k U_{\mp}(s)\, ds, \qquad 0 \leqslant y \leqslant h, \\ \psi(x, 0) = 0, \qquad \psi(x, h) = k_0 \equiv \int_0^h y^k U_-\, dy = \int_0^h y^k U_+\, dy, \qquad |x| \leqslant l. \end{gathered} \tag{8}$$

The fact that a function ψ with fixed boundary values is unique follows from the maximum principle, which, according to Lemma 3.1 (§3.4), is satisfied by any admissible function without centers (local extrema in the domain of definition). Under the requirement that at least one of the functions $U_{\mp}(y)$ should not identically vanish on the interval $0 < y < h$ and the velocity component different from identical zero should have in it a finite number of zeros, we see that ψ satisfies the boundary conditions τ from the first subsection of the previous section (§3.1), that the equality $\sigma = 0$ is valid (ψ is not constant on the boundary), and that there is a finite number of stable regions (isolated arcs of the boundary on which ψ acquires constant values). Since, simultaneously, $q = 0$ (ψ has no centers), $g = 0$ (V is a plane domain), and $\nu = 1$ (V is simply connected), relations (3.1) acquire the form

$$p + s = m - 1, \qquad s - m = \overline{s} - \overline{m}. \tag{9}$$

Further, according to (8), the boundary of the domain V contains at least one stable region (if one of the functions $U_{\mp}(y)$ is identically zero, it is the union of the horizontal sides of the rectangle V and the corresponding vertical line; otherwise, only the horizontal sides of V form such regions); thus, a sector cannot be a chamber of the flow (§3.3). The maximum principle and the simple-connectedness of the domain V exclude circulation. From the admissible chambers (§3.3) there remain only the passing flow and the semivortex (which will sometimes be called also "simple passage" and "reverse flow", respectively) introduced in [**34**]. Following the considerations of [**34**], we will analyze some portraits of potential flows.

1. $U_{\mp}(y) > 0$, $0 < y < h$. In this case, ψ has a unique minimum on the side $y = 0$ and a unique maximum on the side $y = h$. We have a passing flow (Figure 9d, §3.3).

In the following examples, the function $U_{-}(y)$ strictly monotonically *increases* and changes the sign at some point y_0, $0 < y_0 < h$. Only the function $U_{+}(y)$ is being varied.

2. $U_{+}(y) \equiv 0$. The function ψ has a unique minimum at the point $(-l, y_0)$ and a unique maximum on the boundary arc of the rectangle V complementing the interval $x = -l$, $0 < y < h$. The flow is a semivortex (shown in §6, Figure 16a) with the opposite orientation of the streamlines.

3. $U_{+}(y) > 0$, $0 < y < h$. In this case, the trace of ψ on the boundary of V attains a relative minimum at the point $(-l, y_0)$ and a positive maximum on the arc $y = h$, $|x| \leqslant l$. Due to the maximum principle, the numbers $m = \overline{m} = 1$. On the side $y = 0$, $|x| \leqslant l$, the function ψ is equal to zero. The interval $y_0 < y < h$, $x = -l$ contains a point y_1 of the level $\psi = 0$. The structure of the flow is defined by the arc γ_1 of the level $\psi = 0$ emanating from $(-l, y_1)$ and entering some point $(x_1, 0)$, $|x_1| < l$. The arc γ_1 divides the domain V into two chambers: a passing flow adjacent to the side $y = h$ and a semivortex whose base is the arc $0 < y < y_1$, $x = -l$ (Figure 12a).

4. The function $U_{+}(y)$ strictly monotonically decreases and changes the sign at some point y_m, $0 < y_m < h$, the constant $k_0 = \psi(-l, h) > 0$. In this case, the trace of ψ on the boundary of V has a unique minimum point $(-l, y_0)$ and a unique maximum point (l, y_m). Hence, $m = \overline{m} = 1$ and, thus (due to relations (9)), $p = s = \overline{s} = 0$. Simultaneously with γ_1, there is an analogous arc γ_2 of the level $\psi = k_0$ emanating from some point (x_2, h), $|x_2| < l$, and entering (l, y_2), $0 < y_2 < y_m$ (Figure 12b).

4a. $k_0 = 0$. We have the coalescence of the arcs γ_1 and γ_2 into one arc of the level $\psi = 0$ dividing V into two semivortices (Figure 12c).

4b. $k_0 < 0$. The portrait of the flow is obtained from the portrait of Example 4b by mirror reflection in the axis $x = 0$ (Figure 12d).

5. The function $U_{+}(y)$ *increases* strictly monotonically and changes the sign at some point y_0', $0 < y_0' < h$; the constant $k_0 = 0$. In this case, the trace of ψ on the boundary of V attains a negative relative minimum at the points $(-l, y_0)$, (l, y_0'), and zero absolute maximum on the sides $y = 0$ and $y = h$, $|x| \leqslant l$. Hence, $m = 1, 2$ and $\overline{m} = 2$ (the maximum principle).

Let us show that $s = 0$.

Indeed, $s \geqslant 1$ together with (9) and $m \leqslant 2$ implies $p = 0$. It follows from (9) that $s = m - 1 \leqslant 1$; hence, $s = 1$. Only one of the points $(-l, y_0)$ or (l, y_0') can correspond to a saddle boundary continuum (ψ attains the maximum on the horizontal sides of the rectangle, and the interior points of the lateral sides, excluding those mentioned above, are regular), the remaining point is either a minimum point or an inverse-saddle point (due to the restriction $s = 1$). The latter possibility is excluded by the inequality $m \geqslant 1$. But then ψ attains its minimum at the unique (remaining) point; hence $m = 1$, which implies $s = m - 1 = 0$, so that we come to a contradiction with $s = 1$. Hence, $s = 0$.

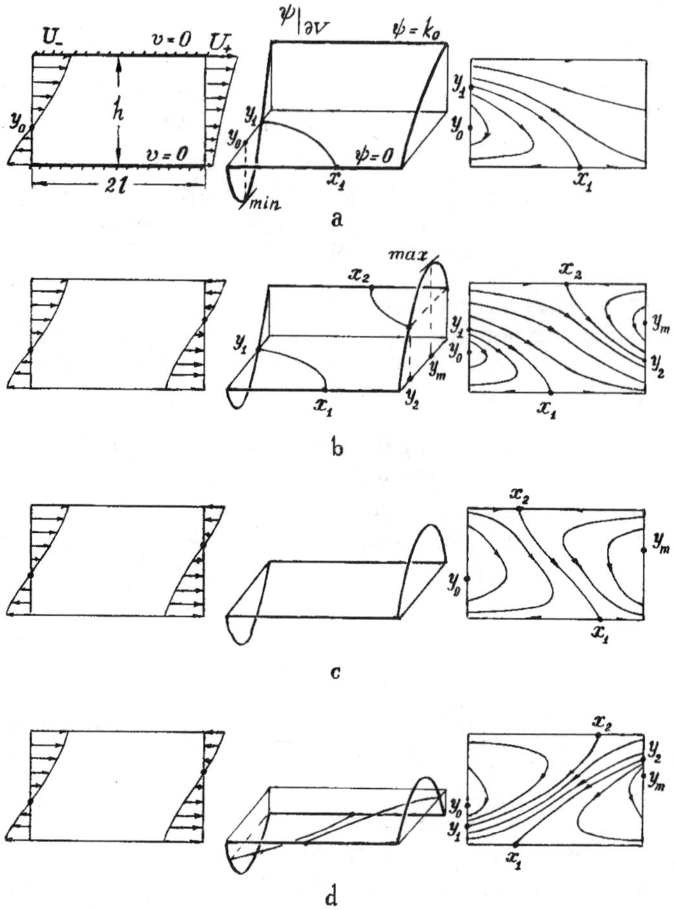

FIGURE 12. The flows from Examples 3 (a), 4 (b), 4a (c), and 4b (d).

Substituting $s = 0$ and $\overline{m} = 2$ in (9), we find

$$p = m - 1, \qquad \overline{s} = 2 - m.$$

Introducing parameters k_1, k_2, k_3 by the equalities

$$k_1 = \psi(-l, y_0) - \psi(l, y_0'), \qquad k_2 = \frac{\partial \psi}{\partial x}(l, y_0'), \qquad k_3 = \frac{\partial \psi}{\partial x}(-l, y_0),$$

for the case $k_1 < 0$, $k_2 \geqslant 0$ we have that $(-l, y_0)$ is a unique minimum point ($m = 1$) and (l, y_0') is an inverse-saddle point of multiplicity $\overline{s} = 2 - 1 = 1$, $p = 1 - 1 = 0$. The portrait of the corresponding flow is shown in Figure 13a.

5a. $k_0 = 0, k_1, k_2 < 0$. In this case, (l, y_0') is a point of local minimum of ψ; hence $m = 2$ and $\overline{s} = 2 - m = 0$, $p = m - 1 = 1$. The inverse-saddle point enters the flow domain (Figure 13b).

5b. $k_0 = 0, k_1, k_3 > 0$. The portrait of the flow does not change.

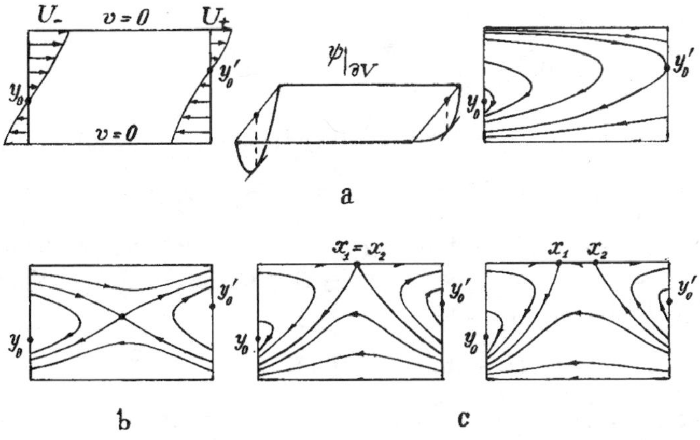

FIGURE 13. The flows from Examples 5 (a), 5a (b), and 5f (c).

5c. $k_0 = 0$, $k_1 > 0$, $k_3 \leqslant 0$. The saddle lies on the boundary, i.e., the points $(-l, y_0)$ and (l, y_0') exchange the roles; the portrait of the flow is obtained from the portrait of Example 5 by reflection in the axis $x = 0$ and reversal of the orientation of the streamlines.

5d. $k_0 = k_1 = 0$. In this case, $(-l, y_0)$ and (l, y_0') are points of minimum $(m = 2)$. The flow has the structure of Example 5a.

The analysis of the remaining cases will be carried out taking into account the signs of the parameters $k_4 = \min_{|x| \leqslant l} u(x,h)$ and $k_5 = \max_{|x| \leqslant l} u(x,0)$.

5e. $k_0 < 0$, $k_4 > 0$. In this case ψ has two maxima, the absolute one on the side $y = 0$ (since $k_0 < 0$) and a relative one on the side $y = h$ (since $k_4 > 0$). Again we come to the portraits of the flows of Examples 5–5d.

In the sequel we shall need the following statement.

PROPOSITION 2. *If ψ attains a relative extremum on the side $y = 0$ (on the side $y = h$, $|x| \leqslant l$) and $U_-(0)U_+(0) > 0$ ($U_-(h)U_+(h) > 0$, respectively), then for all $-l < x < l$*

$$U_-(0)u(x,0) > 0, \qquad (U_-(h)u(x,h) > 0).$$

The validity of Proposition 2 in the case $k = 0$ or $k = 1$ and $y = h$ follows from the classical lemma on the normal derivative **[59]**, since in a neighborhood of the corresponding boundary points the operator L_k from (3) is uniformly elliptic. Only the case $k > 0$ and $y = 0$ (a neighborhood of the degeneration line) requires some additional analysis. In this case, we assume, without loss of generality, that a relative *minimum* of ψ is attained on the side $y = 0$. We have $\psi(x,0) = 0$, $|x| \leqslant 0$ (conditions (8)) and (according to the assumption) there exists a number $0 < y_* < 1$ such that $\psi(x,y) \geqslant 0$ for $|x| \leqslant l$ and $0 < y < y_*$. Hence (since $U_-(0)U_+(0) > 0$),

$$U_-(0) > 0, \qquad U_+(0) > 0. \tag{10}$$

In particular, $\psi \not\equiv \text{const}$; thus (Proposition 1), ψ is admissible in V and has no centers. Due to Lemma 3.1 (§3.4), $\psi(x,y) > 0$ for $|x| < l$ and $0 < y < y_*$ (strict minimum).

As a result, we have (taking into account (10))

$$(11) \qquad m(y) \equiv \min_{|x|\leqslant l} \psi(x,y) > 0, \qquad 0 < y < y_*.$$

Now we introduce an ε-family of "barrier" type functions of the following form:

$$(12) \qquad \Phi_\varepsilon(y) = \varepsilon(y/h)^{k+1}/(k+1), \qquad \varepsilon = \text{const} > 0,$$

and choose sufficiently small constants $\varepsilon, \delta > 0$, $\delta < y_*$ such that on the boundary of the strip $V_\delta : |x| < l, 0 < y < \delta$, the inequality $\psi \geqslant \Phi_\varepsilon$ be satisfied for them.

In order to do this, we put

$$U_0 \equiv \min\{U_-(0), U_+(0)\} > 0$$

and, using the continuity of the functions $U_\mp(y)$ and inequalities (10), we choose a constant $\lambda_0 = \lambda(U_0)$ such that

$$(13) \qquad U_\mp(y) > U_0/2 \quad \text{for } 0 < y < \lambda_0$$

(the functions $U_\mp(y) - U_0/2$ are continuous at $y = 0$; thus, the sign of the inequality

$$U_\mp(0) - U_0/2 > 0$$

is preserved in some neighborhood of this point). On the other hand, ψ is continuous in the closure of the rectangle V and vanishes on its lower side (conditions (8)). Hence, the function $m(y)$ from (11) tends to zero as $y \to 0$; thus, there must exist another constant $\sigma_0 = \sigma(U_0) < \lambda_0$ (sufficiently small) such that

$$m(y) < U_0 h^{k+1}/2 \quad \text{for } 0 < y < \sigma_0.$$

Setting $\delta = \sigma_0/2$ and $\varepsilon = m(\delta)$, we find from the previous inequality that $\varepsilon < U_0 h^{k+1}/2$. Taking into account (13), we obtain

$$\frac{\partial}{\partial y}(\psi(-l,y) - \Phi_\varepsilon(y)) = y^k\left(U_-(y) - \frac{\varepsilon}{h^{k+1}}\right) > 0, \qquad 0 < y < \delta.$$

Since $\Phi(-l,0) - \psi_\varepsilon(0) = 0$, this inequality implies

$$\psi(-l,y) - \Phi_\varepsilon(y) > 0, \qquad 0 < y < \delta.$$

Using (13), by means of analogous considerations we find that

$$\psi(l,y) - \Phi_\varepsilon(y) > 0, \qquad 0 < y < \delta.$$

Since, simultaneously,

$$\psi(x,\delta) - \Phi_\varepsilon(\delta) \geqslant m(\delta) - (\delta/h)^{k+1}\varepsilon/(k+1) \geqslant m(\delta) - \varepsilon = 0,$$

we come to the required inequality $\psi \geqslant \Phi_\varepsilon$ which is valid on the boundary of the strip V_δ.

Since

$$(k+1)L_k\Phi_\varepsilon = -\frac{\varepsilon}{y^k}\frac{d}{dy}\left(\frac{1}{y^k}\frac{d}{dy}\frac{y^{k+1}}{h^{k+1}}\right) = 0 \qquad (y > 0),$$

in the strip V_δ, we have $L_k(\psi - \Phi_\varepsilon) = 0$, i.e., $\psi - \Phi_\varepsilon$ is the stream function of some potential flow. Using the above inequality and the maximum principle, we come to the conclusion that $\psi - \Phi_\varepsilon \geqslant 0$ on all the intervals $x = \text{const}$ of the strip V_δ. Hence, the corresponding derivative with respect to the variable y^{k+1} is $\geqslant 0$, or

$$u(x,0) - \frac{\varepsilon}{h^{k+1}} = \lim_{y \to +0} \frac{1}{y^k} \frac{\partial}{\partial y}(\psi - \Phi_\varepsilon) \geqslant 0, \qquad |x| \leqslant l,$$

which proves the required statement.

Let us return to the analysis of the flow portraits.

5f. $k_0 < 0$, $k_4 \leqslant 0$. The latter inequality together with Proposition 2 implies that the upper side of the rectangle V is a limit continuum of the level $\psi = k_0$ (in the sense of §3.1), and the unique maximum of ψ is attained on the lower side of V, i.e., $\overline{m} = 1$. Using (9), we shall show that $m = 2$.

Indeed, obviously $m \leqslant 2$, i.e., $m = 1$ or 2, (the boundary trace of ψ has two minima and the maximum principle is valid). The equalities $m = 1$ and (9) imply that $p = s = \overline{s} = 0$. Since, simultaneously, the upper side of V is a limit continuum of the level $\psi = k_0$, and, thus, (by statement (i) of Theorem 2 from §3.1) in the domain V there exists a line ω of this level ending at some point (x_1, h) of the upper side. Since $p = 0$, ω is a simple arc with the other endpoint being one of the boundary points of the rectangle V located on the level $\psi = k_0$ but not on the upper side of V (otherwise, ω starts and ends at the mentioned side bounding a domain, which is prohibited by the maximum principle). There are two such points; the point $(-l, y_1)$ with y_1 between 0 and y_0, and the point (l, y_1') with y_1' between 0 and y_0'. Moreover, ω divides V into two domains and on the boundary of one of them the difference $\psi - k_0$ changes the sign (which is obvious from the behavior of ψ on the boundary of V). Hence (due to the maximum principle), this difference changes the sign inside this domain as well, thus, inside the latter there is a new arc ω' of the level $\psi = k_0$ necessarily ending on the upper side of V, i.e., two arcs (ω and ω') of the same level end on this side; hence the sum $s + \overline{s} > 0$, which contradicts the equalities $s = \overline{s} = 0$. The proof is completed.

For $m = 2$ and $\overline{m} = 1$, equalities (9) imply $p = 1 - s$ and $\overline{s} = s - 1$, which is possible only for $p = \overline{s} = 0$ and $s = 1$. The upper side of the rectangle V is, obviously, the corresponding saddle continuum. The structure of the flow is defined by the arcs ω and ω' of the level $\psi = k_0$ emanating from the points (x_1, h) and (l, y_1''), respectively, and ending at the points $(-l, y_1)$ and (x_2, h), $-l < x_1 \leqslant x_2 \leqslant l$. The arcs ω and ω' divide V into a passing flow and two semivortices, as shown in Figure 13c.

5g. $k_0 > 0$, $k_5 \geqslant 0$. In this case, the upper and the lower sides of the rectangle V change their roles. The flow portrait is obtained from the portrait of the previous example by reflection in the axis $y = h/2$ and reversal of the orientation of the streamlines.

5h. $k_0 > 0$, $k_5 < 0$. In this case, we have the same flow portraits as in Examples 5–5d.

Thus, the portraits of the potential flows considered above are uniquely defined by fixing admissible integer numbers in the basic relations (9) and by indicating the sign of the normal derivative of the stream function ψ at those boundary points of the flow domain V where a relative extremum of the trace of ψ is attained on ∂V. It should also be mentioned that the flow portraits from Examples 5f and 5g were not considered in **[34]**.

§5. Imposition of a vortex on a simple passing flow

Although potential flows constitute a rich class of Euler fields, they hardly exhaust the entire class of Euler fields. Singular points of center type are excluded from the consideration in this class. Nonpotential fields (such that $\omega \not\equiv 0$) can be considered as admissible perturbations of potential flows resulting in the generation of vortices. Among the plane-parallel and axisymmetric flows, the simplest class of these perturbations is given by the flows with vorticity of the following type:

$$\omega = \lambda y^k, \qquad \lambda = \text{const}.$$

In this section we shall consider the imposition of this vorticity on the potential flow from Example 1 of the previous section. By setting in this example

$$U_-(y) = U_+(y) = U = \text{const} > 0,$$

for $\lambda = 0$, we obtain a potential flow of the simple passing type with the stream function $\psi = Uy^{k+1}/(k+1)$. For $\lambda \neq 0$, we make the change of variables

$$s = (k+1)\frac{x}{h}, \qquad t = \left(\frac{y}{h}\right)^{k+1},$$

which realizes an analytic diffeomorphic transformation of the rectangle $V : |x| < l, 0 < y < h$, onto the rectangle

$$V' = \{|s| < r = (k+1)\frac{l}{h},\ 0 < t < 1\},$$

and an orientation preserving homeomorphism of the closure of V onto the closure of V'. The flow portrait will be described in terms of the family of level lines of the function

$$\widetilde{\psi}(s,t) = \psi\left(\frac{hs}{k+1}, ht^{\frac{1}{k+1}}\right),$$

which is, obviously, homeomorphic to the family of streamlines under consideration. Proposition 1 from the previous section is valid for the new stream function $\widetilde{\psi}$, if the flow under consideration is potential (which straightforwardly follows from the given properties of the change of variables). Substituting $\omega = \lambda y^k$ and u, v from (4.2) in (4.1), using the following identity for the operator L_k from (4.3):

$$\left(\frac{hy^k}{k+1}\right) L_k\psi(x,y) = T_\alpha\widetilde{\psi}(s,t) \equiv -\frac{\partial^2\widetilde{\psi}}{\partial s^2} - t^\alpha\frac{\partial^2\widetilde{\psi}}{\partial t^2}, \qquad \alpha = \frac{2k}{k+1},$$

and calculating the boundary values of $\widetilde{\psi}$ by means of (4.8), we come to the following problem for the function $\widetilde{\psi}$:

$$\begin{aligned} &T_\alpha\widetilde{\psi} = Mt^\alpha, \qquad |s| < r, \qquad 0 < t < 1, \\ (1) \qquad &\widetilde{\psi}(\mp r, t) = Nt, \qquad 0 \leqslant t \leqslant 1, \\ &\widetilde{\psi}(s,0) = 0, \qquad \widetilde{\psi}(s,1) = N, \qquad |s| \leqslant r. \end{aligned}$$

Here

$$\alpha = \frac{2k}{k+1}, \qquad M = \frac{\lambda h^{2(k+1)}}{(k+1)^2}, \qquad N = \frac{Uh^{k+1}}{k+1}$$

are constants which are temporarily assumed to be arbitrary; the only restriction is imposed on the values of α: $\alpha \geqslant 0$. As proved in Chapter 3, problem (1) has a unique classical solution $\widetilde{\psi}$, twice continuously differentiable in the domain V' and continuous in its closure together with its first-order derivatives. The uniform ellipticity of the operator T_α in neighborhoods of the interior points of the lateral sides $s = \mp r$ and the upper base $t = 1$ of the rectangle V' provides the continuity of the second derivatives of the function $\widetilde{\psi}$ up to the indicated smooth pieces of the domain **[2]**. Taking into account the smoothness properties of the function under consideration, we have the following statement.

PROPOSITION 1. *The solution* $\widetilde{\psi} \equiv \Gamma$ *of problem* (1) *corresponding to* $\alpha = 0, 1$, $M = 1$, *and* $N = 0$ *satisfies the following conditions for* $0 < t < 1$:
(C_1) $-t^\alpha < \Gamma_{ss} < 0, -1 < \Gamma_{tt} < 0$,
(C_2) $\Gamma(-s, t) = \Gamma(s, t)$,
(C_3) $\Gamma_s > 0,\ s < 0$.

Indeed, for $\alpha = 0$ or $\alpha = 1$ the function

$$f = f(s,t) = \Gamma(s,t) + s^2 t^\alpha / 2$$

and the polynomial

$$g = g(s,t) = a_0 + a_1 s + a_2 t + a_3 st, \qquad a_{0,1,2,3} = \text{const}$$

satisfy the identity $T_\alpha f = T_\alpha g = 0$ in V'. Consequently, the same identity holds for their difference $R = f - g$, $T_\alpha R = 0$. As already mentioned, Proposition 4.1 holds for such an R. Since $\Gamma = 0$ on the boundary of the domain V', the function R, coinciding with $s^2 t^\alpha/2 - g$, satisfies the boundary conditions τ from §3.1, simultaneously admitting at most two boundary continua of a relative extremum. Hence, relations (4.9), where $\overline{m} \leqslant 2$, are valid for R, and thus we have

$$p = \overline{m} - \overline{s} - 1 \leqslant \overline{m} - 1 \leqslant 1.$$

The inequality $p \leqslant 1$ ensures the absence of a point $P_0 = (s_0, t_0)$ in the domain V' where the second derivative $f_{ss}(P_0)$ would vanish.

Indeed, otherwise the other second derivative vanishes, $f_{tt}(P_0) = 0$ (the identity $T_\alpha f = 0$ and the inequality $t_0 > 0$), thus the Taylor expansion of f at P_0 up to the terms of second order inclusively can serve as the polynomial g. However, then P_0 is a saddle of the difference $R = f - g$ of multiplicity $n_* \geqslant 2$, which contradicts the previously obtained inequality $p \leqslant 1$ and completes the proof of the absence of zeros of the derivative f_{ss} in the domain V'. Since f_{ss} is continuous in the domain V' up to the interior points of the upper base of the rectangle V' (which was already indicated) and at the indicated points $f_{ss}(s, 1) = 1 > 0$, $|s| < r$, we arrive at the conclusion that this derivative is positive everywhere in V'. Since also $f_{ss} = \Gamma_{ss} + t^\alpha$, we come to the first inequality from (C_1).

Repeating the previous arguments for a new auxiliary function $\phi = \Gamma(s,t) - t(1-t)/2$ (such that $T_\alpha \phi = 0$ in V'), we obtain the inequality $\phi_{tt} > 0$ in V' which is equivalent, due to the identity

$$\Gamma_{ss} = \phi_{ss} = -t^\alpha \phi_{tt} < 0,$$

to the next inequality in (C_1). The remaining inequalities in (C_1) follow from the first two, and from the identity $T_\alpha \Gamma = t^\alpha$.

The conditions of problem (1) are invariant with respect to the change of sign of the variable s. The fact that its classical solution is unique implies in this case the validity of (C_2). Using (C_2), we obtain that $\Gamma_s = 0$ for $s = 0$, which together with the second inequality from (C_2) implies (C_3) and completes the proof of Proposition 1.

We pass to the description of the flow portraits defined by (1). As a consequence of the last inequality from (C_1), the function $\Gamma(s,t)$ is *superelliptic* in V', i.e., the inequality

$$-(\Gamma_{ss}+\Gamma_{tt}) = (t^\alpha - 1)\Gamma_{tt} + t^\alpha \geqslant t^\alpha > 0$$

is valid for this function in the indicated domain (it is taken into account that $0 < t < 1$ and $\alpha \geqslant 0$); thus (**[59]**) the normal derivatives at minimum points satisfy

$$\Gamma_t(s,0) > 0 \qquad \text{and} \qquad \Gamma_t(s,1) < 0 \qquad \text{for} \qquad |s| < r$$

(*the lemma on the normal derivative for superelliptic functions*). Returning to the previous constants M, N, and α from (1), we introduce in this case the constants

$$C_- = -\frac{k+1}{\Gamma_t(0,0)} < 0 \qquad \text{and} \qquad C_+ = -\frac{k+1}{\Gamma_t(0,1)} > 0$$

(depending on $k = 0, 1$ and l/h and having equal absolute values for $k = 0$), and the dimensionless parameter

$$C = \Omega h/U, \qquad \Omega = \lambda h^k,$$

mentioned in the Introduction. By representing the solution $\widetilde{\psi}$ of problem (1) as the sum of functions $Nt + M\Gamma$, taking into account the equality $C = (k+1)M/N$ and the last inequality from (C_1), we find that

(C_4) $\widetilde{\psi}_s = M\Gamma_s$ for any C and $\widetilde{\psi} = Nt$ for $C = 0$,

(C_5) $\widetilde{\psi}_{tt} = C\dfrac{N}{k+1}\Gamma_{tt} < 0$ (> 0) for $C > 0$ (for $C < 0$),

(C_6) $\widetilde{\psi}_t(0,0) = N|\Gamma_t(0,0)|(C - C_-)/(k+1)$,

(C_7) $\widetilde{\psi}_t(0,1) = N|\Gamma_t(0,1)|(C_+ - C)/(k+1)$.

As a consequence of (C_2)–(C_4), the function $\widetilde{\psi}$ is regular ($\widetilde{\psi}_s^2 + \widetilde{\psi}_t^2 \not\equiv 0$) in V', excluding, possibly, the points of the straight line $s = 0$. By varying the parameter C, we come to the following flow portraits:

1. $C_- \leqslant C \leqslant C_+$. In this case, the properties (C_4)–(C_7) imply that $\widetilde{\psi}_t(0,t) > 0$ for $0 < t < 1$, hence $\widetilde{\psi}$ is regular everywhere in V'; the flow preserves the structure of a potential one (the case $C = 0$, a simple passing, Figure 9d, §3.3).

2. $C > C_+$. In this case we find from (C_5)–(C_7) that

$$\widetilde{\psi}_t(0,0) > 0, \qquad \widetilde{\psi}_t(0,1) < 0 \qquad \text{and} \qquad \widetilde{\psi}_{tt}(0,1) < 0$$

for $0 < t < 1$; $\widetilde{\psi}$ has a unique maximum $(0, t_C)$, $0 < t_C < 1$, corresponding to the left center and has no other critical points in V'. Consequently, the function $\widetilde{\psi}$ is admissible, and $p = 0$ and $q = 1$ for it. Moreover, $\widetilde{\psi}$ is superelliptic in V',

$$-(\widetilde{\psi}_{ss} + \widetilde{\psi}_{tt}) = -M(\Gamma_{ss} + \Gamma_{tt}) > 0,$$

and, thus (**[59]**), attains the minimum on the lower side of the rectangle V', where it is attained by the boundary trace of this function (*"unilateral" maximum principle for superelliptic functions*), i.e., for it $m = 1$. Simultaneously, on the upper side of V' the maximum of the boundary trace of $\widetilde{\psi}$ is attained, but the maximum of $\widetilde{\psi}$ is not attained (since $\widetilde{\psi}_t(0,1) < 0$). Hence, this side is a limit boundary continuum of the level $\widetilde{\psi} = N$ (in the sense of §3.1). Taking into account the equalities $\sigma = 0$ and $\chi = 1$ ($g = 0$, $v = 1$), we find from relations (3.1) that

$$s = q - p + m - \sigma - \chi = 1 - 0 + 1 - 0 - 1 = 1$$

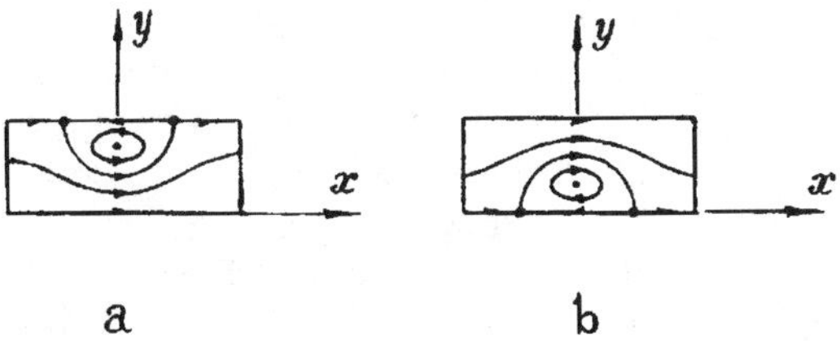

FIGURE 14. The flows from Examples 2 (a) and 3 (b).

and, hence, this continuum is a saddle continuum of multiplicity 1. Taking into consideration statement (i) of Theorem 3.2, we conclude that the structure of the family of level lines of the function $\widetilde{\psi}$ in V' is defined by the arc of the level $\psi = N$ emanating from the point $(-s_C, 1)$, $0 < s_C < 1$, and ending at the point $(s_C, 1)$ (we take into account the property (C_2)), which divides V' into a vortex with center $(0, t_C)$, adjoining the upper side of the rectangle V', and a simple passing flow (Figure 14a); $s_C \to 0$ and $t_C \to 1$ as $C \to C_+$.

3. $C < C_-$. In this case, from (C_5)–(C_7) we find that $\widetilde{\psi}_t(0,0) < 0$, $\widetilde{\psi}_t(0,1) > 0$, and $\widetilde{\psi}_{tt}(0,1) > 0$ for $0 < t < 1$. Replacing $\widetilde{\psi}$ with $-\widetilde{\psi}$, we obtain the flow portrait from the previous example. Consequently, up to a topological equivalence, the flow portrait in this case is obtained from the portrait of the previous one by a mirror reflection in the axis $t = 1/2$ (Figure 14b).

The axisymmetric flow is obtained by rotating the portrait of the plane flow around the axis $y = 0$. In this case, the physical interpretation of the portraits considered, obviously, consists in the statement that for sufficiently large values of the variable parameter C these portraits describe an axisymmetric vortex moving with constant velocity U along the axis of a pipe of circular cross-section (at the points on the central circle of the vortex), whose surface is homeomorphic to a *sphere* (resembling the spherical Hill vortex **[57, 140]**) if $C < C_-$, or homeomorphic to a *torus* if $C > C_+$, as shown in Figure 15.

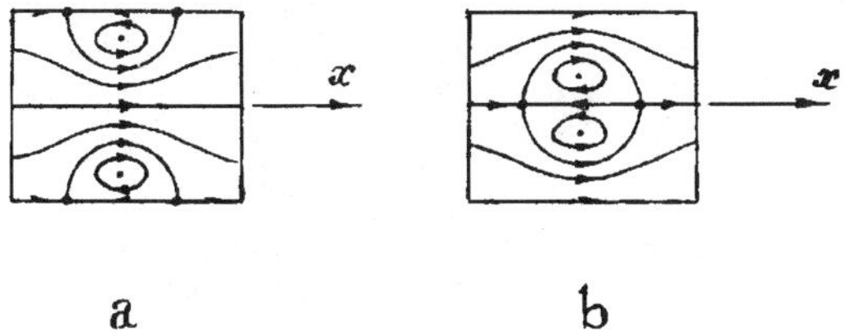

FIGURE 15. Portraits of axisymmetric flows for $C > C_+$(a) and $C < C_-$(b).

§6. Imposition of a vortex on an inverse flow

The problem of perturbation of the plane flow from Example 2 (§4) is connected with the well-known problem of circulation in a trench **[28, 55]**, which was analytically studied in **[34]** for the case of constant vorticity $\omega \equiv \lambda = \text{const}$. As in the case of a simple passing flow, this problem is characterized by the finiteness of the number of centers arising under (unbounded) variation of the parameter λ, the boundedness of the sets of corresponding "critical" values of λ, and the "softness" of the vortex appearance in the sense that at its inception the vortex has arbitrarily small dimensions and only subsequently (with the further increase or decrease of λ) it grows to size comparable with the flow domain. Another alternative possibility is the unbounded increase of the number of centers with the increase of the values of λ for an infinite set of critical values of this parameter and the "stiffness" of the appearance of the vortex characterized by the finiteness of its dimensions at its inception. This possibility is represented in the following example, where $\omega = \lambda y^k \psi$ and the variable parameter λ is the spectral parameter:

$$L_k \psi = \lambda \psi. \tag{1}$$

In the analysis of this example we restrict ourselves to the case of a plane flow with $k = 0$ and the operator $-L_0$ coinciding with the Laplace operator on the plane. Setting

$$U_-(y) = U \cos(\pi y/h), \qquad U_+(y) = 0, \qquad 0 \leqslant y \leqslant h, \qquad U = \text{const} > 0$$

in (4.6), for $\lambda = 0$ we obtain the inverse flow (in the class of potential flows) (Figure 16a).

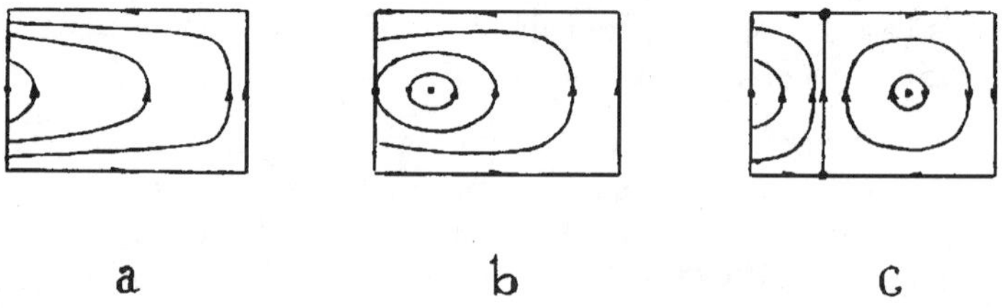

FIGURE 16. Flow portraits for $\Lambda \leqslant \pi/2$ (a), $\pi/2 < \Lambda < \pi$ (b), and $\pi < \Lambda \leqslant 3\pi/2$ (c).

Separating the variables in the boundary problem (1), (4.8) for $k = 0$ and for the admissible values of the parameter λ, or the corresponding dimensionless parameter $\Lambda = |\lambda - \pi^2/h^2|^{1/2} 2l$, which is to be varied in the sequel, we find

$$\psi(x, y) = \frac{Uh}{\pi} A_\Lambda(x) \sin(\pi y/h), \qquad |x| \leqslant l, \quad 0 \leqslant y \leqslant h,$$

$$A_\Lambda(x) = \begin{cases} \dfrac{\sinh[\Lambda(1 - x/l)/2]}{\sinh \Lambda}, & \lambda < \pi^2/h^2, \\[2ex] \dfrac{l - x}{2l}, & \lambda = \pi^2/h^2, \\[2ex] \dfrac{\sin[\Lambda(1 - x/l)/2]}{\sin \Lambda}, & \lambda > \pi^2/h^2, \end{cases}$$

$$\lambda \neq \lambda_n = (\pi/h)^2 + (\pi n/2l)^2 \quad (\Lambda \neq \pi n), \qquad n = 1, 2, \ldots.$$

Varying the parameter Λ within the indicated limits, and analyzing the number of zeros (corresponding to the common boundaries of the rectangular chambers) and the number of points of extremum (which correspond to centers) of the function $A_\Lambda(x)$, we obtain the following discrete series of flows with an indefinite set of critical values

$$\pi/2, \quad \pi, \quad 3\pi/2, \quad 2\pi \dots$$

of this parameter.

1. $\Lambda \leqslant \pi/2$. In this case, ψ is regular in the flow domain and the flow retains the structure of a potential flow (Figure 16a).

2. $\pi/2 < \Lambda < \pi$. As the parameter Λ passes through the first critical value $\pi/2$, there appears, at the point $(-l, h/2)$, a left vortex with center on the straight line $y = h/2$, whose boundary cycle is tangent to the side $x = -l$. In the complement to the vortex, the flow has the structure of a simple passing flow with a multiple level point $(-l, h/2)$ on the boundary. With a further increase of Λ, the vortex increases in volume filling up the whole flow domain for $\Lambda \to \pi - 0$ (Figure 16b).

3. $\pi < \lambda \leqslant 3\pi/2$. When Λ "jumps" through the first "resonance" value π, the flow portrait changes abruptly. The left vortex vanishes, making room for a right vortex with dimensions $(1 - \pi/\Lambda)l < x < l$, $0 < y < h$, which at the moment of inception fills up (for $\Lambda \to \pi + 0$) the half $x > 0$ of the flow domain with its center being located on the axis $y = h/2$ at the position $x = l/2$. As Λ increases, the left part of this vortex chamber, moving to the right, stops for $\Lambda = 3\pi/2$ at the position $x = l/3$. In the complement of the chamber, the flow has the structure of an inverse flow (Figure 16c).

4. $3\pi/2 < \Lambda < 2\pi$. As Λ passes through the fractional critical value $3\pi/2$, there again appears a left vortex evolving in the order described in Example 2. The right vortex is preserved, as it continues to contract.

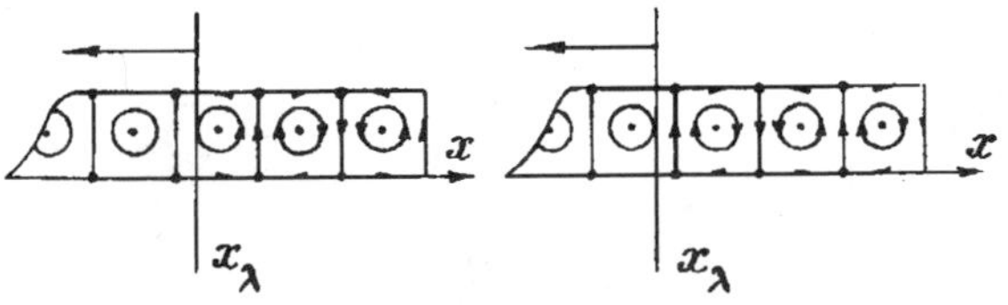

FIGURE 17. The mechanisms of appearance of vortices with the increase of Λ.

5. The sequence of structural rearrangements accompanying the subsequent increase of the values of Λ is a repetition of that described above. It is convenient to study this sequence by shifting to the left the ordinate $x = x_\Lambda$, $x_\Lambda = \text{const}$, which intersects the horizontal chain of intermitting left and right vortices (Figure 17). As Λ passes through the nonresonance (intermediate) value $\pi(n + 1/2)$, $n = 1, 2, \dots$, the vertical straight line $x = x_\Lambda$ intersects the center of the vortex, but the orientation of the flow lines does not change. The latter corresponds to the smooth transformation of the flow portrait (of a homotopic character) and illustrates the mechanism of "soft" rise of the vortex. At the passage of Λ through the resonance (eigen)value πn, the straight line $x = x_\Lambda$ intersects the boundary of the neighboring chambers. Simultaneously, the orientation of the level lines changes for the opposite. The latter corresponds to the transformation of the flow portrait, which is of a jump character (violation of homotopy) and is an illustration of "stiff" rise of the vortex. The total number of chambers becomes infinite as $\Lambda \to \infty$.

An analogous discrete series of flows can be obtained by separating the variables also in the case $k = 1$. The given analysis leads to formal identification of the Euler flow in a bounded domain with a mechanical system that has a countable set of "resonance frequencies" corresponding to the eigenvalues of equation (1) considered in the flow domain under homogeneous boundary conditions for ψ ($\psi = 0$ on the boundary of the domain under consideration) and their "eigenoscillations" corresponding to the eigenfunctions of the indicated homogeneous boundary problem. When the corresponding dimensionless parameter Λ happens to be located in a neighborhood of the appropriate points of the spectrum, the continuous dependence of the solution of the initial problem on Λ is violated. This violation takes place in the form of the above-mentioned "stiff" rise of the vortex. In the example considered above, the number of the arising centers increases with Λ. Whether this dependence is preserved for the case of a general (bounded) flow domain remains unknown. There are examples (cf. **[52]**) in which the number of centers does not increase but the portrait of the plane flow becomes more and more complicated.

CHAPTER 3

A Two-Dimensional Passing Flow Problem for Stationary Euler Equations

§7. Uniqueness of an analytic solution

7.1. Formulation of the problem. We pass to the analysis of solvability of system (4.1) for $k = 0, 1$ in a bounded domain V of the upper half-plane $y \geqslant 0$ with the boundary ∂V consisting of a finite number of sufficiently smooth pieces (e.g., of the class C^3) of arcs intersecting at nonzero angles, some of which may be located on the axis $y = 0$ (where, for $k = 1$, the uniform ellipticity of the operator L_k, generated by system (4.1) and defined as in (4.3), is violated) and also, possibly, consisting of a finite number of isolated cycles (Jordan closed curves) not intersecting the axis $y = 0$, of the same class of smoothness as the arcs (Figure 18).

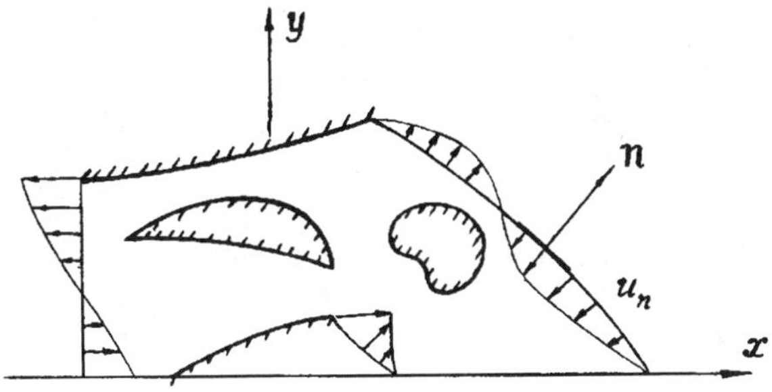

FIGURE 18. The flow domain under consideration.

We assume that the component u_n of the velocity $\mathbf{u} = (u, v)$ in the direction of the outward unit normal n to ∂V at *regular* (nonangular) points of the boundary ∂V is fixed,

$$u_n = \alpha. \tag{1}$$

The fixed real function $\alpha = \alpha(P)$ is defined and continuous at all points $P = (x, y)$ of the boundary ∂V, possibly excluding angular points, and for $k = 1$ vanishes on the axis $y = 0$, i.e., $\alpha(x, 0) = 0$ for $k = 1$ (there are no sources and sinks on the axis of symmetry). We assume that the boundary ∂V is oriented so that each interior point of the domain V that is sufficiently close to the boundary ∂V remains, for the chosen bypass direction of ∂V, to the left of the boundary. We additionally assume the existence of a flow $\mathrm{u} = (a, b)$, defined everywhere in the closure of the domain V, possibly excluding the angular points of the boundary ∂V, that is smooth in V and satisfies system (1) in V for $\omega \equiv 0$ (i.e., is potential in V) under the same boundary conditions as for $\mathbf{u}$,

$$\mathrm{u}_n = u_n = \alpha$$

obeys the requirement of square summability of the following form:

$$\int_V (a^2 + b^2) y^k \, dx \, dy \leqslant \infty,$$

and has a continuous stream function $\Phi = \Phi(P)$ defined in the closure of the domain V.

Let us make some remarks. Formula (1.3) or formula (4.2) for the stream function directly imply that the existence of Φ follows from the fact that α vanishes on each inner cycle J of the boundary ∂V, and the uniqueness of u is obtained additionally fixing the constant value of Φ on J (the consequence of the maximum principle). The existence of the required potential flow u can be established by classical means **[54]**. The corresponding (quite obvious) alternative considerations for flows in a rectangle and a cylinder will be given below. Recall **[55]** that for $k = 0$ the velocity u is infinite at the vertex of an angle of the boundary ∂V directed inside the flow domain V (forming a protruding edge) and vanishes at the vertex of an angle of ∂V directed to the exterior of this domain (forming a cavity). Under the condition that the boundary ∂V has no inner cycles, the existence of Φ is provided by the simple connectedness of the corresponding domain V. For the case when V is not simply connected, the fact that the above-mentioned criterion for the existence of Φ is connected with the integral representation (1.3), is analyzed in detail in the book **[40,** §19.3]. Note that Φ is analytic in V **[54]**. Also note that the boundary conditions (1) correspond to the stationary passing flow problem for an ideal fluid through a given domain V (for $\alpha \not\equiv 0$), which is traditional and well posed for the first pair of equations of system (4.1) for the vorticity $\omega = \omega(x, y)$ **[54]**. By adding the third relation from (4.1) to this pair, we make the problem nonlinear and, simultaneously, come to the problem of obtaining a boundary condition for ω complementing (1), which would distinguish a unique solution. As mentioned in the Introduction, this additional restriction for a nonstationary plane flow is that the vorticity on the influx regions be fixed **[109]**. In the stationary case, the flow is invariant with respect to a change of sign of the time variable, and the values of the vorticity ω can be fixed both on the *influx* and *efflux* regions, i.e. where $\alpha < 0$ and $\alpha > 0$, respectively.

Taking into account the last remark, we complement problem (1) by fixing the values $\beta(x, y)$ of the function $\omega(x, y)/y^k$ on the open arc g of the smooth regions of influx and efflux,

(2) $\qquad \omega(x, y) = y^k \beta(x, y) \quad$ on the smooth boundary arc g where $\alpha \neq 0$.

Note that for $k = 1$ the arc g, *a fortiori*, has no common points with the axis $y = 0$ (since $\alpha(x, 0) = 0$). The function β given on g is assumed to be smooth. Fixing the stream function Φ of the given potential flow u, we introduce the length s of the arc g calculated in the chosen bypass direction of the boundary ∂V and the smooth function $\zeta = \zeta(P)$ defined by the restriction of Φ to g,

$$\zeta = \Phi|_g$$

(the smoothness of ζ is a consequence of the continuity of α). We have

$$y^k \mathrm{u}_n(x, y) = \frac{d\zeta}{ds}(x, y) = y^k \alpha(x, y) \neq 0 \quad \text{on } g;$$

thus, on the interval $\zeta(g)$ defined by the values of the smooth function ζ on the arc g, the smooth inverse function

$$\zeta^{-1} = \zeta^{-1}(t), \qquad \zeta^{-1}(\zeta(P)) \equiv P$$

is defined. Introducing on $\zeta(g)$ the smooth real function

$$\mu_g(t) \equiv \beta[\zeta^{-1}(t)]$$

(as the superposition of the mapping ζ^{-1} of the interval $\zeta(g)$ onto the arc g and the real function β given on g), we require that it admit a continuation to the whole real axis R, and the extended function $\mu = \mu(t)$ defined on R and coinciding with μ_g on g be smooth and satisfy the following *spectral condition*:

The values of the derivative $\mu'(t)$ of the function $\mu(t)$ must not exceed the limits of an interval of the real axis located at a positive distance ε from the set of eigenvalues of equation (6.1), *considered in the domain V under homogeneous boundary conditions* ($\psi = 0$ *on* ∂V).

This restriction is justified by the fact that a stationary Euler flow in a bounded domain has "resonance" frequencies mentioned in §6. The existence of the required complete set of eigenfunctions of the spectral problem under consideration, as well as the countability of the set of its eigenvalues, and the absence of finite limit points in it will be proved below. We pass to the formulation of the main statements concerning system (4.1), considered under these restrictions.

The boundary problem (1), (2) for the system of equations (4.1) will be called *preliminary*. By a *smooth solution* of the preliminary problem we mean a triple of continuous functions u, v, ω, defined on the closure of the domain V, possibly excluding its angular boundary points, and satisfying in V the system of equations (4.1) and the conditions (1) and (2) on the boundary of V, subject to the square integrability of the form

$$\int_V (u^2 + v^2 + (\omega/y^k)^2) y^k \, dx \, dy < \infty. \tag{3}$$

A smooth solution of u, v, ω is *infinitely smooth* (*analytic*), if the functions u, v, ω are infinitely differentiable (analytic) in the domain V.

PROPOSITION 1. *A flow $\mathbf{u} = (u, v)$, defined by any smooth solution u, v, ω of the preliminary problem, has a stream function ψ defined and smooth everywhere in the closure of the domain V, possibly excluding angular boundary points.*

The validity of this proposition is intuitively sufficiently obvious. With accuracy up to an arbitrary additive constant, the required ψ is given by the equality

$$\psi = \Phi + \varphi,$$

where Φ is the stream function of the potential flow $\mathfrak{u}$, whose existence is assumed beforehand, and φ is some classical solution of the Dirichlet problem

$$L_k \varphi = \omega(x, y)/y^k \quad \text{in } V, \qquad \varphi = C \quad \text{on } \partial V. \tag{4}$$

The existence and uniqueness of the solution φ for the indicated right-hand part of equation (4) are established in §9.

THEOREM 1. *The preliminary problem has at least one smooth solution u, v, ω infinitely smooth (analytic) for an infinitely smooth (analytic) function μ and such that the flow $\mathbf{u} = (u, v)$ generated by it is a μ-field in the sense that everywhere in V the following identity is satisfied:*

$$\omega(x,y) = y^k \mu[\psi(x,y)]. \tag{5}$$

We restrict ourselves to the exposition of the general scheme of the proof of this theorem. The necessary details will be given in §9. The required solution is given by equalities (4.2) and (5), and ψ is defined as the solution of the boundary problem of the following form:

$$L_k\psi = \mu(\psi) \quad \text{in } V, \qquad \psi = \Phi \quad \text{on } \partial V. \tag{6}$$

The spectral restriction on μ ensures the existence and uniqueness of the solution ψ with the properties given in the theorem. General theorems concerning weakly nonlinear elliptic equations of the type (6) connected with these restrictions are given in **[96, 98, 99, 131, 132, 151, 153]**, and in §9.

Note that, at least for the analytic continuation μ of the (analytic) function μ_g, the right-hand side of equality (6) does not depend on the additive constant C up to which the stream function ψ is defined. Indeed, the substitution of $\psi + C$ for ψ is equivalent to the substitution of $\Phi + C$ for Φ in the equality $\psi = \Phi + \varphi$ and corresponds to the change of the function $\zeta|_g \equiv \Phi$ to $\zeta + C$ in the arguments given above concerning the construction of the function $\mu_g(t)$. The latter substitution, obviously, does not influence the values of the corresponding inverse function, $\zeta^{-1}(t + C) = \zeta^{-1}(t)$. Consequently,

$$\mu_g(t + C) \equiv \beta[\zeta^{-1}(t + C)] = \beta[\zeta^{-1}(t)] = \mu_g(t).$$

The uniqueness of the analytic continuation $\mu(t)$ of the function $\mu_g(t)$ completes the proof.

The subsequent constructions of this and the following sections are dedicated to the clarification of additional conditions distinguishing a unique solution of problem A.

7.2. The uniqueness theorem. The preliminary problem will be called *problem A* if the functions μ_g and μ defined above are analytic. Simultaneously, the corresponding restriction on μ_g is the requirement that its analytic continuation μ to the whole real axis exist and satisfy the spectral condition. Due to the uniqueness of this continuation, the function μ in problem A is uniquely defined by the right-hand sides given in conditions (1) and (2). Moreover, the dimensions of the region g where the vorticity ω is prescribed are not essential.

PROPOSITION 2. *All analytic solutions of problem A are μ-fields with the same function μ.*

Indeed, ψ is regular on the arc g,

$$\psi_x^2 + \psi_y^2 = y^{2k}(u^2 + v^2) \geqslant y^{2k}\alpha^2 > 0 \quad \text{on } g$$

(identities (4.2) are used), and is continuously differentiable in some relative neighborhood of this arc, including the points of the latter (Theorem 1). As a result of the implicit function theorem, in some curvilinear quadrangle G adjoining g (as shown in Figure 19), the flow defined by ψ has the structure of a simple passing flow, while all the level arcs of the function ψ are smooth and emanate from (or end on) the side of G belonging to g. The domain G is necessarily simple with respect to ψ (in the sense of §2.4).

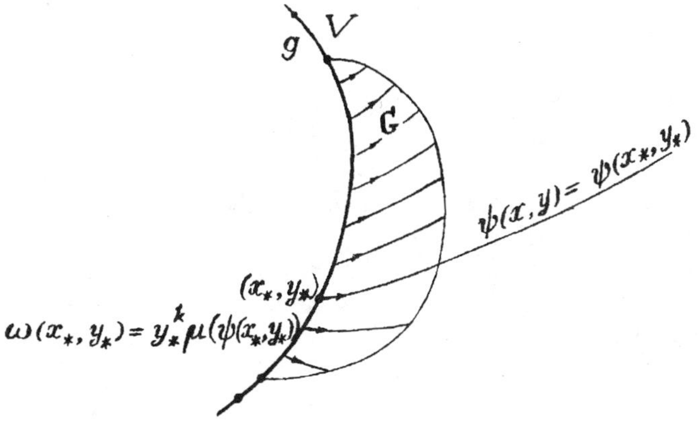

FIGURE 19. A neighborhood of the region of prescribed vorticity.

The third relation from (4.1) implies that the identity

$$\frac{\partial\psi}{\partial x}\frac{\partial}{\partial y}\left(\frac{\omega}{y^k}\right)-\frac{\partial\psi}{\partial y}\frac{\partial}{\partial x}\left(\frac{\omega}{y^k}\right)=0 \tag{7}$$

is valid everywhere in V. Consequently (Theorem 2.1), for some real function $\eta(t)$, the dependence

$$\omega(x,y)=y^k\eta[\psi(x,y)]$$

is defined everywhere in G. On the other hand, as it was already mentioned, every point (x,y) of the domain G belongs to the smooth arc γ of some level $\psi=C$ lying on G, with one of its ends being a point (x_*,y_*) of the arc g. Due to (7), the function ω/y^k is constant on the arc γ; thus,

$$\eta[\psi(x,y)]=\omega(x,y)/y^k=\omega(x_*,y_*)/y_*^k=\beta(x_*,y_*)=\mu[\psi(x_*,y_*)]=\mu[\psi(x,y)],$$

i.e., identity (5) holds everywhere in G. The proof of Proposition 2 is completed by Theorem 2.2 from §2.5.

THEOREM 2. *An analytic solution of problem A is unique.*

Indeed, let u_1, v_1, ω_1 and u_2, v_2, ω_2 be analytic solutions of problem A and let ψ_1 and ψ_2 be the stream functions associated to them and defined as in Proposition 1. For a fixed normal component (1), the boundary values of the functions ψ_1 and ψ_2 differ by a constant. Adding the latter to one of them, we obtain a new stream function for the previous solution of the problem A. Thus, without loss of generality, we assume that ψ_1 and ψ_2 coincide on the boundary. In this case, these functions satisfy the conditions of problem (6) with the same Φ and the same $\mu(t)$ (Proposition 2). Since the solution of problem (6), as it was already mentioned, is unique, $\psi_1=\psi_2$ everywhere in V, which immediately implies the coincidence of our analytic solutions.

§8. Analysis of the uniqueness of a smooth flow

8.1. The choice of the region of prescribed vorticity. In order to clarify the uniqueness of an analytic solution of problem A in the class of smooth (possibly, nonanalytic) solutions, we refine the requirements on the region g of the prescribed vorticity ω in condition (7.2). Let us add to the conditions of the preliminary problem (§7.1) the requirement that this region be the union of all smooth arcs g' of the boundary ∂V, where $\alpha < 0$, or the union of all arcs of the boundary with $\alpha > 0$, i.e., *g is the complete influx or efflux region.* In this case, on each interval $\Phi(g')$, the smooth function $\mu_{g'}(t)$ giving on g' the dependence $\beta(x, y) = \mu_{g'}[\Phi(x, y)]$, uniquely obtained by means of the functions α and β in conditions (7.1) and (7.2), is again defined. However, now the collection of all indicated functions $\mu_{g'}(t)$ generally speaking, does not define the required unique dependence between β and Φ on the complete region g. Therefore we require that conditions (7.1) and (7.2) be *consistent* in the sense that the graphs of all the functions $\mu_{g'}(t)$ lie on the graph of some smooth function $\mu(t)$, defined on the whole real axis $-\infty < t < \infty$ and satisfying the spectral condition formulated in §7.1. The preliminary problem considered under these conditions will be called *problem* B. An additional requirement of the analyticity of the functions $\mu_{g'}(t)$ and $\mu(t)$ (the existence for the functions $\mu_{g'}(t)$ of a common analytic continuation $\mu(t)$ to the real axis satisfying the spectral condition) will define *problem* C.

Note that problem C can be simultaneously considered as being obtained by the restriction of the conditions of problem A from the previous section to the conditions of problem B. As in problem A, the analytic function $\mu(t)$ in problem C is uniquely defined by the functions α and β given in conditions (7.1) and (7.2). In problem B, this function is nonanalytic and, thus, it is uniquely defined only on the range of values of the function Φ taken on the closure of the arc g (i.e., it admits an arbitrarily large number of smooth nonanalytic continuations to the exterior of this set). Problem C is introduced into consideration as a particular case of problem B in order to distinguish a unique analytic solution of problem B. The main goal of the subsequent constructions is to elucidate the conditions necessary for the smooth solution of the latter to be unique.

8.2. Uniqueness of an open flow. For a smooth (infinitely smooth) solution of problem B and an analytic solution of problem C, the definitions from the previous section formulated for the preliminary problem still hold. Problems B and C are obtained by restricting the conditions of the preliminary problem. Therefore, Proposition 1 and Theorem 1 from the previous section are valid for them. For problem C, which is obtained by restricting the conditions of problem A, Proposition 2 and Theorem 2 (§7) hold. In particular, an analytic solution of problem C is unique. The uniqueness of this analytic solution in the class of smooth solutions of the corresponding problem B is of essential interest to us.

We introduce a special class of vector fields $\mathbf{u} = (u, v)$ generated in the domain V by smooth solutions of problem B (or by analytic solutions of problem C) and *open* in the following sense: the set of points $(\overline{x}, \overline{y})$ of the domain V such that through each point $(\overline{x}, \overline{y})$ passes a smooth streamline (arc) γ of this flow $\mathbf{u}$ emanating from an interior point of the influx region and ending at an interior point of the efflux region is dense in V (each point of the domain V is the limit of a sequence of such points). Among the flows with isolated singular points considered in the previous chapters, those without centers of vortices are open and those complementing them are, obviously, not. In particular, the flows from §§5, 6 with $C_- \leqslant C \leqslant C_+$ and $\Lambda \leqslant \pi/2$ (with $C < C_-$

(or $C > C_+$) and $\Lambda > \pi/2$) are open (are not open), respectively. In what follows the stream function ψ of the vector field $\mathbf{u}$ is also assumed to be *normalized* by the condition $\psi = \Phi$ on ∂V ($\psi = \Phi$ on the smooth parts of the boundary ∂V, i.e., everywhere on ∂V, possibly excluding the angular points where, generally speaking, ψ is not defined according to Proposition 7.1; however, Φ is assumed to be continuous on ∂V and the trace of ψ on ∂V can be extended by continuity to the angular points of the boundary).

PROPOSITION 1. *The normalized stream function ψ and the vorticity ω of any open flow* $\mathbf{u}$ *are connected in V by relation* (7.5) (*an open flow is a μ-field*).

Indeed, the function $\omega(x, y)/y^k$ is constant on each indicated arc γ passing in V through the corresponding point $(\overline{x}, \overline{y})$ and starting or ending at an interior point (x_*, y_*) of the interval g where the values $\beta(x_*, y_*)$ of this function are fixed. Since γ is simultaneously a level line of the function ψ, and (according to the normalization) $\psi(x_*, y_*) = \Phi(x_*, y_*)$, we have

$$\omega(\overline{x}, \overline{y})/\overline{y}^k = \omega(x_*, y_*)/y_*^k = \beta(x_*, y_*) = \mu[\Phi(x_*, y_*)] = \mu[\psi(x_*, y_*)] = \mu[\psi(\overline{x}, \overline{y})].$$

The density in V of the points $(\overline{x}, \overline{y})$ at which this identity holds, and the continuity of the functions $\omega(x, y)/y^k$ and $\mu[\psi(x, y)]$, complete the proof.

PROPOSITION 2. *The stream function ψ of an open flow* $\mathbf{u}$ *satisfies the maximum principle*:

$$\min_{\partial V} \psi \leqslant \psi(P) \leqslant \max_{\partial V} \psi \quad \textit{for all } P \textit{ from } V. \tag{1}$$

The validity of this proposition follows from the equality $\psi(x_*, y_*) = \psi(\overline{x}, \overline{y})$ (the notation of previous considerations is used), which guarantees the validity of (1) on the dense set of points $(\overline{x}, \overline{y})$ in V, and from the continuity of ψ.

THEOREM 1. *A smooth solution of problem B is unique in the class of open flows.*

Indeed, as a consequence of Proposition 1, the normalized stream functions ψ_1 and ψ_2 of the smooth solutions u_1, v_1, ω_1 and u_2, v_2, ω_2 of problem B are solutions of the boundary problem (7.6) with some right-hand sides $\mu = \mu_1(t)$ and $\mu = \mu_2(t)$ respectively, satisfying the spectral condition from §7.1 and obtained by smooth continuations to the real axis $-\infty < t < \infty$ of the smooth function $\mu_g(t)$, uniquely defined on the set of values of the function Φ on the closure of a complete interval g of monotonic increase or decrease of Φ on ∂V. Due to the continuity of Φ, this set coincides with the interval

$$\min_{\partial V} \Phi \leqslant t \leqslant \max_{\partial V} \Phi$$

of values of Φ on ∂V. Due to Proposition 2, the values of the functions ψ_1 and ψ_2 on V lie inside this interval, so that ψ_2 is simultaneously a solution of problem (7.6) with $\mu = \mu_1$. As it was already mentioned (and will be established in §9), the spectral restriction imposed on μ_1 implies that the required (classical) solution of the corresponding problem (7.6) is unique. Consequently, $\psi_1 \equiv \psi_2$, which implies the coincidence of the indicated solutions of problem B.

Let us make a few remarks concerning the *existence* of an open flow. The existence of a flow of the simple passing type (example from §4) is ensured by the requirement that the function β be sufficiently small for a fixed function α **[4]**. An example of globalization of this local condition is the restriction $C_- \leqslant C \leqslant C_+$ from §5, which is valid for all the open flows in the case under consideration. We note also that a local statement analogous to the indicated result from **[4]** can be easily obtained by also using the arguments of this paper for the case of an inverse flow, including the axisymmetric Euler fields ($k = 1$).

8.3. Nonuniqueness of a developed flow. Let us consider other classes of Euler fields $\mathbf{u} = (u, v)$ defined by smooth solutions of problem B. The flow $\mathbf{u}$ will be called *weak* (*developed*) if, for its stream function ψ, the maximum principle (1) is valid (violated). The physical interpretation of a weak flow consists in the fact that the discharge of the fluid being in motion defined by the function ψ is limited, i.e., the discharge must not exceed that measured on the boundary of the flow domain. On the contrary, in the domain of a developed flow there must necessarily exist a point P_* of a local extremum of the function ψ. The isolated point P_* coincides with a vortex center (Theorem 3.1). The fact that the point P_* is isolated is provided, as it was already mentioned (§1.1), by P_* being nondegenerate, which is a characteristic feature of the flow. Thus, the existence of a nondegenerate vortex center in the domain of a developed flow is also a characteristic feature, e.g., the developed flows from §§5 and 6 possess this property. Hence, an open (consequently, weak) flow from §5 becomes a developed one with a nondegenerate center P_* (which can easily be proved with the help of Proposition 5.1, straightforwardly verifying the validity of the inequality

$$-\psi_{xy}^2(P_*) + \psi_{xx}(P_*)\psi_{yy}(P_*) > 0)$$

during the passage of the varying parameter C through the critical value C_- or C_+. Similarly, the flow from §6 is rearranged when the varying parameter Λ passes through the critical value $\pi/2$.

THEOREM 2. *Let u_0, v_0, ω_0 be a smooth (arbitrarily smooth) solution of problem B or an analytic solution of problem C (hence, of problem A) from Theorem 7.1, which is a μ_0-field in the domain V, i.e.,*

$$\omega_0(x, y) = y^k \mu_0[\psi_0(x, y)],$$

with a smooth (infinitely smooth) or analytic function $\mu_0 = \mu_0(t)$, $-\infty < t < \infty$, satisfying the spectral condition and having the normalized stream function ψ_0 (such that $\psi_0 = \Phi$ on ∂V). If the flow $\mathbf{u}_0 = (u_0, v_0)$ generated by this solution is developed, i.e., in the domain V there exists a point P_* where one of the inequalities of system (1) is violated, then, together with u_0, v_0, ω_0, problem B (the boundary problem (7.1), (7.2) for the system of equations (4.1) in the domain V considered under the additional restrictions from §§7.1, 8.1) has a family of smooth solutions $u_\sigma, v_\sigma, \omega_\sigma$, $0 \leqslant \sigma \leqslant \sigma_*$, $\sigma_* = \text{const} > 0$, which are infinitely smooth for an infinitely smooth or analytic function μ_0 and pairwise different in the following sense:

$$(2) \quad \int_V \left[(u_\sigma - u_{\sigma'})^2 + (v_\sigma - v_{\sigma'})^2\right] y^k \, dx \, dy \neq 0 \quad \text{for } \sigma \neq \sigma',\ 0 \leqslant \sigma, \sigma' \leqslant \sigma_*.$$

Let us sketch a general scheme of the proof of the formulated theorem. The complete proof will be given in §9.2. Setting

$$m_0 = \min_{\partial V} \psi, \qquad M_0 = \max_{\partial V} \psi,$$

we introduce the infinitely smooth nonanalytic function

$$f(t) = \begin{cases} \exp \dfrac{1}{M_0 - t}, & \text{if} \quad t > M_0, \\ 0, & \text{if} \quad m_0 \leqslant t \leqslant M_0, \\ \exp \dfrac{1}{t - m_0}, & \text{if} \quad t < m_0. \end{cases}$$

The required family of solutions will be defined as the family generated by the solutions ψ_σ of problem (7.6) with

$$\mu(t) = \mu_\sigma(t) \equiv \mu_0(t) + \sigma f(t) \tag{3}$$

(by μ_σ-fields with normalized stream functions ψ_σ). The upper boundary $\sigma_* > 0$ of the corresponding interval $0 \leqslant \sigma \leqslant \sigma_*$ is determined by the presence of a point P_* for ψ_0. In the absence of this point, the values acquired by the function ψ in the domain V do not leave the interval $m_0 \leqslant t \leqslant M_0$, where $\mu_\sigma(t) = \mu(t)$ (equalities $f(t) = 0$ and (3)). As it was shown while proving Theorem 1, in this case the corresponding solutions of problem (7.6) coincide, $\psi_\sigma = \psi_0$ (since they satisfy the maximum principle (1)). If there is a point P_* in the flow domain (which, without loss of generality, can be, obviously, identified with the point of a local extremum of the function ψ_0), then the values of ψ_0 can leave the indicated interval. It follows from $f(t) = 0$ and (3) that for the corresponding $t > M_0$ or $t < m_0$, $\mu_\sigma(t) \neq \mu_{\sigma'}(t)$ for $\sigma \neq \sigma'$. The requirement from §9.2 that the constants σ and σ' are small (the derivatives of the functions $\mu_\sigma(t)$ and $\mu_{\sigma'}(t)$ are sufficiently close to the derivative of the function $\mu_0(t)$ uniformly in all $-\infty < t < \infty$ (which justifies the choice of the function $f(t)$) and, consequently, also satisfy the spectral condition), provides the existence and uniqueness of the corresponding solutions ψ_σ and $\psi_{\sigma'}$ of problem (7.6), and a certain continuous dependence of ψ_σ on the varying (within small limits) parameter σ. It will be shown, in particular, that the values of the function ψ_σ in V leave the indicated interval. The inequalities $\mu_\sigma(t) \neq \mu_{\sigma'}(t)$ (for $\sigma \neq \sigma'$) guarantee that inequalities (2) are valid.

Let us verify that the conditions of problem B for the solutions $u_\sigma, v_\sigma, \omega_\sigma$ are satisfied. The boundary condition from (7.6) for $\mu = \mu_\sigma$ together with the smoothness of the corresponding solution $\psi = \psi_\sigma$ in V up to the smooth parts of the boundary ∂V (established in §9.2) guarantee that the boundary condition (7.1), i.e., $u_n = \mathrm{u}_n = \alpha$, is satisfied for $\mathbf{u} = (u_\sigma, v_\sigma)$. The validity of the equation from (7.6) for the stream function ψ_σ guarantees, as it was already mentioned, that the equations from system (4.1) and the identities $\omega_\sigma(x, y) = y^k \mu_\sigma[\psi_\sigma(x, y)]$ are satisfied for the corresponding $u = u_\sigma$, $v = v_\sigma$, and $\omega = \omega_\sigma$ in V. The latter implies the second boundary condition (7.2).

Indeed, for any sequence of points (x_i, y_i), $i = 1, \ldots,$ of the domain V converging for $i \to \infty$ to an arbitrary fixed interior point (x_*, y_*) of the interval g (necessarily belonging to the smooth part of the boundary ∂V) we have

$$\begin{aligned} \omega_\sigma(x_i, y_i) = y_i^k \mu_s[\psi_\sigma(x_i, y_i)] \to y_*^k \mu_\sigma[\psi_\sigma(x_*, y_*)] &= y_*^k \mu_\sigma[\Phi(x_*, y_*)] \\ &= y_*^k \mu_\sigma[\psi_0(x_*, y_*)]; \end{aligned}$$

and, simultaneously,

$$y_*^k \mu_\sigma[\psi_0(x_*, y_*)] = y_*^k \mu_0[\psi_0(x_*, y_*)] = y_*^k \mu_0[\Phi(x_*, y_*)] = \beta(x_*, y_*)$$

(since for the boundary point (x_*, y_*) we have $m_0 \leqslant \psi_0(x_*, y_*) \leqslant M_0$ and $f[\psi_0(x_*, y_*)] = 0$). The continuity of ω_σ in V up to the smooth parts of the boundary of ∂V completes the proof.

The square integrability (7.3) of the solutions of this family is established in §9.1.

Note that the smooth solutions of problem B with $\sigma > 0$ described in Theorem 2 are, obviously, nonanalytic (a consequence of Theorem 7.2).

Let us make a few remarks concerning the "physical" contents of the above theorems on uniqueness of a solution of the system of equations (4.1) in a bounded domain. Theorem 1 is intuitively sufficiently evident, i.e., according to the third equation of system (4.1), the function $\omega(x, y)/y^k$ is preserved along a smooth streamline and, if the flow domain V does not contain stagnation vortex zones filled out by closed trajectories, then all the information with respect to this function stipulated on the complete influx (or efflux) interval is transferred by these lines to all (or to almost all) the points of the domain V; the remaining two equations of this system are considered in V for a given right-hand side of the second one. On the contrary, the existence of stagnation zones in the flow domain makes it impossible to obtain the information with respect to the given right-hand side with the help of "natural" means (using the third relation from system (4.1)), which implies that the corresponding stationary flow is nonunique (Theorem 2). The restriction of the solutions to holomorphic flows, due to the uniqueness of an analytic continuation, also makes it possible to transfer the information "across" the streamlines. In this case it penetrates inside a stagnation zone along the transversal trajectories, and this guarantees that the corresponding analytic solution is unique (Theorem 7.2).

8.4. Problems on simple passing and inverse flows. We shall illustrate the statements of the above theorems on uniqueness by means of the analysis of holomorphic flows from §§5, 6. First, we show the existence of the required potential flow u from §7.1 for these cases. The flow domain is again the rectangle V from §4, on whose boundary the normal velocity component is fixed in accordance with conditions (4.6), where the smooth functions $U_\mp(y)$ given on the interval $0 \leqslant y \leqslant h$ obey the restriction (4.7). As already mentioned in §4, for $k = 1$ it is additionally required that the derivatives of the functions $U_\mp(y)$ divided by y^k be smooth up to the left end of the indicated interval (which is equivalent to the requirement of twice continuous differentiability of these dependencies as functions of the argument y^{k+1} for $0 \leqslant y \leqslant h$). We shall assume the flow $\mathrm{u} = (a, b)$ to be defined in the domain V by a smooth function $\Phi = \Phi(x, y)$,

$$a(x,y) = \frac{1}{y^k}\frac{\partial \Phi}{\partial y}, \qquad b(x,y) = -\frac{1}{y^k}\frac{\partial \Phi}{\partial x},$$

which admits an extension by continuity together with the functions $a(x, y)$, $b(x, y)$ to the closure of the domain V. The possibility of this extension is provided by the existence of the required properties for the corresponding solution $\psi = \Phi$ of the boundary problem (4.3), (4.8). The existence of a function φ with the indicated properties that satisfies the conditions of the homogeneous boundary problem (7.4) with right-hand side $\omega(x, y)/y^k$, a smooth function in the closure of the domain V, is established in §9.1. This requirement is satisfied, e.g., by a right-hand side with

$$\frac{\omega(x,y)}{y^k} = \frac{l+x}{2l}\frac{1}{y^k}\frac{dU_+}{dy} + \frac{l-x}{2l}\frac{1}{y^k}\frac{dU_-}{dy}$$

(if the restrictions for $U_{\mp}(y)$ are satisfied). The required Φ is connected in this case with φ by the following identity:

$$\Phi(x, y) = \varphi(x, y) + \frac{l+x}{2l}\Phi(l, y) + \frac{l-x}{2l}\Phi(-l, y)$$

(the functions $\Phi(\pm l, y)$ are calculated for $\psi = \Phi$ according to formulas (4.8)); hence it has the same smoothness properties as the function φ. Thus, the existence of the required potential flow u is established. Now, we shall clarify the conditions for the uniqueness of the flows from §§5, 6.

The condition fixing any analytic flow from §5, in accordance with Theorem 7.2, is that the vorticity $\omega = \lambda y^k$ be given on any (arbitrarily small) interval of one of the vertical sides of the rectangle V (and not everywhere in the closure of the given domain, as it was assumed in §5). The corresponding analytic function $\mu(t) \equiv \lambda$ is constant on the axis $-\infty < t < \infty$ and satisfies the required spectral condition since the derivative $\mu'(t) \equiv 0$ and the corresponding points of the spectrum of the homogeneous problem for equation (6.1), as it is proved in §10.4, lie to the right of 0. Due to the conditions of problem B, in order that this analytic solution be unique in the class of smooth (possibly nonanalytic) flows, it is necessary that ω be defined at all interior points of one of the mentioned sides (i.e., on the whole influx or efflux region). However, for large values of the corresponding dimensionless parameter C (§5), there appears a vortex center in the domain of an analytic flow; the flow becomes developed (the maximum principle (1) is violated for it when $C < C_-$ or $C > C_+$, where the corresponding critical values $C_{\mp}$ are defined as in §5). By Theorem 2, the latter implies that there simultaneously exists an infinite number of arbitrarily smooth nonanalytic solutions of this boundary problem. For small C (for all $C_- \leqslant C \leqslant C_+$), the analytic flow under consideration is open, hence it is unique in the class of smooth open flows satisfying the conditions of this problem.

By passing to the dimensionless variables according to the formulas

$$(x, y, u, v, \omega) \to (hx, hy, Uu, Uv, U\omega/h),$$

we come to the same equations (4.1) considered under the corresponding "dimensionless" boundary conditions. For the case when ω is prescribed on the influx region, the latter will acquire the following form:

$$\text{(4)} \qquad \begin{gathered} u(\mp\tau, y) = 1, \qquad v(x, 0) = v(x, 1) = 0, \qquad \omega(-\tau, y) = Cy^k, \\ |x| < \tau = l/h, \qquad 0 < y < 1. \end{gathered}$$

The previous considerations lead to the following statement.

PROPOSITION 3. *The boundary problem* (4) *for the system of equations* (4.1) *in the rectangle* $|x| < \tau$, $0 < y < 1$, *has a unique analytic solution for any values of the varying parameter* $-\infty < C < \infty$ *and an uncountable set of arbitrarily smooth nonanalytic solutions for* $C < C_-$ *or* $C > C_+$. *For* $C_- \leqslant C \leqslant C_+$, *the flow defined by this analytic solution is open and unique in the class of open (possibly nonanalytic) flows, generated by the conditions for the boundary problem under consideration.*

As already mentioned, this proposition follows from Theorems 1, 2, and 7.2, and the considerations from §5.

Further, according to Theorem 7.2, any analytic flow from §6 is uniquely defined by setting on the boundary the normal velocity component indicated in §6 and the vorticity

$$\beta(y) = \pi\Omega \sin(\pi y/h),$$

with the "characteristic" value $\Omega = \text{const}$ on some (arbitrarily small) influx or efflux region (lying on the side $x = -l$), if the additional conditions of the corresponding problem A are satisfied. Passing to the conditions of problem B, we choose g as the influx region. In terms of the introduced dimensionless variables, the corresponding conditions will be written in the following form:

$$(5) \quad \begin{gathered} u(-\tau, y) = \cos \pi y, \quad u(\tau, y) = v(x, 0) = v(x, 1) = 0, \qquad |x| < \tau, \quad 0 < y < 1, \\ \omega(-\tau, y) = \pi C \sin \pi y, \quad 0 < y < 1/2. \end{gathered}$$

The dimensionless parameter C, $-\infty < C < \infty$, is defined by the same equality as in §5, i.e., $C = \Omega h/U$, where Ω, h, and U are characteristic vorticity, length, and velocity, respectively ($h, U > 0$).

PROPOSITION 4. *The boundary problem* (5) *for the system of equations* (4.1) *with* $k = 0$ *in the rectangle* $|x| < \tau$, $0 < y < 1$ *has a unique analytic solution if*

$$C \neq m^2 + n^2/4\tau^2, \qquad m, n = 1, 2, \ldots,$$

and an uncountable set of arbitrarily smooth nonanalytic solutions if, simultaneously, $C > 1 + 1/16\tau^2$. *For* $C \leqslant 1 + 1/16\tau^2$ *the flow defined by this analytic solution is open and unique in the class of open (possibly nonanalytic) flows, generated by the conditions of the problem under consideration.*

Indeed, calculating the boundary values of the normalized stream function ψ according to the formula from (4.8), under the condition that the last equality from (5) is satisfied, we find

$$\omega(-\tau, y) = \pi^2 C \int_0^y \cos(\pi s)\, ds = \pi^2 C \psi(-\tau, y),$$

and the corresponding analytic function

$$\mu(t) = \pi^2 C t.$$

On the other hand,

$$\pi^2 m^2 + \pi^2 n^2/4\tau^2, \qquad m, n = 1, 2, \ldots,$$

is the known total set of eigenvalues of the corresponding homogeneous boundary problem for equation (6.1) considered for $k = 0$ in the rectangle $|x| < \tau$, $0 < y < 1$. Hence, the spectral condition for the derivative $\mu'(t)$ is reduced to the inequality for C given above. Comparing the parameter Λ from §6 with C, we obtain that for $C \leqslant 1 + 1/16\tau^2$ (or for $\Lambda \leqslant \pi/2$) the analytic flow defined by (5) is open and for $C > 1 + 1/16\tau^2$ (or for $\Lambda > \pi/2$) is developed (§6). Theorems 7.2, 1 and 2 complete the proof.

An analogous statement can be easily obtained for the case $k = 1$.

§9. Existence theorems

9.1. Auxiliary boundary problem. Now we shall prove the existence of a smooth solution of the preliminary problem from §7.1 (or analogous solutions of problems A, B, C obtained by means of the above-mentioned restriction of the conditions of the preliminary problem in order to distinguish a unique solution). The following *auxiliary boundary problem* will play the main role:

$$(1)\qquad \begin{gathered} L_k\varphi \equiv -\frac{1}{y^k}\left(\frac{\partial}{\partial x}\left(\frac{1}{y^k}\frac{\partial\varphi}{\partial x}\right)+\frac{\partial}{\partial y}\left(\frac{1}{y^k}\frac{\partial\varphi}{\partial y}\right)\right)=\eta(x,y,\varphi(x,y)) \\ \text{in the domain } V \text{ from §7.1 and } \varphi=0 \text{ on the boundary } \partial V, \\ k=\text{const}\geqslant 0. \end{gathered}$$

Here $\eta=\eta(x,y,t)$ is a fixed real function defined on the set of all the points (x,y) and t lying in the domain V and the real axis R, respectively. The function η for $t=0$ is assumed to be square integrable in the domain V with the weight y^k,

$$\int_V \eta^2(x,y,0)y^k\,dy<\infty,$$

and to satisfy for all (x,y) from V a Lipschitz condition with respect to t,

$$|\eta(x,y,t)-\eta(x,y,t')|\leqslant M|t-t'|,\qquad -\infty<t,t'<\infty,\quad M=\text{const}\geqslant 0.$$

Now we introduce into consideration the generalized solutions of problem (1) which are of interest for us. We denote by $H_k=H_k(V)$ the Hilbert space obtained by the closure (completion) of the linear manifold (set) $C(\overline{V})$ of the continuous functions $f,g,\ldots$, defined in the *closure* $\overline{V}$ of the domain V supplied with the scalar product and the norm

$$(f,g)_k\equiv\int_V fgy^k\,dx\,dy,\qquad \|f\|_k=(f,f)_k^{1/2}.$$

On the other hand, let $C_0^\infty=C_0^\infty(V)$ be the linear manifold of infinitely differentiable real functions defined in V and vanishing together with partial derivatives of all orders on the boundary ∂V of the domain V. For any $k\geqslant 0$, the fact that the function $f(x,y)$ belongs to the set C_0^∞ obviously implies that the function $y^{-k}f(x,y)$ also belongs to this set. Using this, let us prove the following lemma.

PROPOSITION 1. *The set C_0^∞ is dense in H_k (i.e., any element of the space H_k is the limit in H_k of a sequence of functions from C_0^∞).*

Indeed, the transformation of the halfplane $y\geqslant 0$ provided by the change of variables

$$s=x/(k+1),\qquad t=y^{k+1}$$

is a homeomorphic mapping of $\overline{V}$ onto the closure of some bounded domain V' of the halfplane $t\geqslant 0$ which generates a linear isometry of the space $H_k(V)$ on the standard Hilbert space $L_2(V')\equiv H_0(V)$ of the following form:

$$f(x,y)\to\tilde{f}(s,t)\equiv f\left((k+1)s,t^{\frac{1}{k+1}}\right),\qquad \|f\|_k=\|\tilde{f}\|_0.$$

Due to the above, this transformation of the space $C_0^\infty(V)$ is a one-to-one mapping of this space onto $C_0^\infty(V')$. Then the known density of the set $C_0^\infty(V')$ in $L_2(V')$ **[35]** provides the density of $C_0^\infty(V)$ in $H_k(V)$.

By Proposition 1, the Hilbert space H_k may simultaneously be seen as the completion of the linear manifold C_0^∞ with respect to the norm introduced above. Setting

$$\nabla\varphi \cdot \nabla\psi \equiv \frac{\partial\varphi}{\partial x}\frac{\partial\psi}{\partial x} + \frac{\partial\varphi}{\partial y}\frac{\partial\psi}{\partial y}$$

for an arbitrary pair of functions φ and ψ from C_0^∞, we introduce in C_0^∞ a new scalar product

$$(\nabla\varphi, \nabla\psi)_{-k} \equiv \int_V \frac{\nabla\varphi \cdot \nabla\psi}{y^k}\, dx\, dy.$$

We denote by $H_{-k}^1 = H_{-k}^1(V)$ the Hilbert space obtained by the completion of $C_0^\infty(V)$ in the norm

$$\|\nabla\varphi\|_{-k} \equiv (\nabla\varphi, \nabla\varphi)_{-k}^{1/2}.$$

Introducing the distance h from the axis $y = 0$ to the most remote point of the compact $\overline{V}$, directly from the identities given above we obtain the inequalities

$$\|\nabla\varphi\|_{-k} \geqslant h^{-k/2}\|\nabla\varphi\|_0 \geqslant C_0 h^{-k/2}\|\varphi\|_0 \geqslant C_0 h^{-k}\|\varphi\|_k, \tag{2}$$

which realize the chain of embeddings of the Hilbert spaces

$$H_{-k}^1 \subset H_0^1 \subset H_0 \subset H_k$$

(the existence of the *embedding* $H_{-k}^1 \subset H_0^1$ (of the *space* H_{-k}^1 *in* H_0^1) implies, as usual, that the corresponding inequality for the norms (the first inequality from (2)) is valid on the set C_0^∞, dense in H_{-k}^1 and simultaneously in H_0^1). The second embedding of this chain is the well-known *Poincaré-Steklov embedding*.

PROPOSITION 2. *The embedding $H_{-k}^1 \subset H_k$ is compact (i.e., any sequence of functions from C_0^∞ bounded in H_{-k}^1 contains a subsequence converging in H_k).*

This proposition immediately follows from the fact that the Poincaré-Steklov embedding is compact **[35]** and inequalities (2).

For a given function $\eta(x, y, t)$ we define the *superposition operator* $\widehat{\eta} = \widehat{\eta}[\varphi]$ that maps a function $\varphi \in C_0^\infty$ to the real-valued function $\widehat{\eta}[\varphi(x, y)] \equiv \eta(x, y, \varphi(x, y))$ defined in V. Then, from the Lipschitz condition for $\eta(x, y, t)$, for all φ, ψ from C_0^∞ we obtain

$$\|\widehat{\eta}[\varphi] - \widehat{\eta}[\psi]\|_k \leqslant M\|\varphi - \psi\|_k. \tag{3}$$

Since (as it was agreed above) $\widehat{\eta}[0]$ is an element of the space H_k, for any $\varphi(x, y)$ from C_0^∞ the function $\widehat{\eta}[\varphi(x, y)]$ is also an element of this space (and thus is square integrable in V with the weight y^k),

$$\|\widehat{\eta}[\varphi]\|_k \leqslant \|\widehat{\eta}[0]\|_k + M\|\varphi\|_k$$

(a consequence of inequality (3)). Using this, we shall consider the operator $\widehat{\eta}$ as acting from H_k to H_k with the set C_0^∞ as its domain of definition. Then (3) implies the continuity of the operator $\widehat{\eta}$ (with respect to the natural topology of the Hilbert space H_k). Since the domain of definition C_0^∞ of the operator $\widehat{\eta}$ is dense in H_k (Proposition 1), this operator admits a closure (or extension by continuity) to all the elements of the space H_k. As it is known **[35, 44]**, the closure of a continuous operator defined on a dense set of a Hilbert space is unique (this fact is also established in §10.1). Therefore, without loss of generality, $\widehat{\eta}$ will be uniquely identified with its closure in H_k.

Multiplying both sides of equality (1) by an arbitrary function w from C_0^∞, integrating with the weight y^k over the domain V, and using the identity

$$(L_k\varphi, w)_k = (\nabla\varphi, \nabla w)_{-k}$$

(which can be verified straightforwardly), we obtain

$$(\nabla\varphi, \nabla w)_{-k} = (\widehat{\eta}[\varphi], w)_k. \tag{4}$$

Using the continuity of the scalar product **[35]** and the embedding $H_{-k}^1 \subset H_k$, we extend by continuity both sides of equality (4) to the elements φ, w of the space H_{-k}^1. An element φ of the space H_{-k}^1 that satisfies identity (4) for any w from H_{-k}^1 will be called a *generalized solution* of problem (1).

In the particular case of a linear and homogeneous dependence of the form

$$\widehat{\eta}[\varphi] \equiv \lambda\varphi \qquad (\eta(x, y, t) \equiv \lambda t), \qquad \lambda = \text{const},$$

relation (4) leads to a spectral problem of finding an *eigenvalue* λ and an *eigenelement* $\varphi \neq 0$ in H_{-k}^1 satisfying the identity

$$(\nabla\varphi, \nabla w)_{-k} = \lambda(\varphi, w)_k \tag{5}$$

for any w from H_{-k}^1.

As it follows from the lemma in §10.4 and Proposition 2, the spectral problem (5) has a countable set of eigenvalues

$$\lambda_1 < \lambda_2 < \cdots < \lambda_m < \lambda_{m+1} < \cdots \qquad (\lambda_1 > 0), \tag{6}$$

with no finite limit points (in particular, this set is unbounded); each eigenvalue λ_m has finite multiplicity and possesses a finite set of linearly independent eigenelements, while the whole set of the latter (corresponding to all the numbers of the sequence (6)) forms an orthonormal basis of the space H_k and an orthogonal basis of the space H_{-k}^1 (recall that a given system of elements of a Hilbert space forms its *basis* if any element of this space can be uniquely represented by a linear combination of elements of this system (i.e., with uniquely defined coefficients of this linear expansion); a basis is *orthogonal* if its constituting elements are pairwise orthogonal (with respect to the introduced scalar product); an orthogonal basis is *orthonormal* if the norms of its elements are equal to 1; for the case of a countable basis, the above-mentioned representation of an element as a linear combination of elements of the basis means the convergence in the norm of the sequence of the corresponding partial sums to the given element). The validity of the following *variational Rayleigh inequalities* for the numbers (6) will be also proved (§10.4): if $H_{(m)}^1$ is the linear span of the system of eigenfunctions of the spectral problem (5) corresponding to the first m numbers of the series (6) (for the case when $m = 0$ the corresponding subspace $H_{(0)}^1$ coincides with the zero element of the space H_{-k}^1), and $H_{-k}^1 \ominus H_{(m)}^1$ is the orthogonal complement of the subspace $H_{(m)}^1$ in H_{-k}^1, then

$$\begin{aligned} &\|\nabla\omega\|_{-k}^2 \leqslant \lambda_m \|\omega\|_k^2 && \text{for all } \omega \text{ in } H_{(m)}^1 && (m \geqslant 1); \\ &\|\nabla\omega\|_{-k}^2 \geqslant \lambda_{m+1} \|\omega\|_k^2 && \text{for all } \omega \text{ in } H_{-k}^2 \ominus H_{(m)}^1 && (m \geqslant 0). \end{aligned} \tag{7}$$

Setting

$$\Delta t \equiv t - t', \qquad \Delta\eta \equiv \eta(x, y, t) - \eta(x, y, t'),$$

we require that the following *spectral condition* be satisfied for the function η:

There exists an interval $\lambda_- \leqslant \lambda \leqslant \lambda_+$ *of the real axis located at a positive distance* ε *from the set* (6) *such that for all* (x, y) *from* V *and* $-\infty < t, t' < \infty$

$$\lambda_- \Delta t^2 \leqslant \Delta t \Delta \eta \leqslant \lambda_+ \Delta t^2. \tag{8}$$

Note that this spectral condition is analogous to the one formulated in §7.1 and unites the *cases of Hammerstein* **[141]** (additional requirement that $\lambda_+ < \lambda_1$) and *Dolph* **[131, 132]** (alternative restrictions $\lambda_- > \lambda_m$ and $\lambda_+ < \lambda_{m+1}$ for some $m = 1, 2, \dots$). Under the assumption that η is differentiable with respect to t, inequalities (8) acquire the following form:

$$\lambda_- \leqslant \frac{\partial \eta}{\partial t}(x, y, t) \leqslant \lambda_+, \tag{9}$$

and are assumed to hold for all (x, y) in V and all real t. The spectral condition for a weakly nonlinear elliptic equation was also used in this form by Landesman and Lazer **[151, 153]**. Note also that it follows from (8) that the Lipschitz inequality given above is valid for the function η. Recall also that elements of an abstract (infinite-dimensional) Hilbert space (in particular, the elements of the space H^1_{-k}) are equivalence classes **[35]**. The continuity of a generalized solution of problem (1) as an element of H^1_{-k} means that among the representatives of the corresponding equivalence class there exists a continuous function. The concepts of smoothness and analyticity of a generalized solution are to be understood analogously.

The following theorem concerns the existence and uniqueness of a generalized solution of problem (1) and its smoothness depending on the smoothness of η. It is also established in the theorem that a slight variation of η does not violate the conditions under which the corresponding generalized solution φ_η exists and is unique (the solution is stable with respect to small perturbations of η). Simultaneously, the latter depends continuously on the varying right-hand side.

THEOREM 1. *For any function* $\eta = \eta(x, y, t)$ *belonging to the space* $H_k(V)$ *for* $t = 0$ *and satisfying the spectral condition, there exists a unique generalized solution* φ_η *of problem* (1). *If* η *is replaced by a new function* $\xi = \xi(x, y, t)$ *belonging to the space* $H_k(V)$ *for* $t = 0$ *and such that for some constant* $\delta = \delta(\eta) < \varepsilon$ *(possibly depending on* η*) and all* (x, y) *in* V *and* $-\infty < t, t' < \infty$

$$|\Delta \xi - \Delta \eta| \leqslant \delta |\Delta t|, \qquad \Delta \xi \equiv \xi(x, y, t) - \xi(x, y, t'), \qquad \Delta t \equiv t - t', \tag{10}$$

problem (1) *again has a unique generalized solution* φ_ξ, *and*

$$\|\nabla(\varphi_\xi - \varphi_\eta)\|_{-k} \leqslant \frac{4\lambda_{m+1}}{\varepsilon \lambda_1^{1/2}} \rho(\xi, \eta), \tag{11}$$

where the distance $\rho(\xi, \eta)$ *is defined by the least upper bound,*

$$\rho(\xi, \eta) \equiv \sup_{-\infty < t < \infty} \left(\int_V [\xi(x, y, t) - \eta(x, y, t)]^2 y^k \, dx \, dy \right)^{1/2}.$$

For $k = 0$ *or* 1, *under an additional assumption on the smoothness of the given function* η *on the set of points* (x, y) *of the domain* V *and the variable* $-\infty < t < \infty$, *the function* $\varphi_\eta = \varphi_\eta(x, y)$ *is twice continuously differentiable in the domain* V. *If, moreover,* η *is an infinitely differentiable (analytic) function, then the function* φ_η *is infinitely differentiable (analytic, respectively) in* V. *For the case of a smooth function* $\eta(x, y, t)$ *everywhere up* *to the smooth points of its domain of definition (i.e., the points (x, y) belonging to the regular part of the boundary ∂V and $-\infty < t < \infty$), the function φ_η, together with the components*

$$u(x, y) = \frac{1}{y^k}\frac{\partial \varphi_\eta}{\partial y}, \qquad v(x, y) = -\frac{1}{y^k}\frac{\partial \varphi_\eta}{\partial x}$$

of the vector field $\mathbf{u}_\eta = (u(x, y), v(x, y))$ generated by φ_η in V, admits an extension by continuity to the regular points (x_0, y_0) of the smooth parts of the boundary ∂V. Moreover, φ_η and the normal component of the extended field $\mathbf{u}_\eta$ vanish at (x_0, y_0). When $k = 1$ and $y_0 = 0$, the component $v(x_0, 0) = 0$, and $v(x_0, y) = O(y)$ ($|v(x_0, y)| \leqslant$ const $\cdot y$) as $y \to 0$.

Indeed, associating to problem (1) the real functional

$$\text{(12)} \qquad \underset{\eta}{E}[\varphi] \equiv \frac{1}{2}\|\nabla\varphi\|^2_{-k} - \int_V \left[\int_0^{\varphi(x,y)} \eta(x, y, t)\, dt\right] y^k\, dx\, dy$$

generated by the given function η and defined on functions φ of the linear manifold C_0^∞ of the space H^1_{-k}, by a straightforward verification we establish that it is continuously differentiable (in the sense of Frechét **[44]**) everywhere in C_0^∞ (necessary explanations are given in §10.1) and that the identity

$$\text{(13)} \qquad \underset{\eta}{E'}[\varphi]w = (\nabla\varphi, \nabla w)_{-k} - (\widehat{\eta}[\varphi], w)_k$$

for the corresponding derivative of this functional at an arbitrary function φ from C_0^∞ is valid. The right-hand side of identity (13) admits extension by continuity to the elements φ, w of the space H^1_{-k}. On the other hand, as it will be shown in §10.1, this functional (together with the derivative) admits closure to the elements of the space H^1_{-k}; the closure of the derivative (13) coincides with the derivative of the closure (which again turns out to be a continuously differentiable functional). Taking this into account, we identify the functional (12) and its derivative (13) with their corresponding closures to the elements of H^1_{-k}. Comparing (4) with (13), we come to the conclusion that an element φ of the space H^1_{-k} is a generalized solution of problem (1) if and only if φ corresponds to a *critical point* of the functional (12), i.e.,

$$\text{(14)} \qquad \underset{\eta}{E'}[\varphi]w = 0 \quad \text{for all } w \text{ in } H^1_{-k}.$$

Further, by setting for an arbitrary real s

$$\delta\widehat{\eta}[\varphi; w; s] \equiv \widehat{\eta}[\varphi + sw] - \widehat{\eta}[\varphi - sw]$$

for the operator of superposition $\widehat{\eta}$, as a consequence of inequalities (8) we have

$$\text{(15)} \qquad 2\lambda_- sw^2 \leqslant w\delta\widehat{\eta}[\varphi; w; s] \leqslant 2\lambda_+ sw^2, \qquad s > 0.$$

Setting

$$\Delta^2 \underset{\eta}{E}[\varphi](w) \equiv \underset{\eta}{E}[\varphi + w] - 2\underset{\eta}{E}[\varphi] + \underset{\eta}{E}[\varphi - w],$$

we find from (12) and (15), respectively, that

$$\|\nabla w\|^2_{-k} - \Delta^2 \underset{\eta}{E}[\varphi](w) = \left(w, \int_0^1 \delta\widehat{\eta}[\varphi; w; s]\, ds\right)_k$$

and

$$\lambda_-\|w\|_k^2 \leqslant \left(w, \int_0^1 \delta\widehat{\eta}[\varphi; w; s]\,ds\right)_k \leqslant \lambda_+\|w\|_k^2.$$

By means of the relations obtained and inequalities (7), taking into account that $\lambda_+ \leqslant \lambda_{m+1} - \varepsilon$ if the number $m \geqslant 0$ in the spectral condition for η and $\lambda_- \geqslant \lambda_m + \varepsilon$ for $m \geqslant 1$ $(\lambda_{m+1} > \lambda_m)$, we find that

$$(16)\qquad \begin{aligned} &\Delta^2_{\eta} E[\varphi](w) \geqslant \frac{\varepsilon}{\lambda_{m+1}}\|\nabla w\|_{-k}^2 &&\text{for all } w \text{ in } H^1_{-k} \ominus H^1_{(m)} &&(m \geqslant 0),\\ &\Delta^2_{\eta} E[\varphi](w) \leqslant -\frac{\varepsilon}{\lambda_{m+1}}\|\nabla w\|_{-k}^2 &&\text{for all } w \text{ in } H^1_{(m)} &&(m \geqslant 1). \end{aligned}$$

As it is proved in §§10.2, 10.3, inequalities (16) imply the existence and uniqueness of the required generalized solution of problem (1) as an element φ_η of the space H^1_{-k} satisfying identity (14) (or (4)), and the estimate

$$(17)\qquad \|\nabla(\varphi_\xi - \varphi_\eta)\|_{-k} \leqslant \frac{4\lambda_{m+1}}{\varepsilon}\|E'_{\eta}[\varphi_\xi]\|,$$

where $\|E'_{\eta}[\varphi_\xi]\|$ is the least upper bound of the ratio $|E'_{\eta}[\varphi_\xi]w|/\|\nabla w\|_{-k}$, considered on the set of all the nonzero elements w of the space H^1_{-k}. Using (7), (13), and (17) we come to (11):

$$\begin{aligned} \|\nabla w\|_{-k} &\geqslant \lambda_1^{1/2}\|w\|_k,\\ E'_{\eta}[\varphi_\xi]w &= (\nabla\varphi_\xi, \nabla w)_{-k} - (\widehat{\eta}[\varphi_\xi], w)_k\\ &= (\widehat{\xi}[\varphi_\xi], w)_k - (\widehat{\eta}[\varphi_\xi], w)_k \leqslant \|\widehat{\xi}[\varphi_\xi] - \widehat{\eta}[\varphi_\xi]\|_k\|w\|_k,\\ \|\widehat{\xi}[\varphi_\xi] - \widehat{\eta}[\varphi_\eta]\|_k &= \left(\int_V [\xi(x,y,\varphi_\xi) - \eta(x,y,\varphi_\xi)]^2 y^k\,dx\,dy\right)^{1/2} \leqslant \rho(\xi,\eta). \end{aligned}$$

The existence of a solution φ_ξ follows immediately from (8) and (10); they ensure that the right-hand side ξ of equation (1) again satisfies the spectral condition, where λ_-, λ_+, and ε are changed to $\lambda_- - \delta$, $\lambda_+ + \delta$, and $\varepsilon - \delta$, respectively, while the number m is the same as for η. Passing to the analysis of the smoothness of a generalized solution φ_η, we shall omit the known case when the second-order differential operator is uniformly elliptic [**2, 17, 56, 59, 187–189**]. The operator L_k from (1) is uniformly elliptic if $k = 0$, or $k = 1$ and $y \geqslant \text{const} > 0$. For the case when $k = 1$ and the operator L_k is considered in a neighborhood of a boundary point of the domain V located on the axis $y = 0$ (i.e., in a neighborhood where the uniform ellipticity of L_k is violated), we use the arguments from [**140**]. We associate to the cylindrical coordinates x, y, θ the Cartesian coordinates,

$$x_1 = x, \qquad x_2 = y\cos\theta, \qquad x_3 = y\sin\theta \qquad (y > 0, \quad 0 \leqslant \theta \leqslant 2\pi),$$

and to the real function $f(x, y)$, the complex-valued function

$$\widehat{f}(x_1, x_2, x_3) = e^{i\theta}\frac{f(x,y)}{y}, \qquad i^2 = -1.$$

If $k = 1$, equation (1) after this change of the variables acquires the form

$$\frac{\partial^2 \widehat{\varphi}}{\partial x_1^2} + \frac{\partial^2 \widehat{\varphi}}{\partial x_1^2} + \frac{\partial^2 \widehat{\varphi}}{\partial x_3^2} = -ye^{i\theta}\eta(x, y, \varphi(x, y)).$$

Thus, the operator L_k becomes uniformly elliptic and the boundary points of the domain V lying on the intervals of the axis $y = 0$ become interior for the corresponding domain of rotation (obtained by rotating the domain V around the axis $y = 0$). These facts make it possible to obtain the required estimates concerning the smoothness of the generalized solution φ_η. The necessary details are given in **[140]**. The analyticity of φ_η for an analytic $\eta(x, y, t)$ is ensured by the analyticity of the coefficients of the operator L_k in the domain V **[15, 169]**. The proof of Theorem 1 is completed.

COROLLARY 1. *If the assumptions from* §7.1 *are satisfied, then Proposition* 7.1 *is valid.*

Indeed, as in §7.1, the required ψ is given by equality $\psi = \Phi + \varphi$, where φ is a generalized solution of problem (7.4) with a given right-hand side of the equation. The latter does not depend on φ, and, hence, *a fortiori*, satisfies the spectral condition. The smoothness properties of the function $\omega(x, y)$ in V and restriction (7.3) provide the validity of the assumptions of the theorem.

COROLLARY 2. *The assertion of Theorem* 7.1 *is valid under the same assumptions.*

Indeed, the substitution $\psi = \Phi + \varphi$ reduces the weakly nonlinear problem (7.6) investigated here to (1) with

$$\eta(x, y, t) = \mu[\Phi(x, y) + t]. \tag{18}$$

The spectral condition for μ from §7.1 implies that the spectral condition stated above with respect to η holds. The assumed properties of the function Φ guarantee that the remaining restrictions on η are satisfied. Condition (7.2) is also satisfied:

$$\omega(x, y)|_g = y^k \mu[\psi(x, y)]|_g = y^k \mu[\Phi(x, y)]|_g = y^k \beta(x, y).$$

The validity of (7.3) follows from the inequalities

$$\begin{aligned}
&\left(\int_V (u^2 + v^2 + (\omega/y^k)^2) y^k \, dx \, dy\right)^{1/2} \\
&\qquad \leqslant \left(\int_V (u^2 + v^2) y^k \, dx \, dy\right)^{1/2} + \left(\int_V (\omega/y^k)^2 y^k \, dx \, dy\right)^{1/2} \\
&\qquad \leqslant \left(\int_V (a^2 + b^2) y^k \, dx \, dy\right)^{1/2} + \|\nabla\varphi\|_{-k} + \|\mu[\Phi + \varphi]\|_k,
\end{aligned}$$

$$\|\nabla\varphi\|_{-k}^2 = (\varphi, \mu[\Phi + \varphi])_k \leqslant \|\varphi\|_k \|\mu[\Phi + \varphi]\|_k,$$

$$\|\varphi\|_k \leqslant \lambda_1^{-1/2} \|\nabla\varphi\|_{-k},$$

$$|\mu[\Phi + \varphi]| \leqslant \left|\mu[\Phi] + \varphi \int_0^1 \mu'[\Phi + s\varphi] \, ds\right| \leqslant |\mu[\Phi]| + M|\varphi|,$$

$$M = \sup_{-\infty < t < \infty} |\mu'(t)| < \infty.$$

The continuity of Φ in $\overline{V}$ (the closure of V) completes the proof,

$$\sup_V |\Phi| < \infty, \qquad \sup_V |\mu[\Phi]| < \infty.$$

9.2. The vortex catastrophe. We pass to a detailed proof of Theorem 8.2 which states that a smooth nonanalytic stationary Euler flow is nonunique due to the existence of sufficiently developed (in the sense of the violation of the maximum principle (8.1)) vortex zones in the flow domain. As we have already seen (§§5, 6, 8.4), vortex zones of a stationary flow arise due to the variation of the boundary data for the corresponding problem on a passing flow. Nonuniqueness of the stationary flow connected with their appearance resembles the nonuniqueness of the solutions for the equations of gas dynamics due to the appearance of shock waves (in both cases, the nonuniqueness is connected with qualitative rearrangement of the solution). In analogy with the corresponding gradient catastrophe in gas dynamics, the nonuniqueness of the stationary Euler flow brought about by the appearance of vortices can be called *vortex catastrophe*. Now we turn to the analysis of this phenomenon.

As noted in §8.3, the required σ-family satisfying the assumptions of Theorem 8.2 is assumed to be generated by the stream functions $\psi = \psi_\sigma(x, y)$ which satisfy the conditions of problem (7.6) with $\mu(t)$ from (8.3), where the function $f(t)$ is defined as in §8.3. The substitution

$$\psi_\sigma = \Phi + \varphi_\sigma \tag{19}$$

reduces the problem under investigation to the weakly nonlinear problem (1) with η from (18):

$$L_k \varphi_\sigma = \mu_\sigma[\Phi(x, y) + \varphi_\sigma], \quad \varphi_\sigma|_{\partial V} = 0 \qquad (\mu_\sigma(t) = \mu_0(t) + \sigma f(t)). \tag{20}$$

It is assumed, moreover, that the values taken by the derivative of the function $\mu_0(t)$ on the real axis $-\infty < t < \infty$ do not leave a fixed interval of the real axis,

$$\lambda_- \leqslant \mu_0'(t) \leqslant \lambda_+,$$

located at a positive distance ε from the set (6) (the spectral condition from §7.1). Since

$$|f'(t)| \leqslant (2/e)^2, \qquad -\infty < t < \infty,$$

(which is easily seen by direct verification, using §8.3), the values of the derivatives of the corresponding functions $\mu_\sigma(t)$ for

$$0 \leqslant \sigma \leqslant \sigma_1 = \varepsilon e^2/8 \tag{21}$$

belong to the interval

$$\lambda_- - \varepsilon/2 \leqslant \mu_\sigma'(t) \leqslant \lambda_+ + \varepsilon/2,$$

which is again located at a positive distance $\varepsilon/2$ from the set (6), and this (according to Theorem 1) guarantees the existence, for each given σ, of a generalized solution φ_σ of problem (20) (the required corresponding inequalities for finite differences analogous to (8) are obtained by direct integration of the indicated inequalities).

Further, we have

$$(22)\qquad \begin{gathered} \|\nabla(\varphi_\sigma - \varphi_{\sigma'})\|_{-k} \leqslant \frac{\theta}{\varepsilon}|\sigma - \sigma'|, \qquad 0 \leqslant \sigma, \sigma' \leqslant \sigma_1, \\ \theta = 4\lambda_{m+1}\left(\frac{1}{\lambda_1}\int_V y^k\,dx\,dy\right)^{1/2}, \end{gathered}$$

since, by the estimate (11), we have

$$\begin{aligned} &\|\nabla(\varphi_\sigma - \varphi_{\sigma'})\|_{-k} \\ &\quad \leqslant \frac{4\lambda_{m+1}}{\varepsilon\lambda_1^{1/2}} \sup_{-\infty<t<\infty} \left(\int_V [\mu_\sigma[\Phi(x,y)+t] - \mu_{\sigma'}[\Phi(x,y)+t]]^2 y^k\,dx\,dy\right)^{1/2} \\ &\quad \leqslant \frac{4\lambda_{m+1}}{\varepsilon\lambda_1^{1/2}} \left(\int_V y^k\,dx\,dy\right)^{1/2} \sup_{-\infty<t<\infty} |\mu_\sigma(t) - \mu_{\sigma'}(t)| \end{aligned}$$

and (by the boundedness of f, $|f(t)| \leqslant 1$, $-\infty < t < \infty$),

$$|\mu_\sigma(t) - \mu_{\sigma'}(t)| = |\sigma - \sigma'||f(t)| \leqslant |\sigma - \sigma'|.$$

The assumption on the existence of a point P_* in the domain V such that

$$\psi_0(P_*) < m_0 = \min_{\partial V} \psi_0 \qquad \text{or} \qquad \psi_0(P_*) > M_0 = \max_{\partial V} \psi_0$$

is the central condition of Theorem 8.2. Due to the continuity of ψ in V, there exists a sufficiently small constant $\rho = \rho(P_*, \psi_0) > 0$ (depending on P_* and ψ_0) such that the disc $N = N_\rho(P_*)$ of radius ρ with center at P_* lies in V, and for all its points P the sign of one of the following inequalities is preserved:

$$\psi_0(P) < m_0 \qquad \text{or} \qquad \psi_0(P) > M_0 \qquad \text{for all} \quad P \quad \text{from} \quad N.$$

Since $f(t) > 0$ for $t > m_0$ or for $t < M_0$, in any of the indicated cases $f[\psi_0(P)] > 0$ everywhere in N; hence,

$$F_0 = \int_N f[\psi_0(x,y)]y^k\,dx\,dy > 0.$$

On the other hand,

$$f(\psi_\sigma) = f(\psi_0) + (\psi_\sigma - \psi_0)\int_0^1 f'[\psi_0 + (\psi_\sigma - \psi_0)s]\,ds \geqslant f(\psi_0) - (2/e)^2|\psi_\sigma - \psi_0|,$$

and, by (19) and (22), we have

$$\begin{aligned} \int_V |\psi_\sigma - \psi_0| y^k\,dx\,dy &\leqslant \int_V |\varphi_\sigma - \varphi_0| y^k\,dx\,dy \leqslant \left(\int_V y^k\,dx\,dy\right)^{1/2} \|\varphi_\sigma - \varphi_0\|_k \\ &\leqslant \left(\frac{1}{\lambda_1}\int_V y^k\,dx\,dy\right)^{1/2} \|\nabla(\varphi_\sigma - \varphi_0)\|_{-k} = \frac{\theta}{4\lambda_{m+1}}\|\nabla(\varphi_\sigma - \varphi_0)\|_{-k} \leqslant \frac{\theta^2}{4\varepsilon\lambda_{m+1}}\sigma. \end{aligned}$$

As a result of the estimates obtained,

$$\int_V f[\psi_\sigma(x,y)]y^k\,dx\,dy \geqslant \int_N f[\psi_\sigma(x,y)]y^k\,dx\,dy$$
$$\geqslant \int_N f[\psi_0(x,y)]y^k\,dx\,dy - (2/e)^2\int_N |\psi_\sigma - \psi|y^k\,dx\,dy$$
$$\geqslant F_0 - \left(\frac{\theta}{e}\right)^2 \frac{\sigma}{\varepsilon\lambda_{m+1}}.$$

Hence, for

$$\sigma \leqslant \sigma_2 = \frac{\varepsilon F_0}{2}\left(\frac{e}{\theta}\right)^2 \lambda_{m+1} \tag{23}$$

we have

$$\int_V f[\psi_\sigma(x,y)]y^k\,dx\,dy \geqslant F_0/2 > 0. \tag{24}$$

Now let the left-hand side of inequality (8.2) vanish for some $0 \leqslant \sigma, \sigma' \leqslant \sigma_1$. By the identities

$$u_\sigma = y^{-k}\partial\psi_\sigma/\partial y, \qquad v_\sigma = -y^{-k}\partial\psi_\sigma/\partial x,$$

the indicated left-hand side coincides with $\|\nabla(\psi_\sigma - \psi_{\sigma'})\|_{-k}$. Therefore, we have

$$\|\nabla(\psi_\sigma - \psi_{\sigma'})\|_{-k} = 0,$$

or, since ψ_σ and $\psi_{\sigma'}$ are twice continuously differentiable in V (Theorem 1), $\psi_\sigma \equiv \psi_{\sigma'}$. But then

$$\mu_\sigma(\psi_\sigma) - \mu_{\sigma'}(\psi_{\sigma'}) \equiv L_k(\psi_\sigma - \psi_{\sigma'}) \equiv 0$$

as well. Since

$$\mu_\sigma(\psi_\sigma) - \mu_{\sigma'}(\psi_{\sigma'}) = \mu_0(\psi_\sigma) - \mu_0(\psi_{\sigma'}) + \sigma f(\psi_\sigma) - \sigma' f(\psi_{\sigma'}) = (\sigma - \sigma')f(\psi_\sigma)$$

at the same time, we obtain

$$(\sigma - \sigma')\int_V f[\psi_\sigma(x,y)]y^k\,dx\,dy = 0.$$

According to inequality (24), for $\sigma, \sigma' \leqslant \sigma_2$ this inequality is valid only when $\sigma = \sigma'$. Therefore, in the theorem under consideration the required constant $\sigma_* > 0$ can be the lesser of the constants σ_1 and σ_2, defined in accordance with the equalities (21) and (23).

The validity of Theorem 8.2 is established.

§10. Regular functionals

10.1. The closure of a smooth operator. We pass to the consideration of some facts from functional analysis that are necessary to prove the above assertions concerning the existence and uniqueness of a generalized solution of the auxiliary boundary problem. As already mentioned, this problem can be reduced to some variational problem. A critical point of the differentiable functional satisfying the conditions from (9.16) corresponds to the required solution. These conditions generate a special class of abstract functions resembling the simplest quadratic forms of the following form:

$$x_1^2 + x_2^2 + \cdots + x_n^2 \qquad \text{or} \qquad -x_1^2 - \cdots - x_k^2 + x_{k+1}^2 + \cdots + x_n^2$$

(in n real variables), and, in this sense, are "regular". The constructions of this section (briefly presented in [**96, 98, 99**]) are mainly devoted to the corresponding theory.

First, we agree on the terminology and give some statements of general character. Let $\mathbb{B}_1$ and $\mathbb{B}_2$ be Banach (i.e., complete linear normal) spaces with norms $\|\cdot\|_1$ and $\|\cdot\|_2$, respectively, and let

$$F : \mathbb{B}_1 \to \mathbb{B}_2,\ x \to F(x)$$

be some operator acting from $\mathbb{B}_1$ to $\mathbb{B}_2$ with a nonempty *domain of definition* $\mathbb{D}(F) \subset \mathbb{B}_1$ (*belonging to* $\mathbb{B}_1$). A sequence $\{x_i\}$ of elements x_i of the set $\mathbb{D}(F)$ converging in $\mathbb{B}_1$ will be called *admissible*, if the sequence of images $F(x_i)$ converges in $\mathbb{B}_2$ as well. Now let $\lim x_i \in \mathbb{B}_1$ be the limit element belonging to $\mathbb{B}_1$ (the limit as $i \to \infty$ in $\mathbb{B}_1$) of an admissible sequence $\{x_i\}$ and let $\widetilde{\mathbb{D}}(F) \subset \mathbb{B}_1$ be the set of all limit elements of admissible sequences $\{x_i\} \subset \mathbb{D}(F)$. The *strong extension* (or *closure*) of the operator $F : \mathbb{B}_1 \to \mathbb{B}_2$ is introduced (as it is usually done) as the operator $\widetilde{F} : \mathbb{B}_1 \to \mathbb{B}_2 : x \to \widetilde{F}(x)$ defined for those and only those elements $z \in \widetilde{\mathbb{D}}(F)$ (of the set $\widetilde{\mathbb{D}}(F)$) for which the following conditions are satisfied: $x_i \to z$ (the sequence $\{x_i\}$ converges to z) and $y_i \to z$ for arbitrary $\{x_i\}, \{y_i\} \subset \widetilde{\mathbb{D}}(F)$ imply $\lim F(x_i) = \lim F(y_i)$ and acting according to the rule $\widetilde{F}(\lim x_i) \equiv \lim F(x_i)$. As it is easily seen, this extension is indeed an operator, i.e., the values $\widetilde{F}(z)$ are uniquely defined on the indicated limit elements z (do not depend on the choice of an admissible sequence converging to z in $\mathbb{B}_1$). Moreover, the domain of definition $\mathbb{D}(\widetilde{F})$ of the operator $\widetilde{F}$ contains $\mathbb{D}(F)$ and is contained in $\widetilde{\mathbb{D}}(F)$, $\mathbb{D}(\widetilde{F}) \subset \widetilde{\mathbb{D}}(F)$. Note also that the strong extension $\widetilde{F}$ is necessarily a *closed operator*, i.e., $x_i \to z$ and $\widetilde{F}(x_i) \to h$ (convergence in $\mathbb{B}_1$ and $\mathbb{B}_2$, respectively) implies $z \in \mathbb{D}(\widetilde{F})$ and $\widetilde{F}(z) = h$ (which immediately follows from the definition of $\widetilde{F}$).

It is obvious that any continuous operator $F : \mathbb{B}_1 \to \mathbb{B}_2$ (a continuous function) whose domain of definition $\mathbb{D}(F)$ is a linear set dense in $\mathbb{B}_1$ has a closure. Then $\mathbb{D}(\widetilde{F}) = \widetilde{\mathbb{D}}(F) = \mathbb{B}_1$.

The requirement that the domain of definition be linear and dense in $\mathbb{B}_1$ is assumed to be satisfied also for other operators $F : \mathbb{B}_1 \to \mathbb{B}_2$ investigated below.

Any operator $F : \mathbb{B}_1 \to \mathbb{B}_2$ (with the domain of definition $\mathbb{D}(F)$) such that there exists a linear continuous operator $F'(x_0) : \mathbb{B}_1 \to \mathbb{B}_2,\ u \to F'(x_0)u$ defined on $\mathbb{D}(F)$ and such that

$$\|F(x_0 + u) - F(x_0) - F'(x_0)u\|_2 = o(\|u\|_1)$$

is continuous at the point (a given element) $x_0 \in \mathbb{D}(F)$. In this case, the operator F is said to be *differentiable* (*in the sense of Frechét*) at x_0 and the operator $F'(x_0)$ is the *derivative* of F at x_0. From the estimate for the *remainder term*

$$F(x_0 + u) - F(x_0) - F'(x_0)u$$

it immediately follows that the derivative $F'(x_0) : \mathbb{B}_1 \to \mathbb{B}_2$ is uniquely defined.

In the case when the operator $F : \mathbb{B}_1 \to \mathbb{B}_2$ is differentiable at each point of the interval

$$(1 - t)x_0 + t(x_0 + u), \qquad 0 \leqslant t \leqslant 1,$$

the *formula for finite differences with remainder term* [**44**] or the inequality

$$\|F(x_0 + u) - F(x_0) - F'(x_0)u\|_2 \leqslant \|u\|_1 \sup_{0<t<1} \|F'(x_0 + tu) - F'(x_0)\|$$

is valid. The norm $\|A\|$ of the continuous linear operator

$$A\colon \mathbb{B}_1 \to \mathbb{B}_2 \qquad (A \equiv F'(x_0 + tu) - F'(x_0))$$

with the dense domain of definition $\mathbb{D}(A)$ (for A under consideration $\mathbb{D}(A) = \mathbb{D}(F)$) is defined by the standard identity

$$\|A\| \equiv \sup_{u \in \mathbb{D}(A), u \neq 0} \frac{\|Au\|_2}{\|u\|_1}.$$

For an operator $F\colon \mathbb{B}_1 \to \mathbb{B}_2$ *continuously differentiable* at x_0, or for F differentiable for each element from $\mathbb{D}(F)$ that belongs to a fixed neighborhood of $x_0 \in \mathbb{D}(F)$ under consideration (the topology of the Banach space is assumed to be "natural" or generated by the norm introduced) with the derivative continuous at x_0 in the sense that the requirement

$$\|F'(x_0 + u) - F'(x_0)\| \to 0 \quad \text{as } x_0 \to 0$$

is satisfied, the formula for finite differences defines more precisely the right-hand part $o(\|u\|_1)$ of the indicated estimate for the remainder term

$$F(x_0 + u) - F(x_0) - F'(x_0)u.$$

The set of linear continuous operators $A\colon \mathbb{B}_1 \to \mathbb{B}_2$ with a common domain of definition dense in $\mathbb{B}_1$ forms a new linear normed *space* $[\mathbb{B}_1, \mathbb{B}_2]$ with the norm $\|A\|$ defined as above. If the operator $F\colon \mathbb{B}_1 \to \mathbb{B}_2$ is differentiable everywhere in $\mathbb{D}(F)$, then a new operator

$$F'\colon \mathbb{B}_1 \to [\mathbb{B}_1, \mathbb{B}_2],\ x \to F'(x), \qquad \mathbb{D}(F') = \mathbb{D}(F),$$

is defined, i.e., the *derivative of F on $\mathbb{D}(F)$*. The fact that F' is uniquely defined for a given F again follows from the fact that the derivative $F'(x_0)$ at a point is uniquely defined. Then we can ask whether the operator $F'\colon \mathbb{B}_1 \to [\mathbb{B}_1, \mathbb{B}_2]$ is differentiable at the point x_0 and introduce into consideration its derivative at the given point $F''(x_0) \in [\mathbb{B}_1, [\mathbb{B}_1, \mathbb{B}_2]]$, i.e., the *second derivative* at x_0 of the initial operator F. By the continuation of this sequence of inductive constructions, one can introduce the derivatives of higher orders.

The *smoothness* (or continuous differentiability) of a given operator $F\colon \mathbb{B}_1 \to \mathbb{B}_2$ on $\mathbb{D}(F)$ is provided by the existence of the second derivative F everywhere in $\mathbb{D}(F)$. The following lemma concerns smooth operators.

LEMMA 1. *The closure (extension by continuity) of a smooth operator $F\colon \mathbb{B}_1 \to \mathbb{B}_2$ with a dense domain of definition and acting from one Banach space $\mathbb{B}_1$ to another $\mathbb{B}_2$ is a smooth operator and the closure of the derivative of F coincides with the derivative of the closure of F.*

Indeed, using the continuity of the norm and extending by continuity the above-given formula for finite differences with the remainder term to the elements of the space $\mathbb{B}_1$, we come to the differentiability of $\widetilde{F}$ on $\mathbb{B}_1$ and to the identity

$$\widetilde{F}'(x) = \widetilde{F'(x)}, \qquad x \in \mathbb{B}_1,$$

(the derivative of the closure coincides with the closure of the derivative). Then the continuity of F' implies the continuity of $\widetilde{F}'$.

10.2. The conditions for splitting. We pass to the main constructions. Let $\mathbb{B}$ be a Banach space over the field of real numbers R, let $f : \mathbb{B} \to R$, $z \to f(z)$ be a real functional (function) on $\mathbb{B}$, and let

$$\Delta^2 f(z) \colon \mathbb{B} \to R, \ u \to \Delta^2 f(z)(u) \equiv f(z+u) - 2f(z) + f(z-u)$$

be the *second difference* of f at a point $z \in \mathbb{B}$. The functional f is *regular* if the decomposition of the space $\mathbb{B}$ into the direct sum $\mathbb{B}_x \dot{+} \mathbb{B}_y$ of its subspaces $\mathbb{B}_x, \mathbb{B}_y \subset \mathbb{B}$ is valid for it,

$$\mathbb{B} = \mathbb{B}_x + \mathbb{B}_y : \qquad z = x + y \equiv (x, y),$$

and the constant $a > 0$ is defined such that for any $z \in \mathbb{B}$ the following conditions for splitting are satisfied:

$$(1) \qquad \begin{aligned} \Delta^2 f(z)(u) &\geqslant a\|u\|^2 \qquad \text{for} \quad u \in \mathbb{B}_x, \\ \Delta^2 f(z)(u) &\leqslant -a\|u\|^2 \qquad \text{for} \quad u \in \mathbb{B}_y. \end{aligned}$$

Here $\|\cdot\|$ is the norm of the space $\mathbb{B}$.

Additionally, it is assumed that f is differentiable on $\mathbb{B}$.

We investigate some properties of regular functionals. According to the standard terminology, a function $f : \mathbb{B} \to R$ is *convex downward* (*upward*) if for any $z_0, z_1 \in \mathbb{B}$ and any $0 \leqslant t \leqslant 1$

$$\delta_f^2(z_0, z_1, t) \equiv tf(z_1) + (1-t)f(z_0) - f(tz_1 + (1-t)z_0) \geqslant 0 \qquad (\leqslant 0).$$

LEMMA 2. *A continuous function $f : \mathbb{B} \to R$ defined on $\mathbb{B}$ is convex downward (upward) if and only if for any $z, u \in \mathbb{B}$ its second difference $\Delta^2 f(z)(u) \geqslant 0$ ($\leqslant 0$).*

It is evident that this restriction is necessary. Conversely, it implies that for any sum

$$t_s = a_1 2^{-1} + \cdots + a_s 2^{-s}, \qquad s = 1, 2, \ldots, \qquad a_1, \ldots, a_s \in \{0, 1\},$$

and for any $z, u \in \mathbb{B}$ we have $\delta_f^2(z_0, z_1, t_s) \geqslant 0$ ($\leqslant 0$). The fact that the set of such sums on the interval $0 \leqslant t \leqslant 1$ is dense and that the function f is continuous completes the proof.

PROPOSITION 1. *For any fixed $y_0 \in \mathbb{B}_y$ ($x_0 \in \mathbb{B}_x$), the function $x \to f(x, y_0)$ (the function $y \to f(x_0, y)$) generated by the functional $f : \mathbb{B} \to R$ is convex downward (upward) on $\mathbb{B}_x$ (on $\mathbb{B}_y$).*

The proposition immediately follows from Lemma 2 and the splitting conditions (1).

PROPOSITION 2. *For any $x, x_1, x_2 \in \mathbb{B}_x$ and $y, y_1, y_2 \in \mathbb{B}_x$ the inequalities*

$$(2) \qquad f(x_1, y) - f(x_2, y) - f'(x_2, y)(x_1 - x_2) \geqslant \frac{a}{2}\|x_1 - x_2\|^2,$$

$$(3) \qquad -f(x, y_1) + f(x, y_2) + f'(x, y_2)(y_1 - y_2) \geqslant \frac{a}{2}\|y_1 - y_2\|^2$$

are valid for a regular functional $f : \mathbb{B} \to R$.

Indeed, setting for $n = 1, 2, \dots$

$$\omega = \frac{1}{n}(x_1 - x_2), \qquad [k] \equiv (x_2 + k\omega, y), \qquad k = 0, \mp 1, \dots,$$

we obtain from (1) that

$$f[i+1] - 2f[i] + f[i-1] \geqslant a\|\omega\|^2, \qquad i = 1, 2, \dots.$$

Summing the first m of the these inequalities and then the first n of those obtained, we find

$$f[n+1] - 2f[1] + n(f[1] - f[0]) \geqslant a\frac{n(n+1)}{2}\|\omega\|^2.$$

By letting n tend to ∞, we come to (2). Substituting in (2) $-f$ for f and $\mathbb{B}_y$ for $\mathbb{B}_x$, we obtain (3).

As it is usual, a zero z_* of the derivative $f' \colon \mathbb{B} \to [\mathbb{B}, R]$ is called a *critical point* of the differentiable function $f \colon \mathbb{B} \to R$; the element $z_* \in \mathbb{B}$ is such that for any $u \in \mathbb{B}$, $f'(z_*)u = 0$ (in short, $f'(z_*) = 0$). If $f \colon \mathbb{B} \to R$ is a regular functional, then the algebraic dimension $i(f) = \dim \mathbb{B}_y$ (the dimension $i(-f) = \dim \mathbb{B}_x$) of the space $\mathbb{B}_y$ (of the space $\mathbb{B}_x$), if it is finite, is called the *type* (the *cotype*, respectively) of the functional f. In the general case, the space $\mathbb{B}_x$ (the space $\mathbb{B}_y$) will be called the *attracting* (*repelling*) space of the corresponding regular functional f. Regular functionals of zero type or cotype will be called *definite*. For a definite functional, the space $\mathbb{B}$ coincides with $\mathbb{B}_x$ or $\mathbb{B}_y$. Nondefinite functionals will be called *bivalent*. Hence, the attracting and repelling spaces of a bivalent functional are proper subspaces of the space $\mathbb{B}$ (do not coincide with $\mathbb{B}$ and the zero element of the space $\mathbb{B}$).

PROPOSITION 3. *An element $z_* \in \mathbb{B}$ is a critical point of a regular functional $f \colon \mathbb{B} \to R$ if and only if it coincides with a point of extremum of f (if f is a definite functional),*

$$f(z_*) \leqslant f(z) \quad \text{for any } z \in \mathbb{B} \text{ if } i(f) = 0 \ (\text{or } \mathbb{B} = \mathbb{B}_x),$$
$$f(z_*) \geqslant f(z) \quad \text{for any } z \in \mathbb{B} \text{ if } i(-f) = 0 \ (\text{or } \mathbb{B} = \mathbb{B}_y),$$

or with a point of mixed extremum of f (if f is a bivalent functional),

$$z_* = (x_*, y_*), \quad f(x_*, y) \leqslant f(x_*, y_*) \leqslant f(x, y_*) \quad \text{for any } x \in \mathbb{B}_x \text{ and any } y \in \mathbb{B}_y.$$

The fact that these conditions are necessary immediately follows from (2) and (3). It is quite evident that they are sufficient.

COROLLARY. *For a regular functional $f \colon \mathbb{B} \to R$ having a critical point $z_* \in \mathbb{B}$, the following energy inequality is valid:*

$$(\mathrm{E}_*) \qquad \|f'(z)\| \geqslant \frac{a}{4}\|z - z_*\| \qquad \text{for any} \quad z \in \mathbb{B} \qquad (f'(z_*) = 0)$$

(recall that

$$\|f'(z)\| \equiv \sup\{|f'(z)u| : \quad u \in \mathbb{B}, \quad \|u\| = 1\}).$$

Indeed, let $z_* = z_1 = (x_1, y_1)$ and $z = z_2 = (x_2, y_2)$. By assuming in (2) and (3) that $y = y_2$, $x = x_2$ and using the inequalities

$$f(x_1, y_2) \leqslant f(x_1, y_1) \leqslant f(x_2, y_1),$$

which follow from the condition $f'(x_1, y_1) = 0$ and Proposition 3, we find that

$$(2') \qquad f(z_1) - f(z_2) - f'(z_2)u \geqslant \frac{a}{2}\|u\|^2, \qquad u = x_1 - x_2,$$

$$(2'') \qquad -f(z_1) + f(z_2) + f'(z_2)v \geqslant \frac{a}{2}\|v\|^2, \qquad v = y_1 - y_2.$$

Summing the inequalities obtained, we find that

$$f'(z_2)(v - u) \geqslant \frac{a}{2}(\|u\|^2 + \|v\|^2),$$

which, together with the inequalities

$$\|u \mp v\|^2 \leqslant (\|u\| + \|v\|)^2 \leqslant 2(\|u\|^2 + \|v\|^2),$$

leads to (E_*).

PROPOSITION 4. *A regular functional has at most one critical point.*

This proposition immediately follows from (E_*).

Note that from (E_*) there follows a certain *stability of a critical point* of a regular functional expressed by the condition:

$$\|f'(z_n)\| \to 0, \quad n \to \infty, \quad \text{implies } z_n \to z_* \text{ in } \mathbb{B}.$$

Note also that *for a definite* f (2) and (3) imply a stronger, in comparison to (E_*), inequality

$$\|f'(z_1) - f'(z_2)\| \geqslant \frac{a}{2}\|z_1 - z_2\|, \qquad z_1, z_2 \in \mathbb{B}.$$

Finally, the following inequality is worth mentioning:

$$\|f'(z_2) - g'(z_1)\| \geqslant \frac{a}{2}\|z_1 - z_2\|, \qquad f'(z_1) = g'(z_2) = 0,$$

where $g \colon \mathbb{B} \to R$ is a regular functional on $\mathbb{B}$ with the same $\mathbb{B}_x$, $\mathbb{B}_y$, and a as for f. This inequality immediately follows from (2′) and (2″) written for f and separately for g.

On the other hand, this inequality leads to the explicit form of the continuous dependence of the location of the critical point $z_* = z_*(f)$ of a regular functional f on the derivative f',

$$(\mathrm{E}_{**}) \qquad \|z_*(f) - z_*(g)\| \leqslant \frac{4}{a} \sup_{z \in \mathbb{B}} \|f'(z) - g'(z)\|.$$

Indeed, we have

$$\begin{aligned}\|f'(z_2) - g'(z_1)\| &= \|(f'(z_2) - g'(z_2)) + (f'(z_1) - g'(z_1))\| \\ &\leqslant \|f'(z_2) - g'(z_2)\| + \|f'(z_1) - g'(z_1)\| \\ &\leqslant 2 \sup_{z \in \mathbb{B}} \|f'(z) - g'(z)\|,\end{aligned}$$

which immediately implies (E_{**}).

10.3. The existence of a critical point. We shall clarify the conditions for the existence of a critical point for a regular differentiable functional $f: \mathbb{B} \to R$.

PRORPOSITION 5. *If $f: \mathbb{B} \to R$ is a definite functional with index $i(f) = 0$ (with coindex $i(-f) = 0$), i.e., with $\mathbb{B}_x = \mathbb{B}$ (with $\mathbb{B}_y = \mathbb{B}$), then there exists the minimum point (the maximum point, respectively) $z_* \in \mathbb{B}$ of the functional f.*

Indeed, for $\mathbb{B} = \mathbb{B}_x$, inequality (2) implies the boundedness of the functional f from below,

$$f(z) \geqslant f(0) + \frac{a}{2}\|z\|^2 - \|f'(0)\|\|z\| \geqslant f(0) - \frac{1}{2a}\|f'(0)\|^2 > -\infty, \qquad z \in \mathbb{B}.$$

Consequently, the infimum d of the values $f(z)$ of the functional f on $\mathbb{B}$ exists and is attained on some sequence

$$z_k \subset \mathbb{B} : f(z_k) - d \to +0, \qquad k \to \infty \qquad (k = 1, 2, \dots).$$

Setting in (1) $z = z_{km} = (z_k + z_m)/2$ and $u = (z_k - z_m)/2$, $k, m = 1, 2, \dots$, we obtain the convergence of z_k in $\mathbb{B}$:

$$\frac{a}{4}\|z_k - z_m\|^2 \leqslant f(z_k) - 2f(z_{km}) + f(z_m) \leqslant (f(z_k) - d) + (f(z_m) - d), \qquad k, m \to \infty.$$

The continuity of f (provided by the differentiability of f) implies that

$$f(\lim z_k) = \lim f(z_k) = d.$$

Hence, we can set $z_* = \lim z_k$ (convergence in $\mathbb{B}$). The case $i(-f) = 0$ is reduced to that considered above by formally replacing f by $-f$. The proof is completed.

Now let $\mathbb{B}^* = [\mathbb{B}, R]$ be the Banach space of linear continuous functionals

$$l : \mathbb{B} \to R, \ u \to lu$$

given on $\mathbb{B}$ with the norm

$$\|l\| = \sup\{|lu| : u \in \mathbb{B}, \ \|u\| = 1\}$$

(the so-called *dual space*). The set of linear continuous functionals $X: \mathbb{B}^* \to R: l \to X(l)$ defined on $\mathbb{B}^*$ (for each of these functionals there exists an element $u \in \mathbb{B}$ such that for any $l \in \mathbb{B}^*$ the equality $X(l) = lu$ is valid) forms some subspace (linear closed subset) $\mathbb{B}_0^{**}$ of the second dual space $\mathbb{B}^{**}$, the former coinciding up to a linear isometry (i.e., a linear transformation preserving the norm) with $\mathbb{B}$ **[44]**.

By definition, a space $\mathbb{B}$ is *reflexive* if $\mathbb{B}_0^{**} = \mathbb{B}^{**}$. A (*complete*) *criterion of reflexivity* of a given Banach space $\mathbb{B}$ is the requirement that a bounded set of the space $\mathbb{B}$ be compact with respect to the weak topology (or the topology of the dual space) **[31]**. For this case the following proposition is valid.

PROPOSITION 6. *If the space $\mathbb{B}$ is reflexive, then there exists a point $z_* \in \mathbb{B}$ of mixed extremum of a bivalent functional $f: \mathbb{B} \to R$.*

The proposition follows from the differentiability (hence, continuity) of f on $\mathbb{B}$, Proposition 1, *coercitivity* of f on each of the subspaces $\mathbb{B}_x$ and $\mathbb{B}_y$ ($f(x,y) \to +\infty$ ($f(x,y) \to -\infty$) as $\|x\| \to \infty$ (as $\|y\| \to \infty$) for an arbitrary fixed $y \in \mathbb{B}_y$ ($x \in \mathbb{B}_x$, respectively)), which follows from inequalities (2), (3) and Proposition 2.2 from Chapter 6 of monograph **[108]**.

Now we pass to the study of the existence of a point of mixed extremum for a sufficiently smooth bivalent functional given on an arbitrary (possibly, nonreflexive) Banach space. We associate to the bivalent functional $f: \mathbb{B} \to R$ the mapping $S_+: \mathbb{B}_x \to \mathbb{B}_y$ of the space $\mathbb{B}_x$ to the space $\mathbb{B}_y$, and the mapping $S_-: \mathbb{B}_y \to \mathbb{B}_x$ of the space $\mathbb{B}_y$ to the space $\mathbb{B}_x$ which, for any $x \in \mathbb{B}_x$ and $y \in \mathbb{B}_y$, satisfy the inequalities

$$f(S_-(y), y) \leqslant f(x,y) \leqslant f(x, S_+(x)). \tag{4}$$

The existence and uniqueness of the mappings $S_\mp$ immediately follow from Propositions 3–5. It follows from Proposition 3 that a necessary and sufficient condition for the validity of inequalities (4) for the given $S_\mp$ is the validity of the corresponding identities

$$f'_x(S_-(y), y) = 0, \qquad f'_y(x, S_+(x)) = 0, \tag{5}$$

where $f'_x(z_0): \mathbb{B}_x \to R: u \to f'_x(z_0)u$ is the derivative of f at the point $z_0 = (x_0, y_0)$ for a fixed $y_0 \in \mathbb{B}_y$ (*partial derivative with respect to* x). If there is the "complete" derivative $f'_x(z_0)$ (the Frechét derivative defined as above), then this partial derivative, obviously, coincides with the restriction of the mapping $f'(z_0): \mathbb{B} \to R: u \to f'(z_0)u$ to the elements u of the space $\mathbb{B}_x$. The partial derivative $f'_y(z_0): \mathbb{B}_y \to R$ is introduced into consideration analogously. As before, the equality $f'_x(z_0) = 0$ is considered as an abbreviated form of the identity $f'_x(z_0)u = 0$ which holds for any u from $\mathbb{B}_x$.

PROPOSITION 7. *In the case of an arbitrary Banach space* $\mathbb{B}$, *a bivalent functional* $f: \mathbb{B} \to R$ *has a point of mixed extremum if at least one of the mappings* $S_+: \mathbb{B}_x \to \mathbb{B}_y$ *or* $S_-: \mathbb{B}_y \to \mathbb{B}_x$ *is differentiable* (*everywhere on* $\mathbb{B}_x$ *or* $\mathbb{B}_y$, *respectively*).

Indeed, assuming that the mapping $S_+: \mathbb{B}_x \to \mathbb{B}_y$ is differentiable and setting in this case for an arbitrary $x \in \mathbb{B}_x$

$$\psi_+(x) \equiv f(x, S_+(x)),$$

we find from (1) and (4) that

$$\begin{aligned}
\Delta^2\psi_+(x)(u) &\equiv \psi_+(x+u) - 2\psi_+(x) + \psi_+(x-u) \\
&= f(x+u, S_+(x+u)) - 2f(x, S_+(x)) + f(x-u, S_+(x-u)) \\
&\geqslant f(x+u, S_+(x)) - 2f(x, S_+(x)) + f(x-u, S_+(x)) \\
&\geqslant \Delta^2 f_+(x, S_+(x))(u) \\
&\geqslant a\|u\|^2, \qquad u \in \mathbb{B}_x,
\end{aligned}$$

i.e., $\psi_+: \mathbb{B}_x \to R: x \to \psi_+(x)$ is a definite functional of type 0. The differentiability of S_+ implies the differentiability of ψ_+ (as of a composite function). Then (Proposition 5) the functional $\psi_+: \mathbb{B}_x \to R$ has a minimum point $x_* \in \mathbb{B}_x$ coinciding with a critical point of ψ_+ (Proposition 3). Taking into account (5), we find that

$$z_* = (x_*, S_+(x_*))$$

is a critical point of the functional f,

$$f'(z_*) = f'_x(z_*) = \psi'_+(x_*) = 0.$$

Proposition 3 completes the analysis of this case. The case when the mapping $S_-\colon \mathbb{B}_y \to \mathbb{B}_x$ is differentiable is reduced to the one considered above by a formal replacement of f with $-f$. The validity of the proposition is established.

Now we shall study conditions under which the mapping $S_\mp$ is differentiable. Denote by $C^2(\mathbb{B})$ the class of *twice continuously differentiable functionals* $f\colon \mathbb{B} \to R$ defined on $\mathbb{B}$ (which are introduced in analogy to the continuously differentiable functionals [**44**]) and let

$$f''(z)(u,u), \qquad u \in \mathbb{B},$$

be the quadratic form of the second derivative $f''(z) \in [\mathbb{B}, [\mathbb{B}, R]]$, or the *Hessian* of the functional f at the point $z \in \mathbb{B}$, obtained by the restriction of the bilinear continuous (hence, symmetric [**44**]) mapping $\mathbb{B} \times \mathbb{B} \to R$: $f''(z)(u, v)$ induced by the second derivative $f''(z)$ to the diagonal $u = v$ of the Cartesian direct product $\mathbb{B} \times \mathbb{B}$.

LEMMA 3. *A necessary condition for the validity of the splitting conditions* (1) *for a given* $f \in C^2(\mathbb{B})$ *is that the inequalities*

$$(6) \qquad f''(z)(u,u) \geqslant b\|u\|^2 \quad (\leqslant -b\|u\|^2) \quad \textit{for} \quad u \in \mathbb{B}_x \quad (u \in \mathbb{B}_y) \textit{ and } z \in \mathbb{B}$$

with the constant $b = a/2$ *be valid. Conversely, inequalities* (6) *imply* (1) *with the constant* $a = b$.

Indeed, setting

$$Q_f(z,u,v) \equiv f''(z+v)(u,u) - 2f''(z)(u,u) + f''(z-v)(u,u),$$

we find that

$$\Delta^2 f(z)(u) - f''(z)(u,u) = \int_0^1 (1-t)Q_f(z,u,tu)\,dt,$$

which for $\|u\| < \varepsilon$ and small $\varepsilon = \varepsilon(z) = \text{const} > 0$ implies

$$|\Delta^2 f(z)(u) - f''(z)(u,u)| \leqslant \frac{a}{2}\|u\|^2.$$

Then inequalities (6) with $b = a/2$ follow from (1). Due to the homogeneity of inequalities (6) with respect to the variable $u \in \mathbb{B}$, the restriction $\|u\| < \varepsilon$ in the latter can be omitted. Conversely, substituting $z + tu$, $t \in R$, for z in (6), we obtain

$$f''(z+tu)(u,u) = \frac{d^2}{dt^2} f(z+tu).$$

Integrating these inequalities first with respect to t from $-s$ to $s > 0$ and then with respect to s from 0 to 1, we come to inequalities (1) with $a = b$. The proof is completed.

PROPOSITION 8. *If a bivalent functional* $f \in C^2(\mathbb{B})$, *then* $S_\pm$ *are differentiable mappings.*

Let us prove the differentiability of the mapping $S_+ : \mathbb{B}_x \to \mathbb{B}_y$. In order to do this, we show that for any $x \in \mathbb{B}_x$ and

$$z_+ = (x, S_+(x))$$

the repeated partial derivative $f''_{yy}(z_+) \in [\mathbb{B}_y, [\mathbb{B}_y, R]]$ has the inverse continuous operator $[f''_{yy}(z_+)]^{-1} \in [[\mathbb{B}_y, R], \mathbb{B}_y]$, defined on the whole space $[\mathbb{B}_y, R]$. Fixing an arbitrary linear continuous functional $l \in [\mathbb{B}_y, R]$ and setting

$$\psi(y) \equiv \frac{1}{2} f''_{yy}(z_+)(y, y) - ly, \qquad y \in \mathbb{B}_y,$$

we find that

1) $\psi : \mathbb{B}_y \to R : y \to \psi(y)$ is a definite functional of cotype 0:

$$\Delta^2 \psi(y)(v) = f''_{yy}(z_+)(v, v) \leqslant -b\|v\|^2, \qquad v \in \mathbb{B}_y,$$

2) $\psi : \mathbb{B}_y \to R$ is a differentiable functional and

$$\psi'(y)v = f''_{yy}(z_+)(y, v) - lv, \qquad v \in \mathbb{B}_y,$$

(we use the symmetry of the bilinear form $f''_{yy}(z_+)(u, v)$, $u, v \in \mathbb{B}_y$, resulting from the condition $f \in C^2(\mathbb{B})$),

3) $\|y_1 - y_2\| \leqslant \frac{4}{b}\|l_1 - l_2\|$, $f''_{yy}(z_+)(y_i, v) = l_i v$, $v \in \mathbb{B}_y$, $i = 1, 2$ (a consequence of inequality (E_{**})).

As a consequence of properties 1) and 2), the equation

$$f''_{yy}(z_+)(y, v) = lv,$$

considered for an arbitrary $v \in \mathbb{B}_y$, is uniquely solvable with respect to $y \in \mathbb{B}_y$ for any given $l \in [\mathbb{B}_y, R]$ (Propositions 3–5). Consequently, the inverse operator

$$[f''_{yy}(z_+)]^{-1} : [\mathbb{B}_y, R] \to \mathbb{B}_y$$

is defined on $[\mathbb{B}_y, R]$. By property 3), the inverse operator is continuous on $[\mathbb{B}_y, R]$. The proof is completed by the above-mentioned equivalence of conditions (4) and (5) which follows from inequalities (2) and (3), and by the theorem on the differentiability of the function $S_+ : \mathbb{B}_x \to \mathbb{B}_y$ implicitly defined by the second equation from (5) **[44]**.

The differentiability of the mapping $S_- : \mathbb{B}_y \to \mathbb{B}_x$ is established by analogous arguments.

Summing up, we come to the following statement.

THEOREM 1. *A regular differentiable functional $f : \mathbb{B} \to R$ has a unique critical point if at least one of the following restrictions is valid:*

(A) $\mathbb{B}_x = \mathbb{B}$ *or* $\mathbb{B}_y = \mathbb{B}$,

(B) $\mathbb{B}_x$, $\mathbb{B}_y \neq \mathbb{B}$, *and the space* $\mathbb{B}$ *is reflexive,*

(C) $\mathbb{B}_x$, $\mathbb{B}_y \neq \mathbb{B}$, *and the functional* $f \in C^2(\mathbb{B})$.

The theorem follows from Propositions 3–5 for the case (A), Propositions 3, 4, 6 for the case (B), and Propositions 3, 4, 7, 8 for the case (C).

10.4. An abstract version of the eigenvalue problem. Specific versions of the following lemma frequently appear in applications. For Sobolev spaces, it is proved in the monograph **[88]**. We establish it in the general case.

LEMMA 4. *Let* $M = \dots, x, y, u, v, \dots$ *be a linear manifold equipped with scalar products* (x, y) *and* $\{x, y\}$, *and let* $\mathbb{H}$ *and* $\mathbb{W}$ *be the Hilbert spaces obtained by the completion of* M *in the norms* $\|x\| \equiv (x, x)^{1/2}$ *and* $\|x\|_1 \equiv \{x, x\}^{1/2}$ *(generated by the introduced scalar products), respectively. Let the inequality for the norms* $\|x\|_1 \geqslant C\|x\|$ *be valid for all* $x \in M$ *and some constant* $C > 0$ *(independent of* x*), guaranteeing the embedding of the space* $\mathbb{W}$ *in* $\mathbb{H}$, $\mathbb{W} \subset \mathbb{H}$. *Let this embedding be compact (i.e., any subsequence from* M *bounded in the norm of the space* $\mathbb{W}$ *contains a subsequence converging in* $\mathbb{H}$*). Then the abstract spectral problem, generated by the embedding* $\mathbb{W} \subset \mathbb{H}$, *which consists in finding a number* $\lambda \in R$ *and an element* $u \in \mathbb{W}$, $\|u\| = 1$, *(they are an eigenvalue and an eigenelement for this embedding or for the spectral problem under consideration) that satisfy, for any* $v \in \mathbb{W}$, *the identity*

$$\{u, v\} = \lambda(u, v)$$

has a complete system of eigenelements u *forming an orthonormal basis of the space* $\mathbb{H}$ *and an orthogonal basis of the space* $\mathbb{W}$; *each eigenvalue* λ *is positive and finite, and the total set of the indicated eigenvalues has no finite limit point and can be arranged in the increasing order,*

$$(7) \qquad \lambda_1 < \lambda_2 < \dots < \lambda_n < \lambda_{n+1} < \dots \qquad (\lambda_1 > 0).$$

For any $n = 1, 2, \dots$, *the number* λ_{n+1} *is the greatest lower bound of the Rayleigh function* $R(u) \equiv \|u\|_1^2 / \|u\|^2$, *restricted to the elements* $u \neq 0$ *of* M *orthogonal in* $\mathbb{H}$ *to the eigenelements of the spectral problem corresponding to the first* n *eigenvalues of the series* (7), *and the number* λ_1 *is the greatest lower bound of this function for* $u \neq 0$:

$$\lambda_1 = \inf\{R(u) : u \in M,\ u \neq 0\}.$$

Indeed, let $S \subset M$ be the set of elements u from M that satisfy the condition $\|u\| = 1$. Obviously,

$$\inf\{R(u) : u \in M,\ u \neq 0\} = \inf\{\|u\|_1^2 : u \in S\}.$$

Since $\|u\|_1^2 \geqslant 0$, the greatest lower bound

$$d = \inf\{\|u\|_1^2 : u \in S\}$$

exists and is attained on some sequence $u_1, u_2, \dots \in S$:

$$d = \lim d_n \quad (n \to \infty), \qquad d_n = \|u_n\|_1^2, \qquad n = 1, 2, \dots.$$

The convergence of the sequence $\|u_1\|_1^2, \|u_2\|_1^2, \dots$ to d implies that the minimizing sequence $u_1, u_2, \dots$ from M considered is bounded in $\mathbb{W}$; hence (the embedding $\mathbb{W} \subset \mathbb{H}$ is compact), some subsequence of it converges in $\mathbb{H}$. Obviously, the latter is also minimizing, therefore, without loss of generality it can be identified with the initial sequence $u_1, u_2, \dots$. In this case, for any $k, m = 1, 2, \dots$ we have

$$\|(u_k + u_m)/2\|_1^2 \geqslant d\|(u_k + u_m)/2\|^2.$$

Consequently,

$$\begin{aligned} \|(u_k - u_m)/2\|_1^2 &= \|u_k\|_1^2/2 + \|u_m\|_1^2/2 - \|(u_k + u_m)/2\|_1^2 \\ &\leqslant \|u_k\|_1^2/2 + \|u_m\|_1^2/2 - d\|(u_k + u_m)/2\|^2. \end{aligned}$$

Since

$$\|(u_k + u_m)/2\|^2 = \|u_k\|^2/2 + \|u_m\|^2/2 - \|(u_k - u_m)/2\|^2$$

and

$$\|u_k\| = \|u_m\| = 1, \qquad \|u_k\|_1^2 = d_k, \qquad \|u_m\|_1^2 = d_m,$$

we conclude that

$$\|(u_k - u_m)/2\|_1^2 \leqslant (d_k - d)/2 + (d_m - d)/2 + d\|u_k - u_m\|^2/4,$$

and for $d_k, d_m \to d$ we have that $\|u_k - u_m\| \to 0$ implies $\|u_k - u_m\|_1 \to 0$ (the convergence of the sequence $u_1, u_2, \dots$ in $\mathbb{W}$).

Now let $\varphi \in \mathbb{W}$ be the limit element of the sequence $u_1, u_2, \dots$ in $\mathbb{W}$ (and, hence, in $\mathbb{H}$). By the continuity of the norm, we have

$$\|\varphi\|_1^2 = d, \qquad \|\varphi\| = 1.$$

Thus, $\varphi \neq 0$ (here 0 is the zero element of the space $\mathbb{W}$). We show that for any $v \in \mathbb{W}$, $v \neq 0$,

$$\{\varphi, v\} = d(\varphi, v).$$

Indeed, taking an arbitrary $v \in \mathbb{W}$, $v \neq 0$, for any

$$-1/\|v\| < t < 1/\|v\|,$$

the differentiable function

$$f(t) = \frac{1}{2}\|\varphi + tv\|_1^2/\|\varphi + tv\|^2$$
$$(\|\varphi + tv\| \geqslant \|\varphi\| - |t|\|v\| = 1 - |t|\|v\| > 0)$$

attains its smallest value $d/2$ at the interior point $t = 0$ of its interval of definition. Hence, the derivative

$$f'(0) = \{\varphi, v\}/\|\varphi\|^2 - \|\varphi\|_1^2(\varphi, v)/\|\varphi\|^4 = \{\varphi, v\} - d(\varphi, v) = 0,$$

which completes the proof of the required statement.

Thus, the d and φ obtained are solutions of the spectral problem under investigation. It is clear that $d > 0$ (since $d = \|\varphi\|_1^2 \geqslant C^2\|\varphi\|^2$, $\|\varphi\| = 1$, and $C > 0$). Setting $d = \mu_1$ and $\varphi = \varphi_1$, we repeat the above arguments for the orthogonal complements in $\mathbb{W}$ and $\mathbb{H}$ of the line $s\varphi$, $-\infty < s < +\infty$ (for the elements $u \in S$ such that $\{u, \varphi\} = 0$, hence, (since $\{\varphi, u\} = d(\varphi, u)$ and $d > 0$) such that $(\varphi, u) = 0$), etc., and obtain the sequence of eigenvalues

$$\mu_1 \leqslant \mu_2 \leqslant \mu_3 \leqslant \dots \leqslant \mu_n \leqslant \mu_{n+1} \leqslant \dots \qquad (\mu_1 > 0)$$

and the corresponding sequence of eigenfunctions

$$\varphi_1, \varphi_2, \varphi_3, \dots, \varphi_n, \varphi_{n+1}, \dots, \qquad \varphi_k \in \mathbb{W}, \quad \|\varphi_k\| = 1,$$
$$\{\varphi_k, v\} = \mu_k(\varphi, v) \quad \text{for any } v \in \mathbb{W}, k = 1, 2, \dots,$$

such that

$$(\varphi_2, \varphi_1) = 0,$$
$$(\varphi_3, \varphi_1) = (\varphi_3, \varphi_2) = 0,$$
$$\dots\dots\dots\dots\dots$$
$$(\varphi_{n+1}, \varphi_1) = (\varphi_{n+1}, \varphi_2) = \dots = (\varphi_{n+1}, \varphi_n) = 0,$$
$$\dots\dots\dots\dots\dots\dots\dots$$

and

$$\mu_1 = \inf\{\|u\|_1^2 : u \in S\},$$
$$\mu_2 = \inf\{\|u\|_1^2 : u \in S,\ (u, \varphi_1) = 0\},$$
$$\dots\dots\dots\dots\dots$$
$$\mu_{n+1} = \inf\{\|u\|_1^2 : u \in S,\ (u, \varphi_1) = \dots = (u, \varphi_n) = 0\},$$
$$\dots\dots\dots\dots\dots\dots\dots$$

In the case of a finite-dimensional linear manifold M, these sequences, obviously, terminate at a finite step. We show that for the case of an infinite-dimensional M, the sequence $\mu_1, \mu_2, \dots$ has no finite limit points.

Indeed, otherwise some subsequence of the sequence $\mu_1, \mu_2, \dots$ would converge. Then the corresponding (with the same indices) subsequence of the sequence of the eigenelements $\varphi_1, \varphi_2, \dots$ is bounded in $\mathbb{W}$ (since $\|\varphi_k\|_1^2 = \mu_k$, $k = 1, 2, \dots$) and, hence, contains another subsequence converging in $\mathbb{H}$, which is impossible since for all $k, m = 1, 2, \dots$

$$\|\varphi_k - \varphi_m\|^2 = \|\varphi_k\|^2 - 2(\varphi_k, \varphi_m) + \|\varphi_m\|^2 = 1 - 0 + 1 = 2.$$

It follows immediately from the proof that the eigenvalues $\mu_1, \mu_2, \dots$ have finite *multiplicities* (the number of repetitions in the indicated series is finite), and, hence, the equality $\lim(1/\mu_n) = 0$ $(n \to \infty)$.

Now we prove that the limiting equality obtained guarantees the completeness of the orthonormal system of the eigenelements $\varphi_1, \varphi_2, \dots$ in $\mathbb{H}$.

Indeed, setting for an arbitrary $f \in M$

$$\overset{n}{f} \equiv \sum_{k=1}^{n} (f, \varphi_k)\varphi_k, \qquad \underset{n+1}{f} \equiv f - \overset{n}{f},$$

we have

$$(\underset{n+1}{f}, \varphi_1) = \dots = (\underset{n+1}{f}, \varphi_n) = 0,$$

which implies that $\underset{n+1}{f}$ belongs to the orthogonal complement in $\mathbb{H}$ of the linear span of the first n eigenelements $\varphi_1, \dots, \varphi_n$. Taking into account the variational properties of $\mu_1, \mu_2, \dots$ mentioned above, we find that

$$\|\underset{n+1}{f}\|^2 \leqslant \|\underset{n+1}{f}\|_1^2/\mu_{n+1}.$$

At the same time, the sequence $\|f_{n+1}\|_1^2$, $n = 1, 2, \dots$, is bounded:

$$\|\underset{n+1}{f}\|_1^2 = \|f\|_1^2 - 2\{f, \overset{n}{f}\} + \|\overset{n}{f}\|_1^2 = \|f\|_1^2 - 2\sum_{k=1}^{n} \mu_k (f, \varphi_k)^2 + \sum_{k=1}^{n} \mu_k (f, \varphi_k)^2 \leqslant \|f\|_1^2.$$

Since $\lim(1/\mu_n) = 0$, we conclude that

$$\lim \|\underset{n+1}{f}\| = 0 \quad \text{for } f \in M. \tag{8}$$

Now let $f \in \mathbb{H}$. Since M is dense in $\mathbb{H}$, for any $\varepsilon > 0$ there must exist an element $f_\varepsilon \in M$ such that $\|f - f_\varepsilon\| \leqslant \varepsilon$. Choosing for a given ε a number $n_\varepsilon = 1, 2, \ldots$ such that $\|\underset{n+1}{f_\varepsilon}\| \leqslant \varepsilon$ for $n \geqslant n_\varepsilon$ (by means of (8)), we obtain

$$\|\underset{n+1}{f}\| \leqslant \|\underset{n+1}{f_\varepsilon}\| + \|\underset{n+1}{f} - \underset{n+1}{f_\varepsilon}\| \leqslant \varepsilon + \|\underset{n+1}{f - f_\varepsilon}\| \leqslant 2\varepsilon, \qquad n \geqslant n_\varepsilon,$$

which completes the proof of the completeness of the given orthonormal system $\varphi_1, \varphi_2, \ldots$ in $\mathbb{H}$.

We show the completeness of this system in $\mathbb{W}$. Let $f \in M$ and $\{f, \varphi_k\} = 0$, $k = 1, 2, \ldots$. We have

$$(f, \varphi_k) = \frac{1}{\mu_k}\{f, \varphi_k\} = 0, \qquad k = 1, 2, \ldots.$$

It follows from these equalities that the element $f \in M$ is orthogonal to the closure of the linear span of the system of elements $\varphi_1, \varphi_2, \ldots$ in $\mathbb{H}$. The completeness of this system in $\mathbb{H}$ and the lemma on the orthogonal decomposition of a Hilbert space **[35]** imply that $f = 0$ in $\mathbb{H}$ and, hence (since $f \in M$), $f = 0$ in M (and, thus, in $\mathbb{W}$). Therefore, it is shown that the element $f \in M$, orthogonal in $\mathbb{W}$ to the closure in $\mathbb{W}$ of the linear span of the system $\varphi_1, \varphi_2, \ldots$, is necessarily zero. Again by the lemma on orthogonal decomposition, this element $f \in M$ can be approximated (with any prescribed accuracy) in $\mathbb{W}$ by a sequence of finite Fourier sums of elements of the system $\varphi_1, \varphi_2, \ldots$. Since f is an arbitrary element of M and M is dense in $\mathbb{W}$, any element of the space $\mathbb{W}$ can be approximated by the indicated sequences of finite Fourier sums. By identifying the multiple points of the sequence $\mu_1, \mu_2, \ldots$, we come to the required sequence $\lambda_1 < \lambda_2 < \ldots$. The lemma is proved.

CHAPTER 4

The Dissipative Top and the Navier-Stokes Equations

§11. Axioms of the dissipative top

11.1. Plane flows of a viscous incompressible fluid. A stationary plane (or plane-parallel) flow of a viscous incompressible fluid is described by a vector field $\mathbf{u} = (u, v)$ whose components

$$u = u(x, y), \quad v = v(x, y), \qquad u_x + v_y = 0, \qquad u_x \equiv \frac{\partial u}{\partial x}, \quad v_x \equiv \frac{\partial v}{\partial x}$$

satisfy the system of two-dimensional stationary Navier-Stokes equations with a parameter $\nu = \text{const} > 0$ (the coefficient of kinematic viscosity) at the highest derivatives. Introducing characteristic dimensional values of length h and velocity a ($h, a = \text{const} > 0$) and passing to the dimensionless variables

$$x \to hx, \qquad y \to hy, \qquad u \to au, \qquad v \to av,$$

we rewrite the equations in the following form:

$$\begin{aligned} u\frac{\partial u}{\partial x} + v\frac{\partial u}{\partial y} - \frac{1}{\mathrm{Re}}\Delta u + \frac{\partial p}{\partial x} &= w_1(x, y), \\ u\frac{\partial v}{\partial x} + v\frac{\partial v}{\partial y} - \frac{1}{\mathrm{Re}}\Delta v + \frac{\partial p}{\partial y} &= w_2(x, y). \end{aligned} \tag{1}$$

Here $\Delta \equiv \partial^2/\partial x^2 + \partial^2/\partial y^2$ is the Laplace operator, $\mathrm{Re} = ah/\nu$ is the Reynolds number ($\mathrm{Re} = \text{const} > 0$). As in the case of Euler equations (the limiting case $\mathrm{Re} = \infty$), under the assumption that in the flow domain there exists a flow function

$$\psi = \psi(x, y), \qquad u = \psi_y, \qquad v = -\psi_x,$$

the equations can be reduced (by means of the Gromeka-Lamb form, §2.1) to the equation for ψ

$$\frac{1}{\mathrm{Re}}\Delta\Delta\psi - \frac{\partial\psi}{\partial y}\frac{\partial\Delta\psi}{\partial x}\frac{\partial\psi}{\partial x}\frac{\partial\Delta\psi}{\partial y} = f(x, y), \tag{2}$$

which is considered in the flow domain V for a given *vorticity* $f \equiv \partial w_2/\partial x - \partial w_1/\partial y$ of the *field of the mass accelerations* $\mathbf{w} = (w_1, w_2)$.

The flow domain V is understood to be a domain of a two-dimensional smooth Riemannian manifold from §3.1, closed ($V = \overline{V}$) or with a (nonempty) boundary $\partial V = \overline{V} - V \neq \varnothing$ which satisfies the requirements of the same subsection. For the sake of simplicity, the metric is assumed to be Euclidean. In this case, equation (2) can be considered as the coordinate form of the corresponding invariant relation given in the Appendix (§17.7).

For small Reynolds numbers Re $\ll 1$, an approximation to relation (2) is the stationary *Stokes equation*:

$$\frac{1}{\text{Re}}\Delta\Delta\psi = f(x, y). \tag{3}$$

For $\partial V \neq \varnothing$ ($V \neq \overline{V}$), any of the following boundary problems:

$$\psi|_{\partial V} = \Delta\psi|_{\partial V} = 0, \tag{4}$$

or

$$\psi|_{\partial V} = \left.\frac{\partial\psi}{\partial n}\right|_{\partial V} = 0, \tag{5}$$

is correct for the (linear) equation (2). If $\partial V = \varnothing$ ($V = \overline{V}$ is a closed manifold), the correct problem for (3) is defined by the restriction

$$\int_V \psi\, dx\, dy = 0 \qquad (\partial V = \varnothing). \tag{6}$$

The same condition is assumed to hold in this case for $f(x, y)$. In (5), n is the outer normal at a regular (nonangular) point of the boundary ∂V. Restrictions (5) are the well-known *no-slip conditions.* Conditions (4) are the conditions of impermeability and absence of vorticity sources on ∂V valid for the interface of liquid media.

In what follows, we shall consider the following two-dimensional domains:

1. $V = K$ is *the disc* of radius $h = \text{const} > 0$ with center $x = y = 0$, $(x^2+y^2)^{1/2} = r < h$,
2. $V = \Pi$ is *the rectangle* with the sides $l, h = \text{const} > 0$, $0 < x < l$, $0 < y < h$,
3. $V = R$ is *the annulus* obtained by the identification of the opposite sides $x = 0$ and $x = l$ of the rectangle Π, or *the infinite channel* $-\infty < x < \infty$, $0 < y < h$ *under the conditions of periodicity* with respect to x with period l,

$$\psi(x + l, y) = \psi(x, y). \tag{7}$$

4. $V = T$ is *the torus* obtained by the identification of two pairs of opposite sides of the rectangle Π, or the *Euclidean plane factorized by the condition of periodicity with respect to the variables x and y with periods l and h, respectively,*

$$\psi(x + l, y) = \psi(x, y + h) = \psi(x, y). \tag{8}$$

For example, considered in the disc K under the no-slip conditions (5) with an infinitely differentiable in K right-hand part $f = f(x, y)$, both equations (2) and (3) have, as it is known **[56, 154]**, at least one infinitely smooth solution ψ in $\overline{K}$. However, in contrast to (3), the solution of equation (2) remains to be unique only for sufficiently small values of the parameter Re > 0 **[56, 154]**. The uniqueness of the solution of the formulated boundary problem for the nonlinear equation (2) is, therefore, "local" (corresponding to a bounded set of values of the Reynolds number). The fact that the corresponding uniqueness theorem is "local" cannot be explained away by a "misfortune" or defects of the method employed. The fact that the solution of the nonlinear problem under consideration is not unique is a characteristic feature which shows the real state of affairs **[94]**. The study of stationary solutions would make it possible to describe the preturbulent flow conditions mentioned in the Introduction (or secondary flows bifurcating from the basic flow under consideration) **[26, 38, 94]**. A part of this kind of analysis is the investigation of conditions imposed on $f(x, y)$ that ensure the uniqueness and stability (with respect to the corresponding nonstationary equation) of the main flow *for all the numbers* Re > 0. The use of the corresponding *nonlocal uniqueness and stability conditions* could make it possible to maintain the fixed laminar regime of the flow in the whole range of admissible values of the Reynolds number. The well-known example of a global laminar regime is the above-mentioned *Kolmogorov flow* on the two-dimensional torus T with the periods $l = 2\pi/\alpha$, $h = 2\pi$ for $f(x, y) = \text{const}\cos(y)$ **[13, 26, 66, 110]**. A known global uniqueness and stability condition for the corresponding main flow with $\psi(x, y) = \text{const}\,\text{Re}\cos(y)$ is in this case the inequality $\alpha \geqslant 1$ **[66, 110]**. In what follows, analogous conditions are given for some other flows mentioned above, considered under the restrictions (4). Following **[97]**, we give the corresponding general scheme of arguments.

From the point of view of algebra, relation (2) resembles the vector form of the system of three scalar equations:

$$\varepsilon\omega + \omega \times J\omega = \zeta, \tag{9}$$

defining the components of the angular velocity $\omega = (\omega_1, \omega_2, \omega_3)$ of a steady rotation of a *dissipative top* or a solid with one fixed point subject to the moment of external forces ζ and to the moment of friction forces $\zeta_\omega = -\varepsilon\omega$. In the case under consideration, the given moment ζ_ω is proportional to the angular velocity ω with the coefficient of proportionality $-\varepsilon$; $\varepsilon = \text{const} > 0$ is the friction coefficient, the cross $\times$ is the vector product,

$$J = \begin{pmatrix} \lambda_1 & 0 & 0 \\ 0 & \lambda_2 & 0 \\ 0 & 0 & \lambda_3 \end{pmatrix}, \qquad 0 < \lambda_1 < \lambda_2 < \lambda_3,$$

is the matrix of the principal moments of inertia of the top, λ_1, λ_2, and λ_3, reduced to diagonal form. These moments are assumed to be pairwise distinct and indexed in the order of increase. The tops such that the moment of the friction force is proportional to the kinetic moment $J\omega$, $\zeta_\omega = -\varepsilon J\omega$, may also be considered. In this case, instead of (9) we consider the following equation:

$$\varepsilon J\omega + \omega \times J\omega = \zeta. \tag{10}$$

A complete formal analogy between equation (2) and the equation of the dissipative top is attained in the following hypothetical case, when $\zeta_\omega = -\varepsilon JJ\omega$:

$$\varepsilon JJ\omega + \omega \times J\omega = \zeta. \tag{11}$$

In this case, the stream function ψ corresponds to the vector ω, the operation $-\Delta$ to the matrix J, the vorticity $-\Delta\psi$ to the kinetic moment $J\omega$, the Poisson brackets

$$[\varphi, \psi] = \frac{\partial\varphi}{\partial y}\frac{\partial\psi}{\partial x} - \frac{\partial\varphi}{\partial x}\frac{\partial\psi}{\partial y} \tag{12}$$

to the vector multiplication $\times$, the vorticity of the field of mass accelerations f to the moment of external forces, the inverse value of the Reynolds number $1/\text{Re}$ to the friction coefficient ε. Moreover, both equation (2) considered under boundary conditions (4), (5), or (6), and equation (11) (as well as (9) and (10)) admit (as it will be shown below) at least one solution ω for any value of the parameter $\varepsilon > 0$ (the "global" existence theorem), which is unique for sufficiently large ε (the "local" uniqueness theorem). The requirement guaranteeing the corresponding "global" (for all $\varepsilon > 0$) uniqueness of the solution ω turns out to be the (known for the corresponding limiting case $\varepsilon = 0$) requirement that the moment of external forces ζ be directed along the principal axis of inertia of the top (that ζ be the eigenvector of the matrix J) corresponding to the smallest or the greatest principal moment,

(13) $$J\zeta = \lambda\zeta, \qquad \lambda = \lambda_1 \quad \text{or} \quad \lambda = \lambda_3.$$

The validity of the proposition formulated can be established by using the coordinate form of the vector equation of the top [**26**, p. 58–62]. This proposition is proved below by a method that can also be used for the infinite-dimensional case.

As it will be shown later, any of the above boundary problems for equation (2) can serve as an example of an *abstract* (or infinite-dimensional) top. The requirement analogous to condition (13) for f and having the form

(14) $$-\Delta f = \lambda f \quad \text{in the domain } V, \qquad f|_{\partial V} = 0, \quad \lambda = \lambda_1,$$

(where λ_1 is the smallest eigenvalue of the spectral problem under consideration), which guarantees uniqueness and stability of the corresponding solution $\psi(x, y)$ *for all* $\mathrm{Re} > 0$, can reasonably be used only under conditions (4) (if $\partial V \neq \varnothing$) or (6) (if $\partial V = \varnothing$). In these cases, the abstract top under consideration turns out to be "commutative" in the sense of the permutability of those of its operators which, on one hand, correspond to the "inertia" matrix J, and, on the other hand, to the matrix E (identity matrix), J, or JJ (the "dissipative" matrix); the matrix J obviously, commutes with each of the matrices E, J, or JJ. The no-slip conditions lead to the violation of the requirement of permutability for the corresponding operators of "inertia" and "dissipation" of the abstract top defined by equation (2). The top generated by these conditions is noncommutative, in contrast to the commutative top generated by conditions (4) and (6). In what follows, the case of a *commutative* top is the principal one. The case of a *noncommutative* top is not considered in detail and is used only to illustrate some alternative possibilities.

11.2. Quasicompact Lie algebras. We shall give some additional constructions that are necessary for obtaining the above-mentioned nonlocal theorems.

Let $M = \{\dots, x, y, u, v, \dots\}$ be a real linear space equipped with a scalar product (x, y) and also with a *commutator* $[u, v]$ (a bilinear *skew-symmetric* operation $u, v \to [u, v] = -[v, u]$ satisfying the *Jacobi identity*

$$[x, [u, v]] + [u, [v, x]] + [v, [x, u]] = 0, \qquad x, u, v \in M),$$

which is a Euclidean (in the case of finite dimension) or a pre-Hilbert space (in the infinite-dimensional case) with respect to the first operation, and a Lie algebra with respect to the second. The set M_c (the set $M_{c'}$) of the elements $x \in M$ satisfying the identity

(15) $$(u, [v, x]) = (x, [u, v])$$

for all $u, v \in M$ (or, respectively, the identity

(16) $$(u, [v, x]) = (v, [x, u]))$$

will be called *the first* (*the second*) *compact subalgebra* of the algebra M, respectively. The linearity of the sets M_c and $M_{c'}$ immediately follows from identities (15) and (16) (these sets are linear subspaces of the space M). Moreover, (16) is equivalent to the fact that the identity

(17) $$(u, [u, x]) = 0$$

is satisfied for any $u \in M$.

Indeed, by setting $u = v$ in (16), we obtain (17) (the fact that the commutator is skew-symmetric is used here). Substituting $u + v$ for u in (17), we come to (16), q.e.d.

Further, the inclusion

$$M_c \subset M_{c'} \tag{18}$$

is valid (i.e., M_c is a subset of $M_{c'}$). Indeed, by setting $u = v$ in (15), we obtain (17) and then use the equivalence of (16) and (17).

As it will be shown below, (18) is, in general, a proper inclusion. The inclusion (18) and the linearity of the sets M_c, $M_{c'}$ imply that *M_c is a linear subspace of $M_{c'}$, and $M_{c'}$ is a linear subspace of M.*

In fact, M_c and $M_{c'}$ *are algebras,* i.e.,

$$x_1, x_2 \in M_c \quad \text{implies} \quad [x_1, x_2] \in M_c, \tag{19}$$

$$x_1, x_2 \in M_{c'} \quad \text{implies} \quad [x_1, x_2] \in M_{c'} \tag{20}$$

(thus, *M_c is a subalgebra of $M_{c'}$ and $M_{c'}$ is a subalgebra of M*).

Indeed, as a consequence of the Jacobi identity,

$$((u, [u, [x_1, x_2]]) = (u, [[u, x_1], x_2]) + (u, [x_1, [u, x_2]]).$$

If (16) holds for $x = x_1, x_2$, then

$$(u, [[u, x_1], x_2]) = -([u, x_1], [u, x_2]), \quad (u, [x_1, [u, x_2]]) = ([u, x_2], [u, x_1]).$$

Taking into account the symmetry of the scalar product, from these equalities we find that

$$((u, [u, [x_1, x_2]]) = 0,$$

(the equivalence of identities (16) and (17) is additionally used), which proves the validity of (20). Now we prove the validity of the group property (19).

Using the Jacobi identity again, we find that

$$(u, [v, [x_1, x_2]]) = (u, [[v, x_1], x_2]) + (u, [x_1, [v, x_2]]).$$

As a corollary of (15),

$$\begin{aligned} (u, [[v, x_1], x_2]) &= (x_2, [u, [v, x_1]]), \\ (u, [x_1 [v, x_2]]) &= -(x_1, [u, [v, x_2]]). \end{aligned}$$

The right-hand part of the first of these equalities reduces to the form

$$(x_2, [u, [v, x_1]]) = (x_2, [[u, v], x_1]) + (x_2, [v, [u, x_1]]).$$

Simultaneously , it follows from (15) that

$$\begin{aligned} (x_2, [[u, v], x_1]) &= ([u, v], [x_1, x_2]), \\ (x_2, [v, [u, x_1]]) &= ([u, x_1], [x_2, v]), \\ (x_1, [u, [v, x_2]]) &= ([v, x_2], [x_1, u]). \end{aligned}$$

We conclude that

$$\begin{aligned} (u, [v, [x_1, x_2]]) &= (x_2, [u, [v, x_1]]) - (x_1, [u, [v, x_2]]) \\ &= ([u, v], [x_1, x_2]) + ([u, x_1], [x_2, v]) + ([v, x_2], [x_1, u]) \\ &= ([x_1, x_2], [u, v]), \end{aligned}$$

which completes the proof.

In accordance with the standard terminology [**80**, p. 444; **19**, p. 8], the algebra M will be called *compact*, if $M_c = M$ (i.e., equality (15) holds for all $x, u, v \in M$). For example, the subalgebras M_c and $M_{c'}$ are compact, which can be easily proved by a straightforward verification.

Now let $\mathbb{H} = \mathbb{H}(M)$ be the Hilbert space obtained by the completion of M (as a linear manifold) in the norm

$$\|x\| = (x, x)^{1/2}$$

(the space $\mathbb{H}$ will sometimes be called the *Hilbert space of the algebra* M). The algebra M will be called *quasicompact* if the set M_c is dense in $\mathbb{H}$ (which is, obviously, equivalent to equality (15) being satisfied for all $u, v \in M$ and all x from some subset of the algebra M dense in $\mathbb{H}$).

Any compact algebra M is, obviously, quasicompact. On the other hand, a quasicompact finite dimensional algebra M is necessarily compact (since, in this case, $M = \mathbb{H}$). The requirement that M be finite dimensional is essential, which is evident from what follows.

11.3. The equation of the top. A local uniqueness theorem. Let M be a given quasicompact algebra and let $A, B\colon \mathbb{H} \to \mathbb{H}$ be linear symmetric operators, acting in its Hilbert space $\mathbb{H}$, with common domain of definition $\mathbb{D} \subset M_c$ (belonging to the first compact subalgebra of the algebra M) dense in $\mathbb{H}$ and ranges $\mathbb{R}(A), \mathbb{R}(B) \subset M$ (the ranges of the operators A, B belong to the algebra M). It is assumed that $B\colon \mathbb{H} \to \mathbb{H}\colon u \to Bu$ is a positive definite operator,

$$(21) \qquad (Bu, u) \equiv \|u\|_1^2 \geqslant \beta\|u\|^2, \qquad u \in \mathbb{D}, \qquad \beta = \text{const} > 0,$$

with a complete system of eigenelements

$$(22) \qquad \begin{aligned} &e_k \in \mathbb{D}, \qquad Be_k = \mu_k e_k, \qquad (e_k, e_m) = \delta_{km}, \qquad k, m, = 1, 2, \dots, \\ &\mu_1 \leqslant \mu_2 \leqslant \dots \leqslant \mu_k \leqslant \mu_{k+1} \leqslant \dots \qquad (\mu_1 \geqslant \beta), \end{aligned}$$

forming an orthonormal basis of the space $\mathbb{H}$, and a continuous compact inverse operator

$$B^{-1}\colon \mathbb{H} \to \mathbb{H}, \ B^{-1}Bu = u,$$

whose domain of definition $\mathbb{D}(B^{-1}) = \mathbb{R}(B)$ is dense in $\mathbb{H}$ (note that, by the lemma from §10.4, the existence of the basis of eigenelements (22) is provided by the compactness of the embedding, ensured by inequality (21), and the additional requirement that any solution u of the corresponding general (as in §10.4) spectral problem belong to $\mathbb{D}$). Moreover, we assume that there exist constants $\gamma_1, \gamma_2, \gamma_3 > 0$, $\gamma_1 \leqslant \gamma_2$, such that the following inequalities hold:

$$(23) \qquad \gamma_1\|u\|_1 \leqslant \|Au\| \leqslant \gamma_2\|u\|_1, \qquad \|[u, v]\| \leqslant \gamma_3\|u\|_1\|v\|_1.$$

The equation

$$(24) \qquad \varepsilon Bx + [x, Ax] = y, \qquad x \in \mathbb{D}, \quad y \in \mathbb{R}(B), \quad \varepsilon = \text{const} > 0,$$

will be considered with respect to $x \in \mathbb{D}$ for a given $y \in \mathbb{R}(B)$ and a positive constant $\varepsilon > 0$.

Equation (24) will be understood as defining a stationary rotation of some *abstract dissipative top in the agebra* M. The operators A and B will be referred to as *inertia* and *dissipation* (or as the *operators of inertia and dissipation*), respectively (the right-hand part of equation (24) is interpreted as the moment of external forces, the coefficient ε as the friction coefficient, the element Ax as the kinetic moment).

In the case of an ordinary (three-dimensional) top, we have

$$\mathbb{D} = M_c = \mathbb{R}(A) = \mathbb{R}(B) = M = \mathbb{H} = R^3$$

the three-dimensional Euclidean space, $x = \mathbf{x} = (x_1, x_2, x_3)$, $y = \mathbf{y} = (y_1, y_2, y_3) \in R^3$,

$$(x, y) = \mathbf{x} \cdot \mathbf{y} = x_1 y_1 + x_2 y_2 + x_3 y_3$$

is the scalar product of the vectors **x** and **y**,

$$[x, y] = \mathbf{x} \times \mathbf{y} = (x_2 y_3 - x_3 y_2, x_3 y_1 - x_1 y_3, x_1 y_2 - x_2 y_1)$$

is the vector product of **x** and **y**. In this case, the matrix of the moments of inertia J is "inertia" A, "dissipation" is one of the following three matrices: E (equation (9)), J (equation (10)), and JJ (equation (11)),

$$A = J, \qquad B = E, \quad J \quad \text{or} \quad JJ.$$

Now let $M = M(V)$ be the linear manifold $C^\infty(\overline{V})$ of infinitely differentiable functions $\dots, \varphi, \psi, \dots$, defined in the closure $\overline{V}$ of the two-dimensional domain V (defined as above) equipped with the scalar product of the space of square integrable functions,

$$(\varphi, \psi) = \int_V \varphi(x, y)\psi(x, y)\, dx\, dy$$

(here (x, y) is the variable in the domain V), and with the Poisson brackets (12). In the case of a closed manifold $V = \overline{V}$, the functions from M are assumed to be subject to the additional requirement of *orthogonality to constants* (6) (M is the restriction of $C^\infty(\overline{V})$ under the given condition). As it will be shown below, for this algebra M the first compact subalgebra M_c for $\partial V \neq \varnothing$ is the linear space $\dot{C}^\infty(\overline{V})$ of the functions $\psi \in C^\infty(\overline{V})$ that vanish on the boundary ∂V of the domain V, and for $\partial V = \varnothing$, the algebra $M = M(V)$ is compact ($M_c = M$). The Hilbert space $\mathbb{H}$ of the algebra M coincides with the space of square integrable real functions. In this case the requirement that M_c be dense in $\mathbb{H}$ is valid at least for the above Examples 1–4 of the space V from §11.1 **[35]**. The "inertia" A and "dissipation" B for the top under consideration are the restrictions of the operations $-\Delta$ and $\Delta\Delta$, respectively, to the linear manifold $\mathbb{D} = \mathbb{D}(V)$ formed by the stream functions $\psi \in \dot{C}^\infty(V)$ which satisfy conditions (4) or (5) if $\partial V \neq \varnothing$. In the case $\partial V = \varnothing$, we assume $\mathbb{D} = M$. As it was already mentioned, the space M is defined by the restriction of the manifold $C^\infty(V)$ under condition (6). When $\partial V \neq \varnothing$, the density of $\mathbb{D}$ in $\mathbb{H}$ is obvious (at least for the domains V from Examples 1–4). Moreover, $\mathbb{D} \subset M_c$. The general equation for the top (24) becomes the relation (2). In what follows, we shall verify whether the other axioms of an abstract top are satisfied for the case under consideration.

We shall obtain some corollaries which immediately follow from general axioms mentioned above. First, we note that *for $y = 0$ the trivial solution $x = 0$ is the only solution of equation* (24).

Indeed, multiplying both parts of equation (24) scalarly in $\mathbb{H}$ by x and using the identity $(Bu, u) \equiv \|u\|_1^2$ from (21), we obtain

$$(x, [x, Ax]) = (Ax, [x, x]) = 0$$

(since $x \in \mathbb{D} \subset M_c$, $Ax \in R(A)$, and $[x, x] = 0$). Hence,

$$\varepsilon \|x\|_1^2 = (x, y), \tag{25}$$

which for $y = 0$ immediately implies $x = 0$ (as a result of the operator $B\colon \mathbb{H} \to \mathbb{H}$ being positive definite).

When $y \neq 0$, the equation of the top (24) contains an "implicit" parameter $\|B^{-1}y\|$ along with the explicit dependence on the parameter ε. The equation can be brought to a "canonical" form or reduced to the form that, instead of the two mentioned parameters, contains one parameter of the form

$$\mu = \|B^{-1}y\|/\varepsilon^2. \tag{26}$$

The parameter $\mu = \text{const} > 0$ will be called the *characteristic parameter* of equation (24). Passing to new variables $\overline{x}, \overline{y}$ according to the formulas

$$\overline{x} = (\varepsilon/\|B^{-1}y\|)x, \qquad \overline{y} = (1/\|B^{-1}y\|)y \qquad (y \neq 0), \tag{27}$$

we obtain the required *canonical form of the equation of the top,*

$$B\overline{x} + \mu[\overline{x}, A\overline{x}] = \overline{y}, \qquad \|B^{-1}\overline{y}\| = 1. \tag{28}$$

Then relation (25) acquires the following form:

$$\|\overline{x}\|_1^2 = (\overline{x}, \overline{y}). \tag{29}$$

The requirements on the operator B imply that the bilinear form

$$\{u, v\} \equiv (Bu, v) = (u, Bv), \qquad u, v \in \mathbb{D}, \tag{30}$$

satisfies the axioms of the scalar product. The Hilbert space obtained by the completion of the manifold $\mathbb{D}$ in the norm

$$\|u\|_1^2 = \{u, u\}^{1/2} \tag{31}$$

is denoted by $\mathbb{W}$. As a corollary of inequality (21), the embedding $\mathbb{W} \subset \mathbb{H}$ (of the space $\mathbb{W}$ in $\mathbb{H}$) takes place. Obviously, the inequality

$$\|x\| \leqslant \mu_1^{-1/2}\|x\|_1, \qquad x \in \mathbb{D}, \tag{32}$$

is a sharp bound for this embedding. As a corollary of the fact that the inverse operator $B^{-1}\colon \mathbb{H} \to \mathbb{H}$ is compact, the embedding $\mathbb{W} \subset \mathbb{H}$ is compact.

In terms of the notation introduced above, relation (29) has the following form:

$$\|\overline{x}\|_1^2 = (BB^{-1}\overline{y}, \overline{x}) = \{B^{-1}\overline{y}, \overline{x}\},$$

or

$$\|\overline{x} - B^{-1}\overline{y}/2\|_1^2 = \|B^{-1}\overline{y}/2\|_1^2. \tag{33}$$

As a corollary of identity (33), all the solutions $\overline{x}$ of equation (28) are located on the sphere of the space $\mathbb{W}$ of radius $\|B^{-1}\overline{y}/2\|_1^2$ with the center at $B^{-1}\overline{y}/2$.

We shall give the *local* (i.e., corresponding to comparatively small values of the characteristic parameter $\mu > 0$) theorem on the uniqueness of solution of equation (24).

THEOREM 1. *For*

$$\mu < 1/\gamma_2\gamma_2 \tag{34}$$

equation (24) *admits no more than one solution* $x \in \mathbb{D}$; *the estimate*

$$\|x - x'\|_1 \leqslant \frac{1}{\varepsilon}\frac{1}{1-\gamma_2\gamma_3\mu}\|B^{-1}(y-y')\|_1, \tag{35}$$

providing a continuous dependence of x *on* y, *is valid for the difference of the solutions* x *and* x' *of this equation corresponding to the right-hand sides* y *and* y' *from* $\mathbb{R}(B)$, *respectively.*

Indeed, setting $x' - x = u$, (24) leads directly to the equality

$$\varepsilon Bu + [x, Au] + [u, Ax + Au] = B\xi, \qquad \xi = B^{-1}(y' - y). \tag{36}$$

Multiplying both sides of the latter scalarly by u in $\mathbb{H}$ and taking into account the identities

$$(u, [u, Ax + Au]) = (Ax + Au, [u, u]) = 0, \qquad (u, [x, Au]) = (Au, [u, x])$$

(we use the conditions $x, u \in \mathbb{D} \subset M_c$ and $Ax, Au \in \mathbb{R}(A) \subset M$, and identity (15)), we come to the equality

$$\varepsilon\|u\|_{\bar{1}}^2 + (Au, [u, x]) = \{u, \xi\}. \tag{37}$$

As a corollary of inequalities (23), we have

$$|(Au, [u, x])| \leqslant \|Au\|\|[u, x]\| \leqslant \gamma_2\gamma_3\|x\|_1\|u\|_{\bar{1}}^2.$$

Together with (37), these inequalities imply

$$\varepsilon\|u\|_{\bar{1}}^2 - \gamma_2\gamma_3\|x\|_1\|u\|_{\bar{1}}^2 \leqslant \{u, \xi\} \leqslant \|u\|_1\|\xi\|_1.$$

The inequality

$$\varepsilon\|x\|_{\bar{1}}^2 = (x, BB^{-1}y) = \{x, B^{-1}y\} \leqslant \|x\|_1\|B^{-1}y\|_{\bar{1}}^2$$

follows simultaneously from (25). Consequently,

$$\|x\|_1 \leqslant \|B^{-1}y\|_{\bar{1}}^2/\varepsilon = \varepsilon\mu;$$

thus,

$$\varepsilon(1 - \gamma_2\gamma_3\mu)\|u\|_{\bar{1}}^2 \leqslant \varepsilon\|u\|_{\bar{1}}^2 - \gamma_2\gamma_3\|x\|_1\|u\|_{\bar{1}}^2 \leqslant \|\xi\|_1\|u\|_1,$$

which, when (34) is valid, implies (35). As a corollary of (35), the solution x is unique: $y' = y$ implies $x' = x$. The theorem is proved.

11.4. Types of dissipative tops. Now we return to the canonical form of the equation of the top, or relation (28), which is now considered for an arbitrary $0 < \mu < \infty$. The equation $B\overline{x} = \overline{y}$ is limiting for (28) as $\mu \to 0$. In terms of the "old" variables x, y, the equation

$$\varepsilon Bx = y \qquad (B\overline{x} = \overline{y}) \tag{38}$$

corresponds to it. The Stokes equation (3) is a hydrodynamic analog of equation (38). According to this hydrodynamic analogy, the elements $x \in \mathbb{D}$ satisfying (38) will be called *Stokes elements*. As $\mu \to \infty$, the equation $[\overline{x}, A\overline{x}] = 0$, preserving its form under the passage to the initial variables,

$$[x, Ax] = 0 \qquad ([\overline{x}, A\overline{x}] = 0), \tag{39}$$

is limiting for (28). Its hydrodynamic analog is the stationary Euler equation for the stream function ψ of a plane-parallel flow of an ideal fluid,

$$\frac{\partial\psi}{\partial y}\frac{\partial\Delta\psi}{\partial x} - \frac{\partial\psi}{\partial x}\frac{\partial\Delta\psi}{\partial y} = 0.$$

Thus, the elements $x \in \mathbb{D}$ satisfying (39) will be called *Euler elements*. It is obvious that both the Stokes elements and the Euler elements $x \in \mathbb{D}$ are necessarily solutions of the corresponding "complete" equation of the top (24). We shall call these solutions of equation (24) *basic* (or *simple*). The basic solutions $x \in \mathbb{D}$ correspond to the "basic" flows of a viscous incompressible liquid usually investigated with respect to stability (the Poiseuille, Couette, Kolmogorov, etc. flows defined for the general (multidimensional) case in the Appendix (see §17.7)). The stream function of a plane-parallel simple flow satisfies in V both Stokes and Euler equations,

$$\frac{1}{\mathrm{Re}}\Delta\Delta\psi = f(x, y), \qquad \frac{\partial\psi}{\partial y}\frac{\partial\Delta\psi}{\partial x} - \frac{\partial\psi}{\partial x}\frac{\partial\Delta\psi}{\partial y} = 0. \tag{40}$$

Note also that for a right-hand side y (or $\overline{y}$) of equation (24) (or (28)) such that the latter admits a simple solution x (or $\overline{x}$), the following equality (which immediately follows from (38) and (39)) must be satisfied:

$$[B^{-1}y, AB^{-1}y] = 0 \qquad ([B^{-1}\overline{y}, AB^{-1}\overline{y}] = 0). \tag{41}$$

A necessary and sufficient condition for a given solution $x \in \mathbb{D}$ of equation (24) *to be simple is the requirement that the corresponding solution $\overline{x}$ of the canonical equation of the top* (28) *be independent of the characteristic parameter μ (defined as in* (26)*).*

It is obvious that this condition is necessary: the equations $[\overline{x}, A\overline{x}] = 0$ and $B\overline{x} = \overline{y}$ do not contain the parameter μ. Let us show the validity of the converse statement. Let $\overline{x}', \overline{x}'' \in \mathbb{D}$ satisfy (28) for a fixed right-hand side $y \in \mathbb{R}(B)$ and some values $\mu' \neq \mu''$ of the parameter μ, respectively. Writing down equation (28) for μ' and μ'' and subtracting one of the obtained equalities from the other, for $\overline{x}' = \overline{x}'' = \overline{x}$ (i.e., when the solution $\overline{x}$ of equation (28) does not depend on μ) we obtain $(\mu' - \mu'')[\overline{x}, A\overline{x}] = 0$, which immediately implies that the solution $\overline{x}$ is Euler, $[\overline{x}, A\overline{x}] = 0$, and, hence (equality (28)), Stokes, $B\overline{x} = \overline{y}$.

The simple solutions of the equation of the top to be considered below are its *principal rotations*, i.e., eigenelements of the inertia operator,

$$Ax = \lambda x, \qquad x \in \mathbb{D}, \quad x \neq 0, \quad \lambda = \text{const} \neq 0, \tag{42}$$

which are necessarily Euler solutions, $[x, Ax] = \lambda[x, x] = 0$. The corresponding eigenvalues λ will be called *the principal moments*. Note also that the fact that the solution $x = \varepsilon^{-1}B^{-1}y$ is necessarily Stokes leads to a restriction in the choice of the right-hand side of the corresponding equation of the top, $AB^{-1}y = \lambda B^{-1}y$ (which guarantees that the general condition (41) imposed on the right-hand side of equation (24) admitting a basic solution x is satisfied). For the case of a *finite-dimensional top* (when the space M is finite-dimensional), the validity of the imposed restriction is ensured by the additional requirement that the operators A and B commute (which is obvious),

$$AB = BA \qquad (M \text{ is finite-dimensional}). \tag{43}$$

For example, the tops (9)–(11) satisfy this additional requirement. For the finite-dimensional case requirement (43) for symmetric operators $A, B\colon \mathbb{H} \to \mathbb{H}$ implies the validity for any $u, v \in \mathbb{D}$ (here $\mathbb{D} = M = \mathbb{H}$) of the identity

$$(Au, Bv) = (Bu, Av), \qquad u, v \in \mathbb{D}, \tag{44}$$

which admits an extension to the infinite-dimensional case (when $\mathbb{D} \neq M$ or $M \neq \mathbb{H}$).

In the general case, symmetric operators $A, B\colon \mathbb{H} \to \mathbb{H}$ satisfying identity (44) for any $u, v \in \mathbb{D}$ will be called *permutable* (or *commuting*), and the corresponding *tops* (equation (24)) are called *commutative.*

As already mentioned, each of the three-dimensional tops (9), (10), and (11) is commutative. A "hydrodynamic" top generated by equation (2) on a closed manifold $V = \overline{V}$ ($\partial V = \varnothing$) or on a manifold with a boundary $\partial V \neq \varnothing$ under conditions (4) is an example of an infinite-dimensional commutative top.

Indeed, in the indicated cases we have

$$\begin{aligned}
(A\varphi, B\psi) - (B\varphi, A\psi) &= \int_V ((-\Delta\varphi)\Delta\Delta\psi - \Delta\Delta\varphi(-\Delta\psi))\, dV \\
&= \int_V \operatorname{div}\left((-\Delta\varphi)\operatorname{grad}(\Delta\psi) + (\Delta\psi)\operatorname{grad}(\Delta\varphi)\right) dV \\
&= \begin{cases} 0, & \text{if } \ \partial V = \varnothing, \\ \int_{\partial V}((-\Delta\varphi)\frac{\partial\Delta\psi}{\partial n} + (\Delta\psi)\frac{\partial\Delta\varphi}{\partial n})\, dS, & \text{if } \ \partial V \neq \varnothing. \end{cases}
\end{aligned} \tag{45}$$

Here $dV = dx\,dy$ is the volume element of V, dS is the length element of a smooth arc of the boundary $\partial V \neq \varnothing$, n is the outward unit normal to the boundary ∂V at a regular point of the latter. Conditions (4) guarantee that the contour integral in (45) vanishes, $\Delta\varphi = \Delta\psi = 0$ on ∂V, which completes the proof.

We shall make some remarks concerning an alternative possibility connected with the boundary condition (5). This restriction generates a "hydrodynamic" top which is not commutative (since the equalities $\partial\varphi/\partial n = \partial\psi/\partial n = 0$ on ∂V do not imply in the general case that the equalities $\Delta\varphi = \Delta\psi = 0$ are satisfied on ∂V). However, for any $u \in \mathbb{D}$ this noncommutative top satisfies the following *isometry condition*:

$$(Bu, u) = (Au, Au) \qquad (\text{or } \|u\|_1 = \|Au\|). \tag{46}$$

Indeed, the equalities

$$\begin{aligned}
(B\psi, \psi) - (A\psi, A\psi) &= \int_V (\psi\Delta\Delta\psi - (\Delta\psi)\Delta\psi)\, dV \\
&= \int_V \operatorname{div}\left(\psi \operatorname{grad}(\Delta\psi) - (\Delta\psi)\operatorname{grad}\psi\right) dV \\
&= \int_{\partial V} \left(\psi\frac{\partial\Delta\psi}{\partial n} - \Delta\psi\frac{\partial\psi}{\partial n}\right) dS
\end{aligned} \tag{47}$$

together with (5) imply that the contour integral in equalities (47) vanishes, q.e.d.

Note that the contour integral from (47) also vanishes when conditions (4) are satisfied or when $\partial V = \varnothing$ (in the latter case it is identically 0). These constructions show that conditions (44) and (46) are independent (which is essential for the case of infinite dimension; *infinite-dimensional tops are, generally speaking, noncommutative* (e.g., the top generated by equation (2) and boundary conditions (5))). In the case of a finite-dimensional algebra M, the isometry condition (46) is reduced to the equality $B = AA$ and, thus, implies the permutability condition (43), equivalent to equality (44) in the case under consideration. Hence, *a finite-dimensional isometric top is commutative.*

In the case of a commutative isometric top, the equation of principal rotations (42) is equivalent to the equation

$$Bu = \lambda A u. \tag{48}$$

Indeed, equation (42) is equivalent to the identity $(Au, h) = \lambda(u, h)$ for any h from a dense set in $\mathbb{H}$ (which is obvious). As it will be proven below (in §11.5), the range $\mathbb{R}(A)$ of the inertia operator $A\colon \mathbb{H} \to \mathbb{H}$ of a commutative top is a dense subspace of the space $\mathbb{H}$. Then, by setting $h = Av$ for an arbitrary fixed $v \in \mathbb{D}$, we obtain the required equivalence taking into account the fact that identity (46) is additionally satisfied and $\mathbb{D}$ is assumed to be dense in $\mathbb{H}$.

The eigenelements u and the corresponding eigenvalues λ of the spectral problem (48) can be associated to the principal rotations and moments of a general dissipative top. For example, for an isometric top generated by conditions (2) and (5), the latter are defined by the solutions ψ, λ of the spectral problem of the following form:

$$\Delta\Delta\psi + \lambda\Delta\psi = 0 \quad \text{in } V, \qquad \psi = \partial\psi/\partial n = 0 \quad \text{on } \partial V. \tag{49}$$

Problem (49) can be studied with the help of a simpler problem, obtained by replacing the no-slip conditions with less restrictive conditions (4) see **[30]**. However, in what follows the problems concerning noncommutative tops are not analyzed in detail.

As already mentioned, the main attention will be paid to the commutative top. By means of one of the additionally introduced restrictions of the form

$$\begin{gathered} (Au, u) \geqslant \sigma\|u\|^2, \qquad \sigma = \text{const} > 0, \qquad \text{for all } u \in \mathbb{D} \\ \text{or} \\ (Au, u) \leqslant -\sigma\|u\|^2, \qquad \sigma = \text{const} > 0, \qquad \text{for all } u \in \mathbb{D} \end{gathered} \tag{50}$$

it will be shown in §11.5 that there exists a complete set of principal rotations forming an orthonormal basis of the Hilbert space $\mathbb{H}$ for a commutative top. In the general case, tops satisfying one of the restrictions (50) will be called *definite*. A definite top is *normal* if the first restriction from (50) is satisfied for it (the inertia of a definite top is by definition a positive definite operator). A dissipative top that is not normal will be called *anomalous* (e.g., a definite top such that the second condition from (50) is satisfied is anomalous).

Finally, note that not every principal rotation of a dissipative top satisfying the conditions of the spectral problem (48) can be a simple solution of equation (24). The corresponding additional requirements were considered in **[97]** and are given in §11.5 (Proposition 4 below). The requirement of extremality of the corresponding principal moment which provides the required global uniqueness of the corresponding rotation of a commutative top (which is considered below) possibly does not hold (in the sense of obtaining the conditions for nonlocal uniqueness of a basic solution). From this point of view, Theorem 2 **[97]** on the global uniqueness of the extremal principal rotation of a noncommutative isometric top should be considered as a hypothesis. In what follows, other differences between commutative and noncommutative tops will be considered.

11.5. The existence and completeness of the system of principal rotations.

PROPOSITION 1. *Any commutative top has a complete system of principal rotations defined by nontrivial solutions of equation* (42). *This system forms an orthonormal basis of the space* $\mathbb{H}$ *and an orthogonal basis of the space* $\mathbb{W}$. *Each principal rotation* x *of this top is simultaneously an eigenelement of the dissipation operator* $B\colon \mathbb{H} \to \mathbb{H}$. *The principal moments of the top are real, different from zero, and of finite multiplicity, and their full set has no limit points and can be ordered in the following way*:

$$\begin{gathered}\lambda_1, \lambda_2, \dots, \lambda_k, \lambda_{k+1}, \dots, \\ 0 < |\lambda_1| \leqslant |\lambda_2| \leqslant \dots \leqslant |\lambda_k| \leqslant |\lambda_{k+1}| \leqslant \dots .\end{gathered} \tag{51}$$

Indeed, let the operator $B\colon \mathbb{H} \to \mathbb{H}$ be defined as in §11.3 and let

$$N_\mu(B) \equiv \ker(B - \mu)$$

be the kernel of the operator $B - \mu\colon u \to Bu - \mu u$ (or the set of zeros of the operator $B - \mu$, which coincides with the set of eigenelements of the operator B) corresponding to a given real number μ. Then $N_\mu(B)$ is a finite-dimensional subspace of the space $\mathbb{H}$, which is reduced to the zero element of this space when μ does not coincide with an eigenvalue of the operator B. Assuming that the system of the eigenelements of B is complete in $\mathbb{H}$ and denoting the (point) spectrum of this operator (the union of all the eigenvalues of B) by

$$P\sigma B = \{\mu : Bu = \mu u,\ u \neq 0\},$$

we decompose $\mathbb{H}$ into the orthogonal sum of the following form:

$$\mathbb{H} = \bigoplus_{\mu \in P\sigma B} N_\mu(B). \tag{52}$$

Using the permutability condition (44) and the positiveness of the set $P\sigma B$ (cf. with (22)), we find that for any $\mu, \nu \in P\sigma B$ and any $u \in N_\mu(B)$, $v \in N_\nu(B)$

$$(Au, v) = \frac{1}{\nu}(Au, Bv) = \frac{1}{\nu}(Bu, Av) = \frac{\mu}{\nu}(u, Av).$$

Then the symmetry of the operator $A\colon \mathbb{H} \to \mathbb{H}$ implies $(Au, v) = 0$ if $\mu \neq \nu$. Taking into account (52), we conclude that $u \in N_\mu(B)$ implies $Au \in N_\mu(B)$, i.e., $N_\mu(B)$ is an invariant subspace of the operator $A\colon \mathbb{H} \to \mathbb{H}$. By restricting the latter to the indicated finite-dimensional subspace $N_\mu(B)$ and again using the symmetry of A, we come to the existence and completeness of the system of eigenelements of that restriction in the space $N_\mu(B)$ **[25]**. On the other hand, since each eigenelement of the operator A obtained by this method necessarily belongs to one of the components $N_\mu(B)$ of decomposition (52), this element is an eigenelement of B. The compactness of the embedding of the space $\mathbb{W}$ into $\mathbb{H}$, realized by inequality (21) on the dense subset $\mathbb{D} \subset \mathbb{W}$ and guaranteed by the assumed compactness of the inverse operator $B^{-1}\colon \mathbb{H} \to \mathbb{H}$, together with the first inequality from (23) implies the compactness of the inverse operator $A^{-1}\colon \mathbb{H} \to \mathbb{H}$, $A^{-1}Au = u$, $u \in \mathbb{D}$, whose existence and continuity follow from (21) and the first equality from (23). The estimate

$$|\lambda_1| \geqslant \gamma_1 \mu_1^{1/2} \geqslant \gamma_1 \beta^{1/2} > 0 \tag{53}$$

also follows from these equalities. As a result of the completeness of the system of principal rotations in $\mathbb{H}$, which follows from the fact that decomposition (52) is valid and that each component of this decomposition is invariant for A, the domain of definition of the operator $A^{-1}: \mathbb{H} \to \mathbb{H}$, coinciding with the range $\mathbb{R}(A)$ of the operator $A: \mathbb{H} \to \mathbb{H}$, is dense in $\mathbb{H}$. This fact, together with the compactness of $A^{-1}: \mathbb{H} \to \mathbb{H}$, implies the finite multiplicity of the principal moments (51) (the Fredholm theorems are used here). The completeness of the system of principal rotations in $\mathbb{W}$ is established with the help of considerations analogous to those used in the similar situation in the proof of Lemma 10.4 (§10.4). The proof is completed.

PROPOSITION 2. *Let an element $y \in \mathbb{D} \cap \mathbb{R}(B)$ (belonging both to $\mathbb{D}$ and to $\mathbb{R}(B)$) be a principal rotation of a commutative top corresponding to the principal moment λ, $Ay = \lambda y$ $(y \in \mathbb{D} \cap \mathbb{R}(A))$. Then the element $B^{-1}y \in \mathbb{D}$ is also a principal rotation of the top corresponding to this λ (which necessarily belongs to $\mathbb{R}(A)$). For this y, equation (24) admits the solution $x = \varepsilon^{-1}B^{-1}y$.*

Indeed, for any $v \in \mathbb{D}$ we have

$$\begin{aligned}(AB^{-1}y, Bv) &= (BB^{-1}y, Av) = (y, Av) = (Ay, v) \\ &= \lambda(y, v) = \lambda(BB^{-1}y, v) = \lambda(B^{-1}y, Bv).\end{aligned}$$

The set $\mathbb{R}(B) = \{Bv : v \in \mathbb{D}\}$ is dense in $\mathbb{H}$; hence,

$$AB^{-1}y = \lambda B^{-1}y.$$

Since $\lambda \neq 0$ (Proposition 1), $B^{-1}y = \lambda^{-1}AB^{-1}y \in \mathbb{R}(A)$. The validity of the last statement of the proposition immediately follows from the above arguments.

REMARK 1. To prove the proposition, we additionally used the identity $BB^{-1}y = y$, $y \in \mathbb{D}$, which follows from the equalities

$$(B^{-1}Bu, Bv) = (u, Bv) = (Bu, v) = (Bu, B^{-1}Bv), \qquad u, v \in \mathbb{D},$$

(the symmetry of the operator B) and

$$(BB^{-1}y, v) = (B^{-1}y, Bv) = (y, B^{-1}Bv) = (y, v), \qquad v \in \mathbb{D}.$$

REMARK 2. The validity of the inclusion $B^{-1}y \in \mathbb{R}(A)$ is established up to the zero element of the space $\mathbb{H}$, i.e., it is proved that among the representatives of the equivalence class of the element $B^{-1}y \in \mathbb{D}$ in $\mathbb{H}$ there exists an element $\lambda^{-1}B^{-1}y \in \mathbb{R}(A)$ coinciding with the former in $\mathbb{H}$. The equality $B^{-1}By = y$ is understood analogously.

As a consequence of the proved propositions, in the case of a commutative top equation (24) admits a rich variety of solutions, given in Proposition 1. These solutions correspond to the given principal rotations y. In order to make the arguments complete, we indicate an analogous class of solutions of this equation considered for the case of a noncommutative top (i.e., when condition (44) is violated). First we prove the following statement.

PROPOSITION 3. *The principal rotations of a definite dissipative (possibly, noncommutative) top defined by the nontrivial solutions of equation* (48) *form an orthogonal basis of the space* $\mathbb{W}$. *The corresponding principal moments satisfy Proposition* 1 *and are of the same sign. It is assumed that if an element* $u \in \mathbb{W}$ *satisfies* (48) *in some "general sense" (in the sense of equalities* (54) *and* (56) *given below), then* $u \in \mathbb{D}$ *(an "abstract" version of the condition of "sufficient smoothness" of the solution of the spectral problem, or of the equivalence of the corresponding "classical" solution (belonging to* $\mathbb{D}$*) to the "generalized" one (belonging to* $\mathbb{W}$*)).*

Indeed, without loss of generality we shall assume the operator $A\colon \mathbb{H} \to \mathbb{H}$ in equation (48) to be strictly positive definite (the case of a negatively defined operator is reduced to the one given above by changing in what follows the signs of A and λ to the opposite ones). We introduce on $\mathbb{D}$ a new scalar product

$$\langle u, v\rangle \equiv (Au, v) = (u, Av), \qquad u, v \in \mathbb{D}, \tag{54}$$

and the corresponding norm

$$\|u\|_{1/2} \equiv \langle u, u\rangle^{1/2}.$$

Denote by $\mathbb{Z}$ the Hilbert space obtained by the completion of $\mathbb{D}$ in the norm $\|\cdot\|_{1/2}$. We find that the embedding $\mathbb{W} \subset \mathbb{Z}$ realized on $\mathbb{D}$ by the inequalities

$$\|u\|_{1/2} = (u, Au)^{1/2} \leqslant \|u\|^{1/2}\|Au\|^{1/2} \leqslant \gamma_2^{1/2}\|u\|^{1/2}\|u\|_1^{1/2} \leqslant \beta^{-1/4}\gamma_2^{1/2}\|u\|_1 \tag{55}$$

holds. The boundedness of a sequence of elements of the set $\mathbb{D}$ in $\mathbb{W}$ implies, according to the last inequality from (55), the convergence of some subsequence of it in $\mathbb{H}$ (the compactness of the embedding $\mathbb{W} \subset \mathbb{H}$) and, consequently, the convergence of this subsequence in $\mathbb{Z}$. Therefore, the embedding $\mathbb{W} \subset \mathbb{Z}$ is compact, hence (by the "spectral" Lemma 10.4), the spectral problem generated by it

$$\{u, v\} = \lambda\langle u, v\rangle, \qquad v \in \mathbb{W}, \tag{56}$$

(analogous to that formulated in the above-mentioned "spectral" lemma) has the required complete set of eigenelements $u \in \mathbb{W}$, the corresponding eigenvalues λ satisfying Proposition 1. The stability of the sign of these numbers is a corollary of (56) (for $u = v$). Simultaneously, (56) is the corresponding generalized notation of equation (48), since

$$\{u, v\} = (Bu, v), \qquad \langle u, v\rangle = (Au, v), \qquad u \in \mathbb{D}.$$

The inequality $\lambda \neq 0$ follows from (55):

$$\|u\|_1^2 = \lambda\|u\|_{1/2}^2 \leqslant \lambda\beta^{-1/2}\gamma_2\|u\|_1^2$$

and $u \neq 0$ imply

$$\lambda \geqslant \beta^{1/2}\gamma_2^{-1} > 0.$$

The proposition is proved.

PROPOSITION 5. *Let* $u \in \mathbb{D}$ *be a principal rotation of a definite dissipative top defined by a solution of equation* (48) *such that* $Au - \lambda u \in M_0$, *where* M_0 *is the center of the algebra* M, *i.e., the set of elements* $\theta \in M$ *satisfying the identity* $[v, \theta] = 0$ *for any* $v \in M$. *Then for* $y = Au$ *the element* $x = \varepsilon^{-1}\lambda^{-1}u$ *is a simple solution of equation* (24) *corresponding to the given* y.

Indeed, we have

$$[x, Ax] = [x, Ax - \lambda x] = \varepsilon^{-1}\lambda^{-1}[x, Au - \lambda u] = 0$$

(since $Au - \lambda u \in M_0$), and $\varepsilon Bx = \lambda^{-1}Bu = Au = y$, q.e.d.

Propositions 3 and 4 describe the set of possible solutions of equation (24) restricted (in comparison with the case of a commutative top considered above) by the additional condition $Au - \lambda u \in M_0$. For the case of an isometric top, the element $Au - \lambda u$ is orthogonal in $\mathbb{H}$ to the element Au. In this case it is necessary to require that *the center M_0 be orthogonal in $\mathbb{H}$ to the range $\mathbb{R}(A)$ of the operator A*, which would guarantee the compatibility of the conditions $Au - \lambda u \in M_0$ and $(Au - \lambda u, Au) = 0$.

11.6. The basic scalar relations. It will be proved in this subsection that *if $x \in \mathbb{D}$ is a solution of equation* (24) *satisfying the requirement $Ax - \lambda x \in M_0$ for some real λ, then for the difference $u = x - x'$ of two solutions of this equation $x, x' \in \mathbb{D}$, which correspond to the same $\varepsilon > 0$ and $y \in \mathbb{R}(B)$, the following relations are valid*:

$$(Au, Bu) = \lambda(u, Bu), \qquad (u, Bu) = \frac{1}{\varepsilon}(x, [u, Au]). \tag{57}$$

Note that for a commutative top, any principal rotation $x \in \mathbb{D}$ (such that $Ax = \lambda x$) satisfies the requirement $Ax - \lambda x \in M_0$ given above. In this case, the principal moment of the top is the corresponding λ and the zero element of the space M is the element $Ax - \lambda x$ (necessarily belonging to the subspace M_0). In connection with this case, it will be shown in the next section that $Au = \lambda u$ follows from the first equality in (57) for the minimal principal moment λ. Then the second equality from (57) implies $u = 0$ (uniqueness of the corresponding solution x for any $\varepsilon > 0$).

Now we shall obtain equalities (57). Rewriting (24) for the new solution $x' = x + u$, we find from the equality obtained that

$$\varepsilon Bu + [x, Au] + [u, Ax] + [u, Au] = 0. \tag{58}$$

By multiplying both parts of the obtained equality scalarly in $\mathbb{H}$ by u and taking into account the identities

$$\begin{aligned}(u, [x, Au]) &= (x, [Au, u]) = -(x, [u, Au]),\\ (u, [u, Ax]) &= (Ax, [u, u]) = 0,\\ (u, [u, Au]) &= (Au, [u, u]) = 0\end{aligned}$$

(we have used the conditions $x, u \in \mathbb{D} \subset M_c \subset M_{c'}$ and $Ax, Au \in \mathbb{R}(A) \subset M$, and identities (15), (16)), we come to the equality

$$\varepsilon(u, Bu) - (x, [u, Au]) = 0, \tag{59}$$

which is equivalent to the second equality from (57). Multiplying both parts of equality (58) scalarly in $\mathbb{H}$ by Au, and taking into account the identities

$$\begin{aligned}(Au, [x, Au]) &= (x, [Au, Au]) = 0,\\ (Au, [u, Ax]) &= (Ax, [Au, u]) = -(Ax, [u, Au]),\\ (Au, [u, Au]) &= (u, [Au, Au]) = 0\end{aligned}$$

(using again the conditions and identities given above), we come to the relation

$$\varepsilon(Au, Bu) - (Ax, [u, Au]) = 0. \tag{60}$$

Multiplying equality (59) by λ and subtracting the equality obtained from (60), we obtain

$$(61)\qquad \varepsilon(Au, Bu) - \varepsilon\lambda(u, Bu) - (Ax - \lambda x, [u, Au]) = 0.$$

For

$$Ax - \lambda x = \theta \in M_0,$$

the third term of equality (61) vanishes,

$$(Ax - \lambda x, [u, Au]) = (\theta, [u, Au]) = (Au, [\theta, u]) = 0$$

(since $[\theta, u] = 0$), which implies the equivalence of this equality to the first relation from (57) when $\varepsilon > 0$.

§12. Stability of a principal rotation

12.1. A nonlocal uniqueness theorem. Assuming that the top (24) is *commutative*, we shall obtain the result globalizing the proposition of the local uniqueness theorem (§11.3). First, we shall prove the validity of the estimate (6) given below.

Let $\lambda \in P\sigma A$ be a fixed principal moment of the commutative top under consideration and let

$$u_\lambda \in N_\lambda(A) = \ker(A - \lambda)$$

be the orthogonal projection in $\mathbb{H}$ of the given element $u \in \mathbb{D}$ onto the invariant space $N_\lambda(A)$ of this principal moment. Since $N_\lambda(A)$ is finite-dimensional (Proposition 1 from the previous section), we have $u_\lambda \in N_\lambda(A) \subset \mathbb{D}$. Hence, $\xi = u - u_\lambda \in \mathbb{D}$. Introducing the quadratic form

$$K_\lambda(u) \equiv (Bu, Au) - \lambda(Bu, u), \qquad u \in \mathbb{D},$$

we establish the validity of the identity

$$(1)\qquad K_\lambda(u) = K_\lambda(\xi), \qquad \xi = u - u_\lambda.$$

Indeed, using again Proposition 11.1, we have $Au_\lambda = \lambda u_\lambda$ and $Bu_\lambda = \mu u_\lambda$ for some $\mu \in P\sigma B$, which implies

$$\begin{aligned}
K_\lambda(u_\lambda) &= (Bu_\lambda, Au_\lambda - \lambda u_\lambda) = 0,\\
(A\xi, Bu_\lambda) &= \mu(A\xi, u_\lambda) = \mu(\xi, Au_\lambda) = \lambda\mu(\xi, u_\lambda) = 0,\\
(B\xi, Au_\lambda) &= (A\xi, Bu_\lambda) = 0,\\
(\xi, Bu_\lambda) &= \mu(\xi, u_\lambda) = 0,\\
(B\xi, u_\lambda) &= (\xi, Bu_\lambda) = 0,
\end{aligned}$$

and, consequently,

$$\begin{aligned}
K_\lambda(u) &= K_\lambda(\xi + u_\lambda)\\
&= K_\lambda(\xi) + K_\lambda(u_\lambda) + (A\xi, Bu_\lambda) + (B\xi, Au_\lambda) - \lambda(\xi, Bu_\lambda) - \lambda(B\xi, u_\lambda)\\
&= K_\lambda(\xi),
\end{aligned}$$

q.e.d.

As a consequence of Proposition 11.1 (§11.5), we have

$$\varkappa = \inf\{|\lambda_k - \lambda| : \lambda_k \in P\sigma A, \quad \lambda_k \neq \lambda\} > 0.$$

We show that

$$\text{(2)}\qquad \begin{gathered}|K_\lambda(\xi)| \geqslant \varkappa\|\xi\|_1^2, \qquad \varkappa = \inf\{|\lambda_k - \lambda| : \lambda_k \in P\sigma A,\ \lambda_k \neq \lambda\} > 0, \\ \text{if } \lambda = \min\{\lambda_1 : i = 1, 2, \dots\} \text{ or } \lambda = \max\{\lambda_i : i = 1, 2, \dots\}.\end{gathered}$$

Let

$$\text{(3)}\qquad e_1, e_2, \dots, e_k, \dots, \qquad Ae_k = \lambda_k e_k, \quad (e_k, e_m) = \delta_{km}, \quad k, m = 1, 2, \dots,$$

be the complete orthonormal system of principal rotations of our commutative top in the space $\mathbb{H}$ corresponding to the ordered sequence of the principal moments (11.51) (the existence and completeness of this system follow from Proposition 11.1). By setting for an arbitrary $u \in \mathbb{H}$

$$\overset{n}{u} = \sum_{k=1}^{n} (u, e_k) e_k, \qquad \underset{n+1}{u} = u - \overset{n}{u},$$

we show that for $u \in \mathbb{D}$

$$\text{(4)}\qquad A\overset{n}{u} = \overset{n}{(Au)}, \qquad B\overset{n}{u} = \overset{n}{(Bu)}.$$

Indeed, using (3) we find

$$A\overset{n}{u} = \sum_{k=1}^{n} (u, e_k) Ae_k = \sum_{k=1}^{n} (Au, e_k) e_k = \overset{n}{(Au)}.$$

Further, by Proposition 11.1,

$$Be_k = \mu_k e_k, \qquad k = 1, 2, \dots, \qquad \mu_k \in P\sigma B$$

(the indexing of the eigenvalues used here corresponds to that for the sequence (11.51); therefore it may differ from the indexing chosen in (11.22)). Therefore, we have again

$$B\overset{n}{u} = \sum_{k=1}^{n} (u, e_k) Be_k = \sum_{k=1}^{n} (Bu, e_k) e_k = \overset{n}{(Bu)},$$

which completes the proof of the validity of the required properties.

Now we shall show that for any $u \in \mathbb{D}$

$$\text{(5)}\qquad A\underset{n+1}{u} = \underset{n+1}{(Au)}, \qquad B\underset{n+1}{u} = \underset{n+1}{(Bu)}.$$

Indeed, by setting P equal to A or B and using (4), we find

$$P\underset{n+1}{u} = Pu - \overset{n}{(Pu)} = \underset{n+1}{(Pu)},$$

which proves (5).

As a consequence of (4) and (5), for any $u \in \mathbb{D}$

$$(A\overset{n}{u}, B\underset{n+1}{u}) = 0, \qquad (\overset{n}{u}, B\underset{n+1}{u}) = 0.$$

Hence, for any $u \in \mathbb{D}$

$$K_\lambda(u) = K_\lambda(\overset{n}{u}) + K_\lambda(\underset{n+1}{u}).$$

Simultaneously, by (5),

$$|K_\lambda(\underset{n+1}{u})| \leqslant \|B\underset{n+1}{u}\|\|A\underset{n+1}{u}\| + |\lambda|\|B\underset{n+1}{u}\|\|\underset{n+1}{u}\| \equiv \varphi_n(u),$$
$$\|u\|_1^2 - \|\overset{n}{u}\|_1^2 = (Bu, \overset{n}{u} + \underset{n+1}{u}) - (B\overset{n}{u}, \overset{n}{u}) = \|\underset{n+1}{u}\|_1^2 \leqslant \|B\underset{n+1}{u}\|\|\underset{n+1}{u}\| \equiv \psi_n(u).$$

The completeness of system (3) in $\mathbb{H}$ implies

$$\varphi_n(u), \psi_n(u) \to 0, \qquad n \to \infty.$$

Simultaneously,

$$\begin{aligned} K_\lambda(\overset{n}{\xi}) &= \sum_{k=1}^{n} (B\overset{n}{\xi}, e_k)\lambda_k(\xi, e_k) - \lambda \sum_{k=1}^{n} (B\overset{n}{\xi}, e_k)(\xi, e_k) \\ &= \sum_{k=1}^{n} (\lambda_k - \lambda)(\overset{n}{\xi}, Be_k)(\xi, e_k) = \sum_{\substack{k=1 \\ \lambda_k \neq \lambda}}^{n} (\lambda_k - \lambda)\mu_k(\overset{n}{\xi}, e_k)^2. \end{aligned}$$

The validity of one of the two restrictions for λ mentioned in (2) implies that the sign of the terms of the sum obtained is constant,

$$\operatorname{sgn}(\lambda_k - \lambda) = \operatorname{sgn}(\lambda_m - \lambda), \qquad k, m = 1, 2, \dots, \qquad \lambda_k, \lambda_m \neq \lambda.$$

Consequently, taking into account that $(\xi, e_m) = 0$, $e_m \in N_\lambda(A)$, we have

$$\begin{aligned} |K_\lambda(\overset{n}{\xi})| &= \sum_{\substack{k=1 \\ \lambda_k \neq \lambda}}^{n} |\lambda_k - \lambda|\mu_k(\overset{n}{\xi}, e_k)^2 \geqslant \varkappa \sum_{\substack{k=1 \\ \lambda_k \neq \lambda}}^{n} \mu_k(\xi, e)^2 \\ &= \varkappa \sum_{k=1}^{n} \mu_k(\xi, e)^2 = \varkappa(B\overset{n}{\xi}, \overset{n}{\xi}) = \varkappa\|\overset{n}{\xi}\|_1^2. \end{aligned}$$

Simultaneously,

$$|K_\lambda(\overset{n}{\xi})| = |K_\lambda(\xi) - K_\lambda(\underset{n+1}{\xi})| \leqslant |K_\lambda(\xi)| + \varphi_n(\xi),$$
$$\|\overset{n}{\xi}\|_1^2 = \|\xi\|_1^2 - \|\underset{n+1}{\xi}\|_1^2 \geqslant \|\xi\|_1^2 - \psi_n(\xi),$$

and we find that for any (sufficiently large) n the inequality

$$|K_\lambda(\xi)| + \chi_n(\xi) \geqslant \varkappa\|\xi\|_1^2,$$

where

$$\chi_n(\xi) = \varphi_n(\xi) + \varkappa\psi_n(\xi) \to 0, \qquad n \to \infty,$$

is valid. Passing in the inequality obtained to the limit as $n \to \infty$, we come to (2). Therefore, (1) and (2) imply the following statement.

PROPOSITION 1. *If $\lambda \in P\sigma A$ is the greatest or the smallest principal moment of inertia of a commutative top, then for the constant $\varkappa > 0$ from (2) and any $u \in \mathbb{D}$ the following inequality is valid:*

$$|(Bu, Au) - \lambda(Bu, u)| \geqslant \varkappa\|u - u_\lambda\|_1^2, \qquad \varkappa = \text{const} > 0, \tag{6}$$

where $u_\lambda \in \mathbb{D}$ is the orthogonal projection in $\mathbb{H}$ of the element u on the invariant space $N_\lambda(A) = \ker(A - \lambda)$ of the given principal moment λ.

The following theorem globalizes the proposition of the local theorem on uniqueness from §11.3 and concerns the solutions whose existence was established in the previous section.

THEOREM 1. *Let the top (11.24) be commutative and let $y \in N_\lambda(A)$ (i.e., $Ay = \lambda y$). Then $x = \varepsilon^{-1}B^{-1}y$ is a solution of equation (11.24), considered for the indicated y. This solution is unique in $\mathbb{D}$ for any $\varepsilon > 0$ and for any $y \in N_\lambda(A)$, if λ is the greatest or the smallest eigenvalue of the operator $A\colon \mathbb{H} \to \mathbb{H}$.*

This theorem follows from Proposition 11.2, relations (11.57) (from §11.6), and inequality (6) (Proposition 1). Thus, according to Proposition 11.3, we have

$$x = \varepsilon^{-1}B^{-1}y \in N_\lambda(A).$$

According to relations (11.57), the difference $u = x - x' = 0$ if $u \in N_\lambda(A)$ (since, in this case,

$$\|u\|_1^2 = \varepsilon^{-1}(x, [u, Au]) = \lambda\varepsilon^{-1}(x, [u, u]) = 0).$$

Due to inequality (6), we have $u = u_\lambda$, which completes the proof.

12.2. A nonlocal theorem on stability. Following the mechanical analogy, we describe the nonstationary rotations of an abstract top by means of the following equation:

$$\frac{d}{dt}Ax + \varepsilon Bx + [x, Ax] = y. \tag{7}$$

Preserving the assumptions from §11.3, we show the *a priori* stability (i.e., without proving its existence) of a stationary principal rotation of a commutative top from Theorem 1.

THEOREM 2. *Any smooth solution $x = x(t) \in \mathbb{D}$ defined on the semiaxis $t \geqslant 0$ tends in $\mathbb{W}$, as $t \to 0$, to the stationary solution $x_* = \varepsilon^{-1}B^{-1}y$, $Ay = \lambda y$, which corresponds to the smallest principal moment $\lambda = \lambda_1 \in P\sigma A$, if the commutative top under consideration is normal,*

$$(Au, u) \geqslant \sigma\|u\|^2, \qquad u \in \mathbb{D}, \quad \sigma = \text{const} > 0,$$

and the operations of inertia A and of differentiation d/dt with respect to time are permutable, i.e., the domain of definition of the operation d/dt belongs to $\mathbb{D}$ and $Ad/dt = d/dtA$.

Indeed, setting $u(t) = x(t) - x_*$, we obtain immediately from (7) that

$$\frac{d}{dt}Au + \varepsilon Bu + [x_*, Au] + [u, Ax_*] + [u, Au] = 0. \tag{8}$$

Multiplying both parts of the equality obtained scalarly by u and then by Au in $\mathbb{H}$, taking into account the permutability of the operations d/dt and A (cf. with §11.3), we obtain

$$\frac{d}{dt}\frac{(Au, u)}{2} + \varepsilon(Bu, u) - (x_*, [u, Au]) = 0,$$
$$\frac{d}{dt}\frac{(Au, Au)}{2} + \varepsilon(Bu, Au) - (Ax_*, [u, Au]) = 0.$$

By introducing additionally to $K_\lambda(u) \equiv (Bu, Au) - \lambda(Bu, u)$ an analogous quadratic form $J_\lambda(u) \equiv (Au, Au) - \lambda(Au, u)$, taking into account the equality $Ax_* = \lambda x_*$, and multiplying the first of the equalities obtained by λ and subtracting the result of the multiplication from the second, we obtain

$$\frac{d}{dt}J_\lambda(u) + 2\varepsilon K_\lambda(u) = 0, \tag{9}$$
$$K_\lambda(u) \equiv (Bu, Au) - \lambda(Bu, u), \qquad J_\lambda(u) \equiv (Au, Au) - \lambda(Au, u).$$

Introducing again the projection of the element $u(t) \in \mathbb{D}$ onto the subspace $N_\lambda(A) \subset \mathbb{D}$ of the eigenvalue $\lambda \in P\sigma A$ (cf. with §12.1), i.e., the element $u_\lambda(t) \in N_\lambda(A)$, we find for the difference $\xi(t) = u - u_\lambda \in \mathbb{D}$ that

$$J_\lambda(u) = J_\lambda(\xi), \qquad K_\lambda(u) = K_\lambda(\xi), \qquad \xi = u - u_\lambda.$$

The second of these identities was proved above (§12.1), and the first follows from the equalities

$$\begin{aligned} J_\lambda(\xi + u_\lambda) &= (\lambda u_\lambda + A\xi, \lambda u_\lambda + A\xi) - \lambda(u_\lambda + \xi, \lambda u_\lambda + A\xi) \\ &= \lambda^2\|u_\lambda\|^2 + 2\lambda(u_\lambda, A\xi) - \lambda^2\|u_\lambda\|^2 - \lambda(u_\lambda, A\xi) - \lambda^2(\xi, u_\lambda) + J_\lambda(\xi), \\ (u_\lambda, A\xi) &= (Au_\lambda, \xi) = \lambda(\xi, u_\lambda) = 0. \end{aligned}$$

Therefore, we have

$$\frac{d}{dt}J_\lambda(\xi) + 2\varepsilon K_\lambda(\xi) = 0, \qquad \xi = u - u_\lambda. \tag{10}$$

Then, as a consequence of (2),

$$K_\lambda(\xi) \geqslant \varkappa\|\xi\|_1^2 \qquad \text{for } \lambda = \lambda_1 \qquad (\varkappa = \text{const} > 0).$$

Simultaneously, by the second inequality from (11.23),

$$J_\lambda(\xi) = \|A\xi\|^2 - \lambda(\xi, A\xi) \leqslant \|A\xi\|^2 \leqslant \gamma_2^2\|\xi\|_1^2 \leqslant (\gamma_2^2/\varkappa)K_\lambda(\xi).$$

Then identity (10) implies

$$\frac{d}{dt}J_\lambda(\xi) + \alpha J_\lambda(\xi) \leqslant 0, \qquad t > 0, \quad \alpha = 2\varkappa\varepsilon/\gamma_2^2.$$

Setting $\varphi(t) \equiv J_\lambda(\xi)\exp(\alpha t)$, we obtain $d\varphi/dt \leqslant 0$, $t > 0$, or

$$\varphi(t) \leqslant \varphi(0) = J_\lambda(\xi(0)), \qquad t \geqslant 0.$$

Hence,

$$(11)\qquad J_\lambda(\xi(t)) \leqslant J_\lambda(\xi(0))\exp(-\alpha t) \leqslant \gamma_2^2\|\xi(0)\|_1^2\exp(-\alpha t), \qquad t \geqslant 0.$$

Now, taking the principal moment $\lambda_{11} > \lambda_1$ (which comes after the smallest λ_1) of the commutative top under consideration and using the additional condition of it being definite, we show that

$$(12)\qquad J_{\lambda_1}(\xi(t)) \geqslant (1-\lambda_1/\lambda_{11})\|A\xi\|^2, \qquad \lambda_{11} = \inf\{\lambda' \in P\sigma A : \lambda' > \lambda_1\} > \lambda_1.$$

Indeed, taking into account (4), we have

$$J_\lambda(\xi) = \|A\xi\|^2 - \lambda(\xi, A\overset{n}{\xi}) - \lambda(\xi, A\underset{n+1}{\xi}), \qquad \lambda = \lambda_1.$$

Simultaneously,

$$(\xi, A\overset{n}{\xi}) = \sum_{\substack{k=1\\ \lambda_k\neq\lambda}}^{n} \lambda_k(\xi, e_k)^2 = \sum_{\substack{k=1\\ \lambda_k\neq\lambda}}^{n} \frac{1}{\lambda_k}(\xi, Ae_k)^2$$
$$\leqslant \frac{1}{\lambda_{11}}\sum_{\substack{k=1\\ \lambda_k\neq\lambda}}^{n}(\xi, Ae_k)^2 = \|A\overset{n}{\xi}\|^2/\lambda_{11} = (\|A\xi\|^2 - \|A\underset{n+1}{\xi}\|^2)/\lambda_{11}$$

(we use $(\xi, e_k) = 0$, $e_k \in N_\lambda(A)$). Hence, for any sufficiently large $n = 1, 2, \dots$

$$J_{\lambda_1}(\xi) \geqslant (1-\lambda_1/\lambda_{11})\|A\xi\|^2 + (\lambda_1/\lambda_{11})\|A\underset{n+1}{\xi}\|^2 - \lambda_1(\xi, A\underset{n+1}{\xi}).$$

Passing to the limit as $n \to \infty$, we obtain (12) (since $\|A\underset{n+1}{\xi}\|^2 \to 0$).

As a consequence of (11.23), (11), and (12), we have

$$\|\xi\|_1^2 \leqslant \gamma_1^{-2}\|A\xi\|^2 \leqslant \gamma_1^2 J_{\lambda_1}(\xi(t))/(1-\lambda_1/\lambda_{11})$$
$$\leqslant (\gamma_2/\gamma_1)^2\|\xi(0)\|_1^2(1-\lambda_1/\lambda_{11})^{-1}\exp(-\alpha t).$$

Taking into account the equalities (cf. (2))

$$\lambda_{11} - \lambda_1 = \inf\{\lambda' - \lambda_1 : \lambda' \in P\sigma A,\ \lambda' > \lambda_1\} = \varkappa,$$

we have

$$(13)\qquad \|\xi(t)\|_1 \leqslant \gamma_2\gamma_1^{-1}(1+\lambda_1/\varkappa)^{1/2}\|\xi(0)\|_1\exp(-\varkappa\varepsilon t/\gamma_2^2), \qquad t \geqslant 0.$$

We obtain an analogous estimate for $u_\lambda(t)$ for $\lambda = \lambda_1$. In order to do this, we multiply both parts of equality (8) scalarly by u_λ in $\mathbb{H}$. Then we have

$$\left(u_\lambda, \frac{d}{dt}Au\right) = \left(u_\lambda, A\frac{d}{dt}u\right) = \left(Au_\lambda, \frac{du}{dt}\right) = \lambda\left(u_\lambda, \frac{du}{dt}\right)$$
$$= \lambda\left(u_\lambda, \frac{du_\lambda}{dt}\right) = \frac{\lambda}{2}\frac{d}{dt}\|u_\lambda\|^2 = \frac{1}{2\lambda}\frac{d}{dt}\|Au_\lambda\|^2$$

(since $(u_\lambda, d\xi/dt) = 0$, which follows from the permutability of the operations A and d/dt,

$$Adu_\lambda/dt = dAu_\lambda/dt = \lambda du_\lambda/dt;$$

consequently, $du_\lambda/dt \in N_\lambda(A)$, and, thus, $(du_\lambda/dt, \xi) = 0$; simultaneously,

$$(du_\lambda/dt, \xi) = d(u_\lambda, \xi)/dt - (u_\lambda, d\xi/dt) = -(u_\lambda, d\xi/dt),$$

which completes the proof of the required property). Then we have

$$(u_\lambda, Bu) = (Bu_\lambda, u) = \mu(u_\lambda, u) = \mu(u_\lambda, u_\lambda) = (Bu_\lambda, u_\lambda) = \|u_\lambda\|_1^2$$

(since, according to Proposition 11.1, $Bu_\lambda = \mu u_\lambda$);

$$(u_\lambda, [x_*, Au]) = (x_*, [Au, u_\lambda]) = (x_*, [A\xi, u_\lambda]) = (A\xi, [u_\lambda, x_*])$$

(since $[Au, u_\lambda] = [\lambda u_\lambda + A\xi, u_\lambda] = [A\xi, u_\lambda]$);

$$(u_\lambda, [u, Ax_*]) = \lambda(u_\lambda, [\xi, x_*]) = (Au_\lambda, [\xi, x_*])$$

(since $u = \xi + u_\lambda$ and $Ax_* = \lambda x_*$);

$$(u_\lambda, [u, Au]) = (u_\lambda, [\xi, A\xi]) = (A\xi, [u_\lambda, \xi])$$

(since

$$[u, Au] = [u_\lambda + \xi, \lambda u_\lambda + A\xi] = [u_\lambda, A\xi] + \lambda[\xi, u_\lambda] + [\xi, A\xi]$$

and $(u_\lambda, [v, u_\lambda]) = 0$ for any $v \in M$). Hence,

$$\frac{1}{2\lambda}\frac{d}{dt}\|Au_\lambda\|^2 + \varepsilon\|u_\lambda\|_1^2 + (A\xi, [u_\lambda, x_*]) + (Au_\lambda, [\xi, x_*]) + (A\xi, [u_\lambda, \xi]) = 0.$$

Using again (11.23), we find

$$\begin{aligned}
&\gamma_2^{-2}\|Au_\lambda\|^2 \leqslant \|u_\lambda\|_1^2,\\
&(A\xi, [u_\lambda, x_*]) \leqslant \|A\xi\|\gamma_3\|u_\lambda\|_1\|x_*\|_1 \leqslant \gamma_2\gamma_3\|\xi\|_1\|Au_\lambda\|\|x_*\|_1/\gamma_1,\\
&(Au_\lambda, [\xi, x_*]) \leqslant \gamma_3\|Au_\lambda\|\|\xi\|_1\|x_*\|_1,\\
&(A\xi, [u_\lambda, \xi]) \leqslant \gamma_2\gamma_3\|\xi\|_1\|Au_\lambda\|\|\xi\|_1/\gamma_1.
\end{aligned}$$

Taking into account (13), we have

$$\begin{aligned}
\frac{d}{dt}\|Au_\lambda\|^2 + 2\lambda\varepsilon\|Au_\lambda\|^2/\gamma_2^2 &\leqslant 2\lambda\gamma_3((1+\gamma_2/\gamma_1)\|x_*\|_1 + \gamma_2\|\xi\|_1/\gamma_1)\|\xi\|_1\|Au_\lambda\|\\
&\leqslant \beta(t)\|Au_\lambda\|,
\end{aligned}$$

where

$$\begin{aligned}
\beta(t) &= M\exp(-\varkappa\varepsilon t/\gamma_2^2),\\
M &= 2\lambda\gamma_3((1+\gamma_2/\gamma_1)\|x_*\|_1 + (\gamma_2/\gamma_1)^2(1+\lambda_1/\varkappa)^{1/2}\|\xi(0)\|_1)\\
&\quad\times(\gamma_2/\gamma_1)(1+\lambda_1/\varkappa)^{1/2}\|\xi(0)\|_1,\\
\lambda &= \lambda_1.
\end{aligned}$$

By setting

$$\|Au_\lambda\| \equiv f(t), \qquad 2\lambda\varepsilon/\gamma_2^2 = \delta \qquad (\lambda = \lambda_1), \tag{14}$$

we come to the inequality

$$\frac{df^2}{dt} + \delta f^2 \leqslant \beta(t)f, \qquad t > 0 \qquad (f(t) \geqslant 0),$$

which for all $t > 0$ such that $f(t) > 0$ implies

$$\frac{df}{dt} + \delta f/2 \leqslant \beta(t)/2,$$

or

$$\exp(-\delta t/2)d(f\exp(\delta t/2))/dt \leqslant \beta(t)/2,$$

or

$$d(f\exp(\delta t/2))/dt \leqslant \beta(t)\exp(\delta t/2)/2.$$

Integrating the obtained inequality over the interval where the inequality $f(t) > 0$ is satisfied (due to the continuity of $f(t)$, the set where this inequality is satisfied is covered by a finite or countable system of intervals; the continuity of the function $f(t)$ follows from the assumed smoothness of the solution of the nonstationary equation of the top (or the smoothness of the function $u = u(t)$ in the natural topology of the space $\mathbb{W}$, and, hence, of the space $\mathbb{H}$) and from the continuity of the norm), we find

$$f(t)\exp(\delta t/2) \leqslant f(0) + \frac{1}{2}\int_0^t \beta(s)\exp(\delta s/2)\,ds = f(0) + \frac{M}{2}\int_0^t \exp(ks)\,ds,$$
$$k = \delta/2 - \varkappa\varepsilon/\gamma_2^2 = \lambda_1/\gamma_2^2 - \varkappa\varepsilon/\gamma_2^2 = \varepsilon(2\lambda_1 - \lambda_{11})/\gamma_2^2.$$

For $k \leqslant 0$ (or for $\lambda_{11} \geqslant 2\lambda_1$) we have

$$\int_0^t \exp(ks)\,ds \leqslant \int_0^t ds = t.$$

For $k > 0$ (or for $\lambda_{11} < 2\lambda_1$) we have

$$\int_0^t \exp(ks)\,ds \leqslant \int_0^t ds\exp(kt) = t\exp(kt).$$

Hence,

$$\int_0^t \exp(ks)\,ds\exp(-\delta t/2) \leqslant t\exp(-\delta t/2) \qquad \text{for} \qquad k \leqslant 0,$$

$$\int_0^t \exp(ks)\,ds\exp(-\delta t/2) \leqslant t\exp((k-\delta/2)t) = t\exp(-\varkappa\varepsilon t/\gamma_2^2) \qquad \text{for} \qquad k > 0.$$

Setting

$$\rho = \min\{\delta/2, \varkappa\varepsilon/\gamma_2^2\} \qquad (\rho > 0),$$

we conclude that *either* $f(t) = 0$ *at the point* $t > 0$, *or* $f(t) > 0$ *and then*

$$f(t) \leqslant f(0)\exp(-\delta t/2) + \frac{Mt}{2}\exp(-\rho t), \qquad \delta, \rho = \text{const} > 0. \tag{15}$$

Since the right-hand side of the inequality is nonnegative, the latter is valid *for all* $t \geqslant 0$. Simultaneously, $\|u_\lambda(t)\|_1 \leqslant f(t)/\gamma_1$ (the estimate from (11.23)). The required *a priori* stability immediately follows from (13)–(15). Theorem 2 is proved.

§13. Generalized solutions of the equation of the top

13.1. Spatial flows of a viscous incompressible fluid. By multiplying both parts of the equation of the top (11.24) scalarly in $\mathbb{H}$ by an arbitrary element $v \in \mathbb{D}$ and taking into account the identities $(v, [x, Ax]) = (Ax, [v, x])$ (we use $v \in \mathbb{D} \subset M_c$ and $Ax \in \mathbb{R}(A) \subset M$, and $(v, Bx) = (Bv, x) = \{v, x\} = \{x, v\}$), we come to the following *generalized form of the equation of the top*:

$$\varepsilon\{x, v\} + (Ax, [v, x]) = (y, v). \tag{1}$$

The introduction of the generalized form (1) allows us to include in the consideration spatial flows of a viscous incompressible fluid. In order to obtain the corresponding hydrodynamic equations from relation (1), we change the requirement that the domain of definition $\mathbb{D}$ of the operators $A, B\colon \mathbb{H} \to \mathbb{H}$ be dense in $\mathbb{H}$, to the condition that $\mathbb{D}$ *is dense in a given subspace* $\mathbb{H}_s$ *of the space* $\mathbb{H}$ *invariant with respect to the operators* A, B *in the sense that the domains of definition of these operators belong to* $\mathbb{H}_s$, $\mathbb{R}(A)$, $\mathbb{R}(B) \subset \mathbb{H}_s$ (*the domains of definition of* A, B *coincide with* $\mathbb{D}$ *as before*).

Now let V be a fixed domain of some connected compact C^∞-smooth oriented three-dimensional Riemannian manifold. The metric of the latter is again assumed to be Euclidean. If V does not coincide with the given manifold (V does not coincide with its closure $\overline{V}$), then the boundary $\partial V = \overline{V} \setminus V \neq \varnothing$ of the domain V is assumed to consist of a finite number of pieces of smooth, of class C^∞, surfaces intersecting at nonzero angles. In this case the algebra M, is the linear manifold $C^\infty(\overline{V})$ of infinitely differentiable fields

$$\ldots, \mathbf{u}(\mathbf{x}) = (u_1, u_2, u_3), \mathbf{v}(\mathbf{x}) = (v_1, v_2, v_3), \ldots,$$

defined on the closure $\overline{V}$ of the domain V. The triple of variables x_1, x_2, x_3 will simultaneously define the system of local coordinates and the varying point $\mathbf{x}$ of the manifold $\overline{V}$. The scalar product $(\mathbf{u}, \mathbf{v})$ and the commutator $[\mathbf{u}, \mathbf{v}]$ of the algebra M are introduced by the equalities

$$(\mathbf{u}, \mathbf{v}) = \int_V \mathbf{u} \cdot \mathbf{v}\, dV, \qquad \mathbf{u} \cdot \mathbf{v} \equiv u_1 v_1 + u_2 v_2 + u_3 v_3, \quad dV \equiv dx_1\, dx_2\, dx_3,$$
$$[\mathbf{u}, \mathbf{v}] = -\mathbf{u} \times \mathbf{v} \equiv -(u_2 v_3 - u_3 v_2, u_3 v_1 - u_1 v_3, u_1 v_2 - u_2 v_1).$$

For any $\mathbf{u}, \mathbf{v}, \mathbf{w} \in M$, we have

$$(\mathbf{u}, [\mathbf{v}, \mathbf{w}]) = -\int_V \mathbf{u} \cdot (\mathbf{v} \times \mathbf{w}) dV = (\mathbf{w}, [\mathbf{u}, \mathbf{v}]),$$

i.e., the algebra M is compact ($M = M_c$). The space $\mathbb{H}$ is the space of vector fields square integrable in V. The operators $A, B\colon \mathbb{H} \to \mathbb{H}$ are defined by the identities

$$A\mathbf{u} = \operatorname{rot} \mathbf{u} = \nabla \times \mathbf{u} = (D_2 v_3 - D_3 v_2, D_3 v_1 - D_1 v_3, D_1 v_2 - D_2 v_1),$$
$$B\mathbf{u} = -\Delta \mathbf{u} = -(D_1^2 + D_2^2 + D_3^2)\mathbf{u},$$
$$D_i \equiv \partial/\partial x_i, \qquad i = 1, 2, 3.$$

We associate the domain of definition $\mathbb{D}$ of the operators A, B with the linear space of infinitely differentiable vector fields $\mathbf{u}$ defined on $\overline{V}$ and *solenoidal* in V,

$$\operatorname{div} \mathbf{u} = \nabla \cdot \mathbf{u} = D_1 u_1 + D_2 u_2 + D_3 u_3 = 0, \qquad \mathbf{x} \in V; \tag{2}$$

if $\partial V = \varnothing$ (V is a closed manifold), then it is additionally required that $\mathbf{u}$ be orthogonal to the constants,

$$\int_V \mathbf{u}\, dV = 0 \qquad \left(\text{or } \int_V u_i\, dV = 0\right), \tag{3}$$

and if $\partial V \neq \varnothing$, then it is required that the *no-slip conditions*

$$\mathbf{u}|_{\partial V} = 0 \qquad (\text{or } u_i|_{\partial V} = 0) \tag{4}$$

hold.

The substitution of the operations defined above in equation (1) and the formal replacement of the abstract elements x and y by the introduced solenoidal fields $x = \mathbf{u}$ and $y = \mathbf{f}$ lead to the following integral identity:

$$\int_V (-\varepsilon\Delta\mathbf{u} + \operatorname{rot}\mathbf{u} \times \mathbf{u} - \mathbf{f}) \cdot \mathbf{v}\, dV = 0 \tag{5}$$

considered for an arbitrary $\mathbf{v} \in \mathbb{D}$. For a domain V of the standard three-dimensional Euclidean space, the integral identity (5) is equivalent to the requirement that the vector field $-\varepsilon\Delta\mathbf{u} + \operatorname{rot}\mathbf{u} \times \mathbf{u} - \mathbf{f}$ be gradient or to the requirement that there exist a smooth scalar function $\gamma = \gamma(\mathbf{x})$ such that everywhere in V

$$-\varepsilon\Delta\mathbf{u} + \operatorname{rot}\mathbf{u} \times \mathbf{u} - \mathbf{f} = -\operatorname{grad}\gamma.$$

This fact is a corollary of the Weyl theorem on the coincidence of the orthogonal complement $\mathbb{H}_g$ of the space $\mathbb{H}_s$ of solenoidal fields introduced above,

$$\mathbb{H} = \mathbb{H}_s \oplus \mathbb{H}_g, \tag{6}$$

with the square integrable gradient fields in the domain V **[23, 56]**. An analogous Weyl decomposition can easily be obtained for a torus. However, omitting the analysis of the corresponding general case, we dwell on corollaries of the above considerations that will be needed.

The equations

$$-\varepsilon\Delta\mathbf{u} + \operatorname{rot}\mathbf{u} \times \mathbf{u} - \mathbf{f} = -\operatorname{grad}\gamma, \qquad \operatorname{div}\mathbf{u} = \operatorname{div}\mathbf{f} = 0, \qquad \mathbf{x} \in V, \tag{7}$$

obtained from (1), are the Gromeka-Lamb form of the complete stationary Navier-Stokes system for the spatial flow of an incompressible fluid. The function γ corresponds to the Bernoulli integral

$$p + |\mathbf{u}|^2/2, \qquad |\mathbf{u}|^2 = (\mathbf{u}, \mathbf{u}).$$

Let us verify whether the remaining axioms of the dissipative top given in §11.3 are satisfied. The identity

$$\operatorname{rot}\mathbf{u} \cdot \mathbf{v} - \mathbf{u} \cdot \operatorname{rot}\mathbf{v} = \operatorname{div}(\mathbf{u} \times \mathbf{v}), \tag{8}$$

or

$$\begin{aligned}(D_2u_3 - D_3u_2)v_1 &+ (D_3u_1 - D_1u_3)v_2 + (D_1u_2 - D_2u_1)v_3 \\ &- (D_2v_3 - D_3v_2)u_1 - (D_3v_1 - D_1v_3)u_2 - (D_1v_2 - D_2v_1)u_3 \\ &= D_1(u_2v_3 - u_3v_2) + D_2(u_3v_1 - u_1v_3) + D_3(u_1v_2 - u_2v_1)\end{aligned}$$

(which can be straightforwardly verified), implies

$$\begin{aligned}(\operatorname{rot}\mathbf{u}, \mathbf{v}) - (\mathbf{u}, \operatorname{rot}\mathbf{v}) &= \int_V (\operatorname{rot}\mathbf{u} \cdot \mathbf{v} - \mathbf{u} \cdot \operatorname{rot}\mathbf{v})\, dV \\ &= \int_V \operatorname{div}(\mathbf{u} \times \mathbf{v})\, dV \\ &= \begin{cases} 0, & \text{if } \partial V = \varnothing, \\ \int_{\partial V} \mathbf{n} \cdot (\mathbf{u} \times \mathbf{v})\, dS, & \text{if } \partial V \neq \varnothing. \end{cases}\end{aligned}$$

Here $\mathbf{n}dS$ is the vector area element of the surface ∂V. As a consequence of condition (4), the operator rot is symmetric on $\mathbb{D}$ (the contour integral from the equality just given vanishes). The operator $-\Delta$ is also symmetric on $\mathbb{D}$:

$$\begin{aligned}(-\Delta\mathbf{u},\mathbf{v})+(\mathbf{u},\Delta\mathbf{v}) &= (\operatorname{rot}\operatorname{rot}\mathbf{u},\mathbf{v})-(\mathbf{u},\operatorname{rot}\operatorname{rot}\mathbf{v})\\ &= (\operatorname{rot}\operatorname{rot}\mathbf{u},\mathbf{v})-(\operatorname{rot}\mathbf{u},\operatorname{rot}\mathbf{v})-(\mathbf{u},\operatorname{rot}\operatorname{rot}\mathbf{v})+(\operatorname{rot}\mathbf{u},\operatorname{rot}\mathbf{v})\\ &= \int_V \operatorname{div}(\operatorname{rot}\mathbf{u}\times\mathbf{v}+\mathbf{u}\times\operatorname{rot}\mathbf{v})\,dV\\ &= \begin{cases} 0, & \text{if } \partial V=\varnothing,\\ \int_{\partial V}\mathbf{n}\cdot(\operatorname{rot}\mathbf{u}\times\mathbf{v}+\mathbf{u}\times\operatorname{rot}\mathbf{v})\,dS, & \text{if } \partial V\neq\varnothing;\end{cases}\end{aligned}$$

under condition (4) this implies

$$(-\Delta\mathbf{u},\mathbf{v})+(\mathbf{u},\Delta\mathbf{v})=0$$

(we have additionally used identities $-\operatorname{rot}\operatorname{rot}\mathbf{u}=\Delta\mathbf{u}$, $\operatorname{div}\mathbf{u}=0$).

The positive definiteness of the operator $-\Delta$ on $\mathbb{D}$ follows from the identities

$$\begin{aligned}(-\Delta\mathbf{u},\mathbf{u}) &= (\operatorname{rot}\operatorname{rot}\mathbf{u},\mathbf{u})-(\operatorname{rot}\mathbf{u},\operatorname{rot}\mathbf{u})+(\operatorname{rot}\mathbf{u},\operatorname{rot}\mathbf{u})\\ &= \int_V \operatorname{div}(\operatorname{rot}\mathbf{u}\times\mathbf{u})\,dV+(\operatorname{rot}\mathbf{u},\operatorname{rot}\mathbf{u}),\end{aligned}$$

$$\int_V \operatorname{div}(\operatorname{rot}\mathbf{u}\times\mathbf{u})\,dV=\begin{cases}0, & \text{if } \partial V=\varnothing,\\ \int_{\partial V}\mathbf{n}\cdot(\operatorname{rot}\mathbf{u}\times\mathbf{u})\,dS, & \text{if } \partial V\neq\varnothing,\end{cases}$$

and the boundary conditions (4). As a consequence of the indicated properties, the isometry conditions

(9) $$(-\Delta\mathbf{u},\mathbf{u})=(\operatorname{rot}\mathbf{u},\operatorname{rot}\mathbf{u}),\qquad \mathbf{u}\in\mathbb{D},$$

are satisfied for the operator $-\Delta$. The unique solvability and continuous dependence of the solution of the classical boundary problem

(10) $$\operatorname{rot}\mathbf{u}=\mathbf{f}(\mathbf{x}),\quad \operatorname{div}\mathbf{u}=0\quad(\operatorname{div}\mathbf{f}=0),\qquad \mathbf{x}\in V,\quad \mathbf{n}\cdot\mathbf{u}|_{\partial V}=0$$

on $\mathbf{f}$ in $\mathbb{H}$ (for the case of a closed manifold $V=\overline{V}$, the impermeability conditions from (10) are replaced by the restrictions (3), which are assumed to be satisfied for both $\mathbf{u}$ and $\mathbf{f}$) guarantee the validity of the inequality

(11) $$\|\operatorname{rot}\mathbf{u}\|^2=\int_V|\operatorname{rot}\mathbf{u}|^2\,dV\geqslant\beta\int_V|\mathbf{u}|^2\,dV=\beta\|\mathbf{u}\|^2$$

for some constant $\beta>0$. The complete continuity of the corresponding inverse operator for problem (10) is provided by the compactness of embedding (11). The validity of embedding (11) and its compactness under boundary conditions (4) can be verified by analyzing the following problem:

(12) $$-\Delta\mathbf{u}=\mathbf{f}(\mathbf{x})+\operatorname{grad}\varphi,\quad \operatorname{div}\mathbf{u}=0\quad(\operatorname{div}\mathbf{f}=0),\qquad \mathbf{x}\in V,\quad \mathbf{u}|_{\partial V}=0$$

(*the Stokes problem*). For the case of a bounded domain of Euclidean space, the necessary constructions were given in **[50, 56]**. Note that in the indicated case, the norm $\|\mathbf{u}\|_1=-(\Delta\mathbf{u},\mathbf{u})^{1/2}$ is isometric to the norm $\|\operatorname{rot}\mathbf{u}\|$ from (11),

$$\|\mathbf{u}\|_1^2=(-\Delta\mathbf{u},\mathbf{u})=\int_V|\nabla\mathbf{u}|^2\,dV\equiv\int_V(|D_1\mathbf{u}|^2+|D_2\mathbf{u}|^2+|D_3\mathbf{u}|^2)\,dV,$$

i.e., the identity

$$\| \operatorname{rot} \mathbf{u}\| = \|\nabla \mathbf{u}\|, \qquad \|\nabla \mathbf{u}\| \equiv \left(\int_V |\nabla \mathbf{u}|^2 \, dV\right)^{1/2} \tag{13}$$

is valid. The latter immediately follows from another known identity, namely,

$$(-\Delta \mathbf{u}, \mathbf{u}) = -\int_V (\operatorname{div}(u_1 \operatorname{grad} u_1 + u_2 \operatorname{grad} u_2 + u_3 \operatorname{grad} u_3)\, dV + \|\nabla \mathbf{u}\|^2$$

and from conditions (4). Analogous constructions are also possible in the case of a closed manifold $V = \overline{V}$, when condition (3) is satisfied.

Condition (9) guarantees that the first two inequalities from (11.23) are satisfied (in the case under consideration $\gamma_1 = \gamma_2 = 1$). The remaining estimate from (11.23) takes the form

$$\|\mathbf{u} \times \mathbf{v}\| \leqslant \gamma_3 \| \operatorname{rot} \mathbf{u}\| \| \operatorname{rot} \mathbf{v}\|. \tag{14}$$

In the case under consideration, the indicated estimate follows from the inequality

$$\int_V u_i^4 \, dV \leqslant (4/3)^{3/2} \left(\int_V u_i^2 \, dV\right)^{1/2} \left(\int_V |\operatorname{grad} u_i|^2 \, dV\right)^{3/2},$$

proved in **[56, Chapter 1, §1, Lemma 2]** and inequality (11).

On the other hand, as it was shown in **[56]**, the more subtle (in comparison with (11)) estimate (15) together with the compactness of the embedding realized by (11) guarantees the existence of a generalized solution of problem (4), (5). An analogous theorem is proved below for the generalized equation of the top (1) under an additional condition corresponding to the indicated inequality (15).

If $\partial V = \varnothing$, then the top generated by (5) is commutative, while if $\partial V \neq \varnothing$ and conditions (4) are satisfied, then it is not commutative:

$$\begin{aligned}(-\Delta \mathbf{u}, \operatorname{rot} \mathbf{v}) + (\operatorname{rot} \mathbf{u}, \Delta \mathbf{v}) &= (\operatorname{rot} \operatorname{rot} \mathbf{u}, \operatorname{rot} \mathbf{v}) - (\operatorname{rot} \mathbf{u}, \operatorname{rot} \operatorname{rot} \mathbf{v}) \\ &= \int_V \operatorname{div}(\operatorname{rot} \mathbf{u} \times \operatorname{rot} \mathbf{v})\, dV \\ &= \begin{cases} 0, & \text{if } \partial V = \varnothing; \\ \int_{\partial V} \mathbf{n} \cdot (\operatorname{rot} \mathbf{u} \times \operatorname{rot} \mathbf{v})\, dS, & \text{if } \partial V \neq \varnothing. \end{cases}\end{aligned}$$

13.2. An extension of the class of admissible right-hand sides. By the second and third inequalities from (11.23), the operator $A\colon \mathbb{H}_s \to \mathbb{H}_s$ and equation (1) admit an extension by continuity to the elements of the new Hilbert space $\mathbb{W}_s$ obtained (as the space $\mathbb{W}$ from §11.33) by completion of $\mathbb{D}$ in the norm $\|u\|_1 = \{u, u\}^{1/2}$, generated by the scalar product $\{u, v\} = (Bu, v) = (u, Bv)$ introduced on $\mathbb{D}$. The class of the right-hand sides of equation (1) can be extended to arbitrary linear continuous functionals $l_y\colon \mathbb{W}_s \to R\colon v \to l_y(v)$ given on $\mathbb{W}_s$ which, by definition, belong to the corresponding adjoint space $\mathbb{W}_s^*$. As it follows from the inequalities

$$|(y, v)| \leqslant \|y\| \|v\| \leqslant \beta^{-1/2} \|y\| \|v\|_1$$

(we use (11.21)), the functional $l_y(v) = (y, v)$ determined by the right-hand side of equation (1) also belongs to the space $\mathbb{W}_s^*$. An element $x \in \mathbb{W}_s$ satisfying (16) for any v from $\mathbb{W}_s$ will be called a *generalized solution* of the corresponding equation (1)

$$\varepsilon\{x, v\} + (Ax, [v, x]) = l_y(v). \tag{16}$$

Equation (16) will be called the *extended generalized form of the equation of the top* (11.24).

THEOREM 1. *If*

$$\mu = \|l_y\|/\varepsilon^2 < 1/\gamma_2\gamma_3, \tag{17}$$
$$\|l_y\| \equiv \sup\{|l_y(v)| : v \in \mathbb{W}_s, \|v\|_1 = 1\},$$

then equation (16) *admits at most one generalized solution* $x \in \mathbb{W}_s$, *the latter continuously depending on* $l_y \in \mathbb{W}_s^*$ *in the following sense*:

$$\|x' - x\|_1 \leqslant \|l_{y'} - l_y\|/(\varepsilon(1 - \gamma_2\gamma_3\mu)), \tag{18}$$

where $x' \in \mathbb{W}_s$ *is a generalized solution of equation* (16) *corresponding to the right-hand side* $l_y(v) = l_{y'}(v)$.

Indeed, by setting in (16) $v = x$ and extending by continuity the identity $(Ax, [x, x]) = 0$ from $\mathbb{D}$ to $\mathbb{W}_s$, we obtain

$$\varepsilon\|x\|_1^2 = l_y(x) \leqslant \|l_y\|\|x\|_1. \tag{19}$$

Changing x in (16) to $x' = x + u$, we find

$$\varepsilon\{u, v\} + (Au, [v, x]) + (Ax, [v, u]) + (Au, [v, u]) = l_{y'}(v) - l_y(v). \tag{20}$$

If $v = u$, then we have $(Au + Ax, [u, u]) = 0$ (since the latter is valid for $x, u \in \mathbb{D}$ and $\mathbb{D}$ is dense in $\mathbb{W}_s$); equation (20) together with (19) implies (18):

$$\begin{aligned}\varepsilon(1 - \gamma_2\gamma_3\mu)\|u\|_1^2 &= \varepsilon\|u\|_1^2 - \gamma_2\gamma_3\|l_y\|\|u\|_1^2/\varepsilon \leqslant \varepsilon\|u\|_1^2 - \gamma_2\gamma_3\|x\|_1\|u\|_1^2 \\ &\leqslant \varepsilon\|u\|_1^2 + (Au, [u, x]) = l_{y'}(u) - l_y(u) \leqslant \|l_{y'} - l_y\|\|u\|_1.\end{aligned}$$

The uniqueness of the solution x immediately follows from (18). The proof is completed.

13.3. The existence of a generalized solution. According to the Riesz lemma, for any $l_y \in \mathbb{W}_s^*$ there exists a unique element $z_y \in \mathbb{W}_s$ satisfying the identity

$$l_y(v) = \{z_y, v\} \tag{21}$$

for any $v \in \mathbb{W}_s$. On the other hand, for any fixed $x \in \mathbb{W}_s$, the second term on the left-hand side of equation (16) defines a certain bounded linear (hence, continuous) functional on $\mathbb{W}_s$,

$$|(Ax, [v, x])| \leqslant \|Ax\|\|[v, x]\| \leqslant \gamma_2\gamma_3\|x\|_1^2\|v\|_1.$$

Hence, the transformation $\mathfrak{F}\colon \mathbb{W}_s \to \mathbb{W}_s : x \to \mathfrak{F}(x)$ of the space $\mathbb{W}_s$ satisfying for any $v \in \mathbb{W}_s$ the identity

$$(Ax, [v, x]) = \{\mathfrak{F}(x), v\} \tag{22}$$

is uniquely defined. Taking into account (21) and (22), we rewrite (16) in the form

$$x + \frac{1}{\varepsilon}\mathfrak{F}(x) = \frac{1}{\varepsilon}z_y \quad \text{in } Bw_s. \tag{23}$$

The identities

$$(y, v) = l_y(v) = \{z_y, v\} \tag{24}$$

used here allow us to indicate the form of a new norm of y,

$$\|y\|_{-1} \equiv \sup\{|(y, v)| : v \in \mathbb{D},\ \|v\|_1 = 1\}, \tag{25}$$

relative to which the extension of the right-hand sides of the equation of the top (11.24) is carried out. The existence of a solution $x \in \mathbb{W}_s$ is established below by proving the fact that the image of the transformation

$$\mathfrak{G}\colon \mathbb{W}_s \to \mathbb{W}_s,\ x \to \mathfrak{G}(x) \equiv x + \frac{1}{\varepsilon}\mathfrak{F}(x) \tag{26}$$

of space $\mathbb{W}_s$ generated by the left-hand side of equation (23) is the entire space $\mathbb{W}_s$. The following three properties of the transformation $\mathfrak{F}\colon \mathbb{W}_s \to \mathbb{W}_s$ are used here.

The first property is the *continuity* of $\mathfrak{F}$ everywhere in $\mathbb{W}_s$, which follows from the estimate

$$\|\mathfrak{F}(x+u) - \mathfrak{F}(x)\|_1 \leqslant \gamma_2\gamma_3(2\|x\|_1\|u\|_1 + \|u\|_1^2). \tag{27}$$

Estimate (27) follows from (11.23) and (22),

$$\begin{aligned}\{\mathfrak{F}(x+u) - \mathfrak{F}(x), v\} &= (Au, [v, x]) + (Ax, [v, u]) + (Au, [v, u]) \\ &\leqslant \gamma_2\gamma_3(\|u\|_1\|v\|_1\|x\|_1 + \|x\|_1\|v\|_1\|u\|_1 + \|u\|_1\|v\|_1\|u\|_1),\end{aligned}$$

and

$$\begin{aligned}\|\mathfrak{F}(x+u) - \mathfrak{F}(x)\|_1 &\leqslant \sup\{|\{\mathfrak{F}(x+u) - \mathfrak{F}(x), v\}| : v \in \mathbb{D},\ \|v\|_1 = 1\} \\ &\leqslant \gamma_2\gamma_3(2\|x\|_1\|u\|_1 + \|u\|_1^2).\end{aligned}$$

The second property is the *neutrality* of the transformation $\mathfrak{F}$ defined by the condition

$$\{\mathfrak{F}(x), x\} = 0, \qquad x \in \mathbb{W}_s. \tag{28}$$

Identity (28) follows from (22) for $v = x$.

For the case of a *finite-dimensional* space $\mathbb{W}_s$, the indicated two properties of the transformation $\mathfrak{F}\colon \mathbb{W}_s \to \mathbb{W}_s$ guarantee the required coincidence of the domain of definition $\mathbb{R}(\mathfrak{G})$ of the transformation $\mathfrak{G}$ from (26) with $\mathbb{W}_s$: $\mathfrak{G}(\mathbb{W}_s) = \mathbb{W}_s$.

Indeed, the mapping $\mathfrak{G}\colon \mathbb{W}_s \to \mathbb{W}_s$ satisfies the identity

$$\{\mathfrak{G}(x), x\} = \|x\|_1^2 \tag{29}$$

for a neutral $\mathfrak{F}$ and is, simultaneously, a continuous transformation of the space $\mathbb{W}_s$. The following well-known lemma is valid for the indicated continuous *transformations* of finite-dimensional spaces.

LEMMA 1. *A continuous mapping $f\colon R^n \to R^n$ of n-dimensional Euclidean space R^n to R^n ($n = 1, 2, \cdots < \infty$) satisfying the additional requirement*

$$(x, f(x))/\|x\| \to +\infty \quad \textit{as} \quad \|x\| \to +\infty, \tag{30}$$

$$x = (x_1, \ldots, x_n) \in R^n, \qquad (x, y) = x_1y_1 + \cdots + x_ny_n, \qquad \|x\| = (x, x)^{1/2},$$

maps R^n onto R^n.

The proof of this lemma is given, e.g., in [**73,** p. 25–26].

Equality (29) guarantees that condition (30) is satisfied for $\mathfrak{G}$; then Lemma 1 implies the validity of the required identity $\mathfrak{G}(\mathbb{W}_s) = \mathbb{W}_s$ in the case of a finite-dimensional $\mathbb{W}_s$.

The third property of the transformation $\mathfrak{F}\colon \mathbb{W}_s \to \mathbb{W}_s$ is related to the case of an infinite-dimensional space $\mathbb{W}_s$. The indicated property is provided by a special additional condition imposed on the commutator of the algebra M. First, let us agree on the terminology. Let $\mathbb{B}_1$, $\mathbb{B}_2$ be Banach spaces and let $F\colon \mathbb{B}_1 \to \mathbb{B}_2\colon x \to F(x)$ be a continuous mapping of the space $\mathbb{B}_1$ into $\mathbb{B}_2$. The mapping F will be called *strong continuous* if from any subsequence $\{x_i\}$ of elements x_i, $i = 1, 2, \dots$, of the space $\mathbb{B}_1$ weakly converging to some element $x_* \in \mathbb{B}_1$ (*weak convergence* means that for any functional $l\colon x \to l(x)$ from the dual space $\mathbb{B}_1^*$ (the space of continuous linear functionals given on $\mathbb{B}_1$) the limit

$$\lim l(x_i) = l(x_*), \qquad i \to \infty)$$

we can choose a subsequence

$$\{x_{i_k}\} \qquad (i_k \to \infty \text{ as } k \to \infty, \quad k = 1, 2, \dots)$$

such that the subsequence $\{F(x_{i_k})\}$ converges to $F(x_*)$ in the norm,

$$F(x_{i_k}) \to F(x_*) \qquad \text{in} \qquad \mathbb{B}_2 \qquad \text{as} \qquad i_k \to \infty$$

(*strong convergence*). Thus, the strong continuity for a compact mapping assumes it to be closed in weak topology.

LEMMA 2. *A continuous transformation $f\colon H \to H$ of a separable Hilbert space satisfying condition* (30) *such that the mapping $x \to f(x) - x$ is strong continuous, maps H onto H.*

Before proceeding with the proof of this lemma, we note that for the case of a finite-dimensional $H = R^n$, the conditions of Lemmas 1 and 2 are equivalent. Indeed, in this case the continuity and strong continuity of the transformation of the space H under consideration are equivalent (which immediately follows from the equivalence, in this case, of the notions of strong and weak convergence). Recall also that *separability* of H means the existence in H of a countable and everywhere dense set. Separability implies the existence of an orthonormal basis $e_1, e_2, \dots, (e_k, e_m) = \delta_{km}$, in H.

Setting for an arbitrary $x \in H$

$$\overset{n}{x} = \sum_{k=1}^{n} x_k e_k, \qquad x_k = (x, e_k),$$

we consider the following finite-dimensional analog of the equation $f(x) = y$,

$$\overset{n}{f}(\overset{n}{x}) = \overset{n}{y}. \tag{31}$$

The transformation defined by the left-hand side of equation (31) obviously satisfies the conditions of Lemma 1. According to this lemma, this equation has a solution $\overset{n}{x}$ for any $n = 1, 2, \dots$. As a corollary of condition (30), the sequence $\{\overset{n}{x}\}$ $(n = 1, 2, \dots)$ is bounded in H.

Indeed, the existence of an unbounded subsequence $\{\overset{n_m}{x}\}$ of the sequence $\{\overset{n}{x}\}$ would have implied, according to (30), the unboundedness of the sequence of numbers

$$\alpha_m = (\overset{n_m}{x}, f(\overset{n_m}{x}))/\|\overset{n_m}{x}\|, \qquad m = 1, 2, \ldots,$$

which immediately contradicts equality (31) since, according to (31),

$$|\alpha_m| = |(\overset{n_m}{x}, f(\overset{n_m}{x}))|/\|\overset{n_m}{x}\| = |(\overset{n_m}{x}, \overset{n_m}{y})|/\|\overset{n_m}{x}\| \leqslant \|\overset{n_m}{y}\| \leqslant \|y\|.$$

The boundedness of the sequence $\{\overset{n}{x}\} \subset H$ implies the existence of a subsequence $\{\overset{n_m}{x}\} \subset \{\overset{n}{x}\}$ weakly converging to some element $x_* \in H$ (we use the criterion of reflexivity of a Banach space **[31]**,

$$(\overset{n_m}{x}, v) \to (x_*, v), \qquad m \to \infty \qquad (n_m \to \infty),$$

for any $v \in H$). Taking into account the assumed strong continuity of the transformation $g(x) = f(x) - x$, $x \in H$, we shall assume, without loss of generality, that the subsequence $\{\overset{n_m}{x}\}$ coincides with its subsequence satisfying the required condition

$$\|g(x_*) - g(\overset{n_m}{x})\| \to 0 \qquad (n_m \to \infty).$$

Since, simultaneously,

$$\|\overset{n_m}{g}(x_*) - \overset{n_m}{g}(\overset{n_m}{x})\| \leqslant \|g(x_*) - g(\overset{n_m}{x})\|$$

and

$$\|f(x_*) - \overset{n_m}{f}(x_*)\|, \ \|y - \overset{n_m}{y}\|, \ \|x_* - \overset{n_m}{x_*}\| \to 0, \qquad n_m \to \infty,$$

(the completeness of the orthonormal system is taken into account), we conclude, by passing to the limit as $n_m \to \infty$ in the identity

$$\begin{aligned}(f(x_*) - y, v) &= (f(x_*) - \overset{n_m}{f}(x_*), v) + (\overset{n_m}{g}(x_*) - \overset{n_m}{g}(\overset{n_m}{x}), v) \\ &\quad + (\overset{n_m}{y} - y, v) + (\overset{n_m}{x_*} - x_*, v) + (x_*, v) - (\overset{n_m}{x}, v), \qquad v \in H,\end{aligned}$$

that $f(x_*) = y$. Lemma 2 is proved.

PROPOSITION 1. *If the additional requirement that the transformation $\mathfrak{F}\colon \mathbb{W}_s \to \mathbb{W}_s$ defined by identity* (22) *be strong continuous is satisfied, then equation* (16) *has at least one solution $x \in \mathbb{W}_s$ for any given $\varepsilon > 0$ and $l_y \in \mathbb{W}_s$.*

Indeed, the transformation $\mathfrak{G}\colon \mathbb{W}_s \to \mathbb{W}_s$ defined in this case by (26), satisfies the conditions of Lemma 2. Hence, in this case equation (23), which is equivalent to (16), is solvable in $\mathbb{W}_s$ for any $\varepsilon > 0$ and $z_y \in \mathbb{W}_s$, which completes the proof.

We shall say that there is an *embedding* of one Banach space $\mathbb{B}_1$ into another $\mathbb{B}_2$ and shall denote this by $\mathbb{B}_1 \subset \mathbb{B}_2$, if the elements of the first space are elements of the second, and if for the corresponding norms $|\cdot, \mathbb{B}_1|$ and $|\cdot, \mathbb{B}_2|$, the inequality

$$|x, \mathbb{B}_2| \leqslant \gamma |x, \mathbb{B}_1|, \qquad x \in \mathbb{B}_1, \quad \gamma = \text{const} \geqslant 0,$$

is satisfied. The existence of the embedding $\mathbb{B}_1 \subset \mathbb{B}_2$, of course, does not imply that the inequality inverse to the indicated one is satisfied. In particular, under the embedding $\mathbb{B}_1 \subset \mathbb{B}_2$, the zero element of the space $\mathbb{B}_2$, generally speaking, is not the zero element of the space $\mathbb{B}_1$. In the case when the latter happens, i.e., when among the representatives of the equivalence class of an element of the space $\mathbb{B}_2$ there exists a representative of the equivalence class of an element of the space $\mathbb{B}_1$, the embedding $\mathbb{B}_1 \subset \mathbb{B}_2$ will be called *faithful*. For example, known embeddings of the Sobolev spaces and, in particular, the classical Poincaré-Steklov embedding are faithful. Further, an embedding $\mathbb{B}_1 \subset \mathbb{B}_2$ is *compact* if (as it is usual) any bounded sequence of the space $\mathbb{B}_1$ is partially convergent in the space $\mathbb{B}_2$ (*partial convergence* means the existence of a convergent subsequence). Finally, the embedding $\mathbb{B}_1 \subset \mathbb{B}_2$ is *strong continuous* if the weak convergence of a sequence from $\mathbb{B}_1$ to some $x \in \mathbb{B}_1$ implies its partial convergence to the same x in $\mathbb{B}_2$. Without loss of generality, the element x in this definition can obviously be changed to the zero element of the space $\mathbb{B}_1$ (which is simultaneously the zero element of the space $\mathbb{B}_2$, or belongs to the equivalence class of the zero element of the space $\mathbb{B}_2$). For a reflexive (in particular, a Hilbert) space $\mathbb{B}_1$ where any bounded sequence is known to be partially weakly converging (the above-mentioned reflexivity criterion **[31]**), the strong continuous embedding is compact. The question of strong continuity of a compact embedding is resolved by the following lemma.

LEMMA 3. *Let for the Banach space $\mathbb{B}$ and Hilbert spaces $\mathbb{W}$ and $\mathbb{H}$ the embeddings $\mathbb{W} \subset \mathbb{B} \subset \mathbb{H}$ be valid, and let $\mathbb{W}$ have a dense subset which is also dense in $\mathbb{H}$. If the embedding $\mathbb{W} \subset \mathbb{B}$ is compact and the embedding $\mathbb{B} \subset \mathbb{H}$ is faithful, then the embedding $\mathbb{W} \subset \mathbb{B}$ is strong continuous.*

Indeed, let the sequence $\{x_i\} \subset \mathbb{W}$ weakly converge to the zero element of the Hilbert space $\mathbb{W}$. Then it is bounded in $\mathbb{W}$ **[44]**. Since, simultaneously, the embedding $\mathbb{W} \subset \mathbb{B}$ is compact, a bounded sequence $\{x_i\} \subset \mathbb{W}$ has a subsequence converging in $\mathbb{B}$ to some element $x \in \mathbb{B}$. Without loss of generality, we identify this subsequence with $\{x_i\}$. To complete the proof, we have to show that x is the zero element of the space $\mathbb{B}$ (hence, of the space $\mathbb{H}$). The compactness of the embedding $\mathbb{W} \subset \mathbb{B}$ and the chain of embeddings $\mathbb{W} \subset \mathbb{B} \subset \mathbb{H}$ implies the compactness of the embedding $\mathbb{W} \subset \mathbb{H}$ (which is obvious). The assumed existence in $\mathbb{W}$ of a set which is simultaneously dense in $\mathbb{W}$ and $\mathbb{H}$, guarantees, according to Lemma 10.4, the existence of a common orthogonal basis of the spaces $\mathbb{W}$ and $\mathbb{H}$, defined by the nontrivial solutions $u \neq 0$ of the corresponding spectral problem $\{u, v\} = \lambda(u, v)$, $v \in \mathbb{W}$ $(\lambda > 0)$; here $\{u, v\}$ and (u, v) are the scalar products in the spaces $\mathbb{W}$ and $\mathbb{H}$, respectively. For any basis element u, we have

$$|(x, u)| \leqslant |(x - x_i, u)| + |(x_i, u)|,$$
$$|(x - x_i, u)| \leqslant \|x - x_i\| \|u\| \qquad (\|u\| = (u, u)^{1/2}),$$
$$(x_i, u) = \{x_i, u\}/\lambda.$$

The convergence of $\{x_i\}$ to x in $\mathbb{B}$ implies the convergence of $\{x_i\}$ to x in $\mathbb{H}$ (we use the embedding $\mathbb{B} \subset \mathbb{H}$), $\|x - x_i\| \to 0$, $i \to \infty$. Then the weak convergence $\{x_i, u\} \to 0$ implies the equality $(x, u) = 0$. Since u is an arbitrary basis element of the space $\mathbb{H}$, it follows from the equality obtained that x is the zero element of the space $\mathbb{H}$. The faithfulness of the embedding $\mathbb{B} \subset \mathbb{H}$ implies that in this case x is the zero element of $\mathbb{B}$. The proof is completed.

COROLLARY. *A compact embedding $\mathbb{W} \subset \mathbb{H}$ of Hilbert spaces $\mathbb{W}$ and $\mathbb{H}$ with a common dense subset is strong continuous.*

Indeed, in the proof of Lemma 10.4 we established the faithfulness of the embedding $\mathbb{W} \subset \mathbb{H}$ (*compact embeddings of Hilbert spaces with a common dense subset are necessarily faithful*). Then, setting $\mathbb{B} = \mathbb{H}$ in the conditions of Lemma 2, we come to the required corollary.

Now we pass to the formulation of the above-mentioned additional restriction on the commutator $[u, v]$, analogous to condition (15). Along with the norms $\|\cdot\|$ and $\|\cdot\|_1$, we assume the existence of a third norm $\|\cdot\|_c$ defined on the elements of $\mathbb{D}$ generating (by completion of the linear manifold $\mathbb{D}$ in the given norm) some Banach space $\mathbb{C}$ faithfully embedded in $\mathbb{H}_s$ and compactly absorbing $\mathbb{W}_s$, thus entering a chain of embeddings of the form

$$\mathbb{W}_s \subset \mathbb{C} \subset \mathbb{H}_s, \tag{32}$$

the first of which is compact and the second is faithful. We also assume that for some constant $\gamma_4 > 0$ and for all $u, v, x \in \mathbb{D}$ the inequalities

$$|(Au, [v, x])| \leqslant \gamma_4 \|u\|_c \|v\|_1 \|x\|_1, \gamma_4 \|u\|_1 \|v\|_c \|x\|_1 \tag{33}$$

are satisfied. Note that, according to the second embedding from (32), for some constant $\gamma_5 > 0$ and for all $u \in \mathbb{D}$, the inequality

$$\|u\|_c \leqslant \gamma_5 \|u\|_1 \tag{34}$$

is valid. The latter, together with (33), guarantees that the third inequality from (11.23) is satisfied when $\mathbb{H} = \mathbb{H}_s$. Therefore, for $\mathbb{H} = \mathbb{H}_s$ inequalities (33) indeed strengthen the restriction indicated above.

PROPOSITION 2. *Under the above assumptions concerning the norm $\|\cdot\|_c$ (the existence of the latter as providing the compactness of the first embedding of the chain* (32) *and the faithfulness of the second), if inequalities* (33) *are satisfied, then the transformation $\mathfrak{F}: \mathbb{W}_s \to \mathbb{W}_s$ of the Hilbert space $\mathbb{W}_s$ defined by identity* (22) *is strong continuous.*

Indeed, according to (22), (33), and (34),

$$\begin{aligned}\{\mathfrak{F}(x+u) - \mathfrak{F}(x), v\} &= (Au, [v, x]) - (Ax, [u, v]) + (Au, [v, u]) \\ &\leqslant \gamma_4(\|u\|_c \|v\|_1 \|x\|_1 + \|x\|_1 \|u\|_c \|v\|_1 + \|u\|_c \|v\|_1 \|u\|_1) \\ &\leqslant \gamma_4(2\|x\|_1 + \|u\|_1)\|u\|_c \|v\|_1,\end{aligned}$$

which implies

$$\|\mathfrak{F}(x+u) - \mathfrak{F}(x)\|_1 \leqslant \gamma_4(2\|x\|_1 + \|u\|_1)\|u\|_c. \tag{35}$$

Now let a given sequence $\{x_i\} \subset \mathbb{W}_s$ weakly converge to $x_* \in \mathbb{W}_s$. Since the conditions of Lemma 3 are satisfied for the embeddings (32), there exists a subsequence $\{x_{i_k}\} \subset \{x_i\}$ converging to x_* in $\mathbb{C}$. By setting in (35) $x = x_*$, $u = x_{i_k} - x_*$, we obtain the required convergence $\|\mathfrak{F}(x_{i_k}) - \mathfrak{F}(x_*)\|_1 \to 0$, $i_k \to \infty$, which completes the proof.

THEOREM 2. *If the additional inequalities* (33) *with the norm $\|\cdot\|_c$ introduced on $\mathbb{D}$ (and providing the compactness of the first embedding from* (32) *and the faithfulness of the second) hold, then equation* (16) *has a solution $x \in \mathbb{W}_s$ for any $\varepsilon > 0$ and $l_y \in \mathbb{W}_s^*$.*

Theorem 2 immediately follows from Propositions 1 and 2.

In conclusion we note that for the stationary Navier-Stokes equations in a bounded domain of the above-mentioned type, when the external mass forces are infinitely differentiable in the closure of the given domain, the generalized and the classical (introduced in a standard way as satisfying the required smoothness conditions) solutions are equivalent **[56, 93]**. Consequently, Theorem 2 can be considered as providing the existence of a classical solution of the equations indicated for any (arbitrarily large) values of the Reynolds number (reciprocal to the parameter ε) and any infinitely smooth mass forces.

§14. Examples of dissipative tops

14.1. Motions on the algebra $\mathfrak{so}(n, R)$. Let us illustrate Theorem 12.1 from §12.1 by the analysis of the following example of a finite-dimensional top. Let M be the linear finite-dimensional manifold $\mathfrak{so}(n, r)$ of square skew-symmetric matrices $x = (x_{ik})$ of order $n \geqslant 3$ $(i, k = 1, 2, \dots)$ with real elements x_{ik}. Skew-symmetry means that $x_{ik} = -x_{ki}$, or, in terms of the transposed matrix $x^* = (x_{ki})$, the equality $x^* = -x$. The commutator in the algebra $\mathfrak{so}(n, R)$ is defined by means of the standard identity $[x, y] = xy - yx$, where $xy = (x_{im} y_{mk})$ is the matrix obtained by multiplying the matrix x by the matrix y from $\mathfrak{so}(n, R)$ (summation is carried over the repeated index). The scalar product in $\mathfrak{so}(n, R)$ will be defined as being generated by the Killing metric,

$$(x, y) = \operatorname{Sp} xy^* = -\operatorname{Sp} xy = -(x_{km} y_{mk}).$$

Then, for any $u, v \in \mathfrak{so}(n, R)$, we have

$$\operatorname{Sp}(uuv) = \operatorname{Sp}(uuv)^* = -\operatorname{Sp}(vuu)$$

(since $(uuv)^* = v^* u^* u^* = -vuu$). Simultaneously,

$$\operatorname{Sp}(uuv) = u_{im} u_{mk} v_{ki} = v_{ki} u_{im} u_{mk} = \operatorname{Sp}(vuu).$$

Hence,

$$\operatorname{Sp}(uuv) = \operatorname{Sp}(vuu) = 0.$$

Since for $u, v \in \mathfrak{so}(n, R)$ we have

$$[u, v] \in \mathfrak{so}(n, R), \qquad ([u, v])^* = -[u, v]$$

(which is established by straightforward verification), we have also

$$\begin{aligned}
(u, [u, v]) &= \operatorname{Sp} u[u, v]^* = \operatorname{Sp} u[v, u] \\
&= \frac{1}{2} \operatorname{Sp} u[v, u] + \frac{1}{2} \operatorname{Sp}(u[v, u])^* \\
&= \frac{1}{2} \operatorname{Sp} u[v, u] + \frac{1}{2} \operatorname{Sp}[v, u]u \\
&= \frac{1}{2} \operatorname{Sp}(uvu - uuv + vuu - uvu) \\
&= -\frac{1}{2} \operatorname{Sp}(uuv) + \frac{1}{2} \operatorname{Sp}(vuu) = 0,
\end{aligned}$$

i.e., *the algebra* $\mathfrak{so}(n, R)$ *is compact*. We introduce into consideration diagonal matrices $a = (a_{ik})$ and $b = (b_{ik})$, $a_{ik} = a_k$ $(b_{ik} = b_k)$ for $i = k$, $a_{ik} = b_{ik} = 0$ for $i \neq k$, such that $a_i + a_k \neq 0$ and $b_i + b_k > 0$, $i, k = 1, 2, \ldots, n$, and we set

$$Ax \equiv ax + xa, \qquad Bx \equiv bx + xb, \qquad x \in \mathfrak{so}(n, R).$$

We have $Ax, Bx \in \mathfrak{so}(n, R)$ (since $(Ax)^* = -xa - ax = -Ax$ and similarly for Bx). The permutability of diagonal matrices ($ab = ba$) guarantees the permutability of the corresponding top defined by equation (11.24) (§11.3). The latter takes in $\mathfrak{so}(n, R)$ the form of the matrix relation

$$\varepsilon(bx + xb) + xxa - axx = y, \qquad x, y \in \mathfrak{so}(n, R),$$

or of the system of $n(n+1)/2$ scalar equations of the form

$$(1) \qquad \varepsilon(b_i + b_k)x_{ik} + (a_k - a_i)\sum_{m=1}^{n} x_{im}x_{mk} = y_{ik}, \qquad 1 \leqslant i < k \leqslant n,$$

with respect to the variables

$$\begin{aligned} x_{12}, x_{13}, \ldots, x_{1n},& \\ x_{23}, \ldots, x_{2n},& \\ \ldots\ldots\ldots,& \\ x_{n-1,n},& \end{aligned}$$

defining the elements of the matrix $x = (x_{ik}) = -(x_{ki})$. It can easily be verified that the eigenvalues of the operator A are given by the series $a_i + a_k$, $1 \leqslant i < k \leqslant n$, and the corresponding eigenelements from $\mathfrak{so}(n, R)$ are given by the skew-symmetric matrices $u^{ik} = (\delta^{ik}_{lm})$ with the elements $\delta^{ik}_{lm} = -\delta^{ik}_{ml}$, $1 \leqslant l < m \leqslant n$, equal to 1 (equal to 0) for $l = i$ and $m = k$ (for $l \neq i$ or $m \neq k$). It can easily be seen that the indicated matrices form an orthogonal basis of the corresponding space $\mathbb{H}$, which, in the finite-dimensional case, coincides with the algebra M. Further, by a straightforward verification we establish that for $y = Cz$, $C = \text{const}$, $z = u^{ik}$, the solution of system (1) is

$$x = \varepsilon^{-1}B^{-1}y = \frac{1}{\varepsilon}\frac{1}{b_i + b_k}u^{ik}$$

(in this equality there is no summation over the repeated index). According to Theorem 12.1, the solution is unique for any $\varepsilon > 0$ and $-\infty < C < \infty$, if $a_i + a_k$ is the smallest or the greatest number of the series

$$\begin{aligned} a_1 + a_2, a_1 + a_2, \ldots, a_1 + a_n,& \\ a_2 + a_3, \ldots, a_2 + a_n,& \\ \ldots\ldots\ldots\ldots,& \\ a_{n-1} + a_n.& \end{aligned}$$

According to Theorem 12.2, under the additional requirement that the elements of this series be positive (i.e., that the corresponding operator $A\colon \mathbb{H} \to \mathbb{H}$ be positive definite), the indicated solution is asymptotically stable (as an equilibrium of the corresponding nonstationary system formally obtained by adding the time derivative $d((a_i + a_k)x_{ik})/dt$ to the left-hand side of equation (1)) if $a_i + a_k$ is the smallest number of that series.

14.2. Flows on two-dimensional manifolds. Systems of the form (1) (or finite-dimensional tops) are obtained as a result of Galerkin approximations of solutions of the Navier-Stokes equations **[26]**. In order to illustrate the corresponding infinite-dimensional case, we assume that M is the linear manifold $C^\infty(\overline{V})$ of infinitely differentiable stream functions $\varphi(x,y)$, $\psi(x,y), \dots$, defined on the closure $\overline{V}$ of the domain V from §11.1 equipped with the scalar product from §11.3 and the commutator (11.12). In the case when $\partial V = \varnothing$ ($V = \overline{V}$ is a closed manifold), we additionally assume that condition (11.6) is satisfied (in this case, M is defined by the restriction of the linear manifold $C(\overline{V})$ to the functions ψ orthogonal to constants).

PROPOSITION 1. *The algebra $M = M(V)$ is compact in the case $\partial V = \varnothing$ and is quasicompact in the case $\partial V \neq \varnothing$. In the latter case, the first (second) compact subalgebra M_c (the subalgebra $M_{c'}$) of the algebra M is the linear subspace $\dot{C}^\infty(\overline{V})$ (the linear subspace $C_c^\infty(\overline{V})$) of the functions $\psi \in C(\overline{V})$ vanishing on the boundary ∂V of the domain V, $\psi|_{\partial V} = 0$ (taking a constant value on each boundary cycle J of the domain V, $\psi|_J = \text{const}$, $J \subset \partial V$, respectively).*

Indeed, we have

$$\varphi_1\left(\frac{\partial\psi}{\partial x}\frac{\partial\varphi_2}{\partial y} - \frac{\partial\psi}{\partial y}\frac{\partial\varphi_2}{\partial x}\right) = \frac{\partial}{\partial x}\left(\varphi_1\psi\frac{\partial\varphi_2}{\partial y}\right) - \frac{\partial}{\partial y}\left(\varphi_1\psi\frac{\partial\varphi_2}{\partial x}\right) + \psi\left(\frac{\partial\varphi_2}{\partial x}\frac{\partial\varphi_1}{\partial y} - \frac{\partial\varphi_2}{\partial y}\frac{\partial\varphi_1}{\partial x}\right);$$

consequently,

$$(\varphi_1, [\varphi_2, \psi]) - (\psi, [\varphi_1, \varphi_2]) = \int_V \left(\frac{\partial}{\partial x}\left(\varphi_1\psi\frac{\partial\varphi_2}{\partial y}\right) - \frac{\partial}{\partial y}\left(\varphi_1\psi\frac{\partial\varphi_2}{\partial x}\right)\right) dx\, dy$$

$$= \begin{cases} 0, & \text{if } \partial V = \varnothing, \\ \int_{\partial V} \varphi_1\psi\, d\varphi_2, & \text{if } \partial V \neq \varnothing, \end{cases}$$

$$d\varphi_2 \equiv \frac{\partial\varphi_2}{\partial x}dx + \frac{\partial\varphi_2}{\partial y}dy.$$

In particular, $M = M_c$ for $\partial V = \varnothing$. For $\partial V \neq \varnothing$, the validity of the equality

$$\int_{\partial V} \varphi_1\psi\, d\varphi_2 = 0$$

for any $\varphi_1, \varphi_2 \in C^\infty(\overline{V})$ implies $\psi|_{\partial V} = 0$. The validity of this equality for any φ_1, $\psi \in C^\infty(\overline{V})$ implies $\varphi_2|_J = \text{const}$. The set $\dot{C}^\infty(\overline{V})$ is dense in $\mathbb{H}$ **[35]**, which completes the proof.

Defining the operators $A, B\colon \mathbb{H} \to \mathbb{H}$ as the restrictions of the operations

$$-\Delta \equiv -\partial^2/\partial x^2 - \partial^2/\partial y^2 \quad \text{and} \quad \Delta\Delta$$

(initially defined on the functions from $C^\infty(\overline{V})$), respectively, to the subset of functionssatisfying the boundary conditions (11.4) in the case $\partial V \neq \varnothing$, we come to problem

(11.4) for equation (11.2). The corresponding top is commutative (cf. (11.45)) and isometric (cf. (11.47)). In the case $\partial V = \varnothing$, the restrictions of the above-mentioned operations defined by condition (11.6) are considered as A and B. The isometricity implies the validity of the first two inequalities from conditions (11.23) (which are satisfied in this case with the constants $\gamma_1 = \gamma_2 = 1$). Leaving aside for the time being the verification of the third inequality from (11.23), we turn to the analysis of equation (11.42) which acquires the form of the boundary problem (11.14). Note that in the case $\partial V = \varnothing$, the latter (as well as the nonlinear equation (11.2)) is considered under the restriction (11.6). Preserving the numbering of the examples of the domain V (1)–(4) given in §11.1, we write out the corresponding solutions λ and ψ of the spectral problem under consideration (which admits separation of the variables). Thus, we have

1. $$\lambda_{0m} = (\mu_m^{(0)}/h)^2, \qquad \psi_{0m} = J_0(\mu_m^{(0)} r/h),$$
$$\lambda_{km} = (\mu_m^{(k)}/h)^2, \qquad \begin{cases} \psi_{km}^{+} = J_k(\mu_m^{(k)})r/h)\cos(k\theta), \\ \psi_{km}^{-} = J_k(\mu_m^{(k)} r/h)\sin(k\theta); \end{cases}$$

here $J_\nu(\mu)$, $\mu \geqslant 0$, is the Bessel function of order $\nu \geqslant 0$, $\mu_m^{(\nu)}$ is the positive root of the equation $J_\nu(\mu) = 0$, $\mu_m^{(\nu)} < \mu_{m+1}^{(\nu)}$, $k,\ m = 1, 2, \dots$, $r = (x^2 + y^2)^{1/2}$, θ is the polar angle,

2. $$\lambda_{km} = (\pi k/l)^2 + (\pi m/h)^2, \qquad \psi_{km} = \sin(\pi k x/l)\sin(\pi m y/h),$$

3. $$\lambda_{0m} = (\pi m/h)^2, \qquad \psi_{0m} = \sin(\pi m y/h),$$
$$\lambda_{km} = (2\pi k/l)^2 + (\pi m/h)^2, \qquad \begin{cases} \psi_{km}^{+} = \cos(2\pi k x/l)\sin(\pi m y/h), \\ \psi_{km}^{-} = \sin(2\pi k x/l)\sin(\pi m y/h), \end{cases}$$

4. $$\lambda_{0m} = (2\pi m/h)^2, \qquad \begin{cases} \psi_{0m}^{+} = \cos(2\pi m y/h), \\ \psi_{0m}^{-} = \sin(2\pi m y/h), \end{cases}$$
$$\lambda_{km} = (2\pi k/l)^2 + (2\pi m/h)^2, \qquad \begin{cases} \psi_{km}^{++} = \cos(2\pi k x/l)\cos(2\pi m y/h), \\ \psi_{km}^{+-} = \cos(2\pi k x/l)\sin(2\pi m y/h, \\ \psi_{km}^{-+} = \sin(2\pi k x/l)\cos(2\pi m y/h), \\ \psi_{km}^{--} = \sin(2\pi k x/l)\sin(2\pi m y/h), \end{cases}$$
$$\lambda_{k0} = (2\pi k/l)^2, \qquad \begin{cases} \psi_{k0}^{+} = \cos(2\pi k x/l), \\ \psi_{k0}^{-} = \sin(2\pi k x/l). \end{cases}$$

These series of numbers do not have the greatest elements. The smallest eigenvalues $\lambda_{\min}$ and the corresponding basis functions $\psi_{\min}$ from those given above are the following:

	$\lambda_{\min}$	$\psi_{\min}$	
1.	$\lambda_{01} = (\mu_1^{(0)}/h)^2,$	$\psi_{01} = J_0(\mu_1^{(0)} r/h),$	
2.	$\lambda_{11} = (\pi/l)^2 + (\pi/h)^2,$	$\psi_{11} = \sin(\pi x/l)\sin(\pi y/h),$	
3.	$\lambda_{01} = (\pi/h)^2,$	$\psi_{01} = \sin(\pi y/h)$	
4.	$\lambda_{01} = (2\pi/h)^2,$	$\begin{cases} \psi_{01}^{+} = \cos(2\pi y/h), \\ \psi_{01}^{-} = \sin(2\pi y/h), \end{cases}$	$h \geqslant l$
	$\lambda_{10} = (2\pi/l)^2,$	$\begin{cases} \psi_{10}^{+} = \cos(2\pi x/l), \\ \psi_{10}^{-} = \sin(2\pi x/l). \end{cases}$	$h \leqslant l$

Using the results from §12, in the case under consideration we have the following statement.

THEOREM 1. *Let the flow domain V be the disc K, the rectangle Π, the annulus R or the torus T from §11.1, and let the right-hand side of equation (11.2) be one of the following functions:*

$$\begin{array}{lllll}
1. & f(x,y) = CJ_0(\mu_1^{(0)}(x^2+y^2)^{1/2}/h) & \text{for} & V = K, & \\
2. & f(x,y) = C\sin(\pi x/l)\sin(\pi y/h) & \text{for} & V = \Pi, & \\
3. & f(x,y) = C\sin(\pi y/h) & \text{for} & V = R, & \\
4. & f(x,y) = C_1\cos(2\pi y/h) + C_2\sin(2\pi y/h) & \text{for} & V = T & \text{if } l < h, \\
 & f(x,y) = C_1\cos(2\pi x/l) + C_2\sin(2\pi x/l) & \text{for} & V = T & \text{if } l > h, \\
 & f(x,y) = C_1\cos(2\pi y/h) + C_2\sin(2\pi y/h) & & & \\
 & \qquad + C_3\cos(2\pi x/h) + C_4\sin(2\pi x/h) & \text{for} & V = T & \text{if } l = h,
\end{array}$$

where $C, C_1, C_2, C_3, C_4 = \mathrm{const}$. Then, for any $\mathrm{Re} > 0$ and $-\infty < C, C_1, C_2, C_3, C_4 < \infty$, the unique solution of problem (11.4) (if $V \neq T$) or of problem (11.6) (if $V = T$) for equation (11.2) is, at least in the class $C^\infty(\overline{V})$, $\psi = \lambda_{\min}^{-2}(\mathrm{Re}) f(x,y)$, where $\lambda_{\min} > 0$ is determined from the identity $-\Delta f = \lambda_{\min} f$ which is valid for $f = f(x,y)$. The corresponding stationary flow is asymptotically stable with respect to the corresponding smooth nonstationary flows of a viscous incompressible fluid which are defined under the indicated restrictions by the solutions of the following equation:

$$\frac{\partial}{\partial t}(-\Delta\psi) + \frac{1}{\mathrm{Re}}\Delta\Delta\psi - \frac{\partial\psi}{\partial y}\frac{\partial\Delta\psi}{\partial x} + \frac{\partial\psi}{\partial x}\frac{\partial\Delta\psi}{\partial y} = f(x,y). \tag{2}$$

This theorem immediately follows from Theorems 12.1 and 12.2 (§12) and from the extremal properties of the indicated eigenvalues considered above.

In conclusion of this subsection, we establish the validity of the above-mentioned third inequality from (11.23) for the case under consideration. Introducing the abbreviated notations $\psi_x \equiv \partial\psi/\partial x$, $\psi_y \equiv \partial\psi/\partial y$, we find

$$\int_V ((\psi_y\psi_{xy})_x - (\psi_y\psi_{xx})_y)\,dx\,dy = \begin{cases} 0, & \text{if } \partial V = \varnothing, \\ \int_{\partial V}\psi_y\,d\psi_x, & \text{if } \partial V \neq \varnothing. \end{cases}$$

Simultaneously,

$$\psi_{xx}^2 + 2\psi_{xy}^2 + \psi_{yy}^2 = (2\psi_y\psi_{xy})_x - 2(\psi_y\psi_{xx})_y + (\psi_{xx}+\psi_{yy})^2.$$

Consequently, for any $\psi \in \dot{C}^{\infty}(\overline{V})$ and $V = K, \Pi, R, T$ we have

$$\|\psi\|_w \equiv \left(\int_V (\psi_{xx}^2 + 2\psi_{xy}^2 + \psi_{yy}^2)\, dx\, dy\right)^{1/2} \leqslant \|\Delta\psi\|$$

where

$$\|\Delta\psi\| \equiv \left(\int_V (\psi_{xx} + \psi_{yy})^2\, dx\, dy\right)^{1/2},$$

$$\int_{\partial V} \psi_y\, d\psi_x = 0 \quad \text{for } V = \Pi, R,$$

$$\int_{\partial K} \psi_y\, d\psi_x = -\frac{1}{2}\int_0^{2\pi} \left(\frac{\partial\psi}{\partial r}\right)^2\bigg|_{r=h} d\theta \leqslant 0, \quad \psi \in \dot{C}^{\infty}(\overline{V})$$

and

$$\begin{aligned} \|[\varphi, \psi]\|^2 &\equiv \int_V (\varphi_y\psi_x - \varphi_x\psi_y)^2\, dx\, dy \\ &\leqslant \int_V (\varphi_x^2 + \varphi_y^2)(\psi_x^2 + \psi_y^2)\, dx\, dy \\ &\leqslant \left(\int_V (\varphi_x^2 + \varphi_y^2)^2\, dx\, dy\right)^{1/2} \left(\int_V (\psi_x^2 + \psi_y^2)^2\, dx\, dy\right)^{1/2}, \end{aligned}$$

or

$$\|[\varphi, \psi]\|^2 \leqslant \||\nabla\varphi|^2\|\, \||\nabla\psi|^2\|,$$

where $|\nabla\varphi| \equiv |\operatorname{grad}\varphi| \equiv (\varphi_x^2 + \varphi_y^2)^{1/2}$. The validity of the required estimate is guaranteed by the inequalities

(3)
$$\begin{gathered} \||\nabla\psi|^2\| \leqslant \gamma_4\|\psi\|_c\|\psi\|_w, \qquad \|\psi\|_c \leqslant \gamma_5\|\psi\|_w, \\ \psi \in \dot{C}^{\infty}(\overline{V}), \qquad \|\psi\|_c \equiv \max_{\overline{V}} |\psi|, \qquad \gamma_4, \gamma_5 = \text{const} > 0, \end{gathered}$$

leading to inequalities (13.33) and (13.34), which provide (as it was established in §13) the existence of a generalized solution of the corresponding equation of the top (equation (11.2)), and, simultaneously, lead to the required (the third) inequality from (11.23). The first inequality from (3) follows, in its turn, from the equalities

$$\psi_x^4 + \psi_y^4 = -3\psi\psi_x^2\psi_{xx} - 3\psi\psi_y^2\psi_{yy} + (\psi\psi_x^3)_x + (\psi\psi_y^3)_y,$$

$$\int_V ((\psi\psi_x^3)_x + (\psi\psi_y^3)_y)\, dx\, dy = 0$$

(since $\psi|_{\partial V} = 0$ or $\partial V \neq \varnothing$). As a corollary of the latter,

$$\begin{aligned}
\||\nabla\psi|^2\|^2 &= \int_V (\psi_x^2 + \psi_y^2)^2 \, dx\, dy \\
&\leqslant 2\int_V (\psi_x^4 + \psi_y^4)\, dx\, dy \\
&\leqslant 6\int_V |\psi|(\psi_x^2|\psi_{xx}| + \psi_y^2|\psi_{yy}|)\, dx\, dy \\
&\leqslant 6\|\psi\|_c \int_V |\nabla\psi|^2(\psi_{xx}^2 + \psi_{yy}^2)^{1/2}\, dx\, dy \\
&\leqslant 6\|\psi\|_c \||\nabla\psi|^2\| \|\psi\|_w,
\end{aligned}$$

and we come to the first inequality from (3) with $\gamma_4 = 6$. The second inequality from (3) can be obtained for the case of a plane domain by means of arguments analogous to those given in **[56, 88]** or, in particular, for the examples of two-dimensional domains, by decomposing ψ in a Fourier series with respect to the corresponding orthogonal systems of eigenfunctions given above. The embedding provided by this inequality turns out to be compact. The details of the corresponding constructions (quite evident) are omitted.

14.3. Flows on a three-dimensional torus. Now we show that for a three-dimensional flow the corresponding operator rot, generating the inertia A of the dissipative top from §13.1, is not, generally speaking, positive or negative definite, i.e., the top under consideration is anomalous in the sense of §11.4. The three-dimensional torus $V = T^3$ is obtained by factorization of the Euclidean space R^3 with respect to the lattice generated by the periods l_1, l_2, $l_3 = \text{const} > 0$ along the axes Ox_1, Ox_2, Ox_3, respectively. Introducing into consideration the corresponding integer-valued vector

$$k = (k_1, k_2, k_3), \qquad k_1, k_2, k_3 = 0, \mp 1, \mp 2, \ldots,$$

and the vector

$$\mathfrak{H}_k = (2\pi k_1/l_1, 2\pi k_2/l_2, 2\pi k_3/l_3),$$

we consider the linear manifold of vector fields on V defined by the sums of the following form:

$$u(x) = \sum_{0<|k|\leqslant N} \mathbf{u}_k \exp[i(\mathfrak{H}_k \cdot \mathbf{x})], \qquad N = 1, 2, 3, \ldots,$$

$$|\mathbf{k}| = (\mathbf{k}\cdot\mathbf{k})^{1/2}, \quad \mathbf{x} = (x_1, x_2, x_3), \quad \mathbf{x}\cdot\mathbf{y} = x_1y_1 + x_2y_2 + x_3y_3, \quad i^2 = -1.$$

Here $\mathbf{u}_k = (u_{1k}, u_{2k}, u_{3k})$ is a complex-valued vector such that $\bar{\mathbf{u}}_k = \mathbf{u}_{-k}$ (or $\bar{u}_{\alpha k} = u_{\alpha(-k)}$, $\alpha = 1, 2, 3$; the bar means the complex conjugation). The latter guarantees that the indicated vector sums are real,

$$\overline{\mathbf{u}_k \exp[i(\mathfrak{H}_k \cdot \mathbf{x})]} = \bar{\mathbf{u}}_k \exp[-i(\mathfrak{H}_k \cdot \mathbf{x})] = \mathbf{u}_{-k} \exp[i(\mathfrak{H}_{-k} \cdot \mathbf{x})],$$

$$\bar{\mathbf{u}}(\mathbf{x}) = \sum_{0<|k|\leqslant N} \mathbf{u}_{-k} \exp[i(\mathfrak{H}_{-k} \cdot \mathbf{x})] = \sum_{0<|k|\leqslant N} \mathbf{u}_k \exp[i(\mathfrak{H}_k \cdot \mathbf{x})] = \mathbf{u}(\mathbf{x}).$$

It is evident that the set M of these sums is dense in the Hilbert space $\mathbb{H}$ obtained by the closure of the linear manifold of infinitely differentiable real vector fields $\mathbf{u}$, $\mathbf{v}$, ...

given on V and orthogonal to constants,

$$\int_V \mathbf{u}\,dV = \int_V \mathbf{v}\,dV = \cdots = 0,$$

with respect to the norm $\|\mathbf{u}\| = (\mathbf{u}, \mathbf{u})^{1/2}$ defined by the scalar product

$$(\mathbf{u}, \mathbf{v}) \equiv \int_V \mathbf{u} \cdot \mathbf{v}\,dV. \tag{4}$$

Additionally endowing M with the commutator

$$[\mathbf{u}, \mathbf{v}](\mathbf{x}) \equiv \mathbf{v}(\mathbf{x}) \times \mathbf{u}(\mathbf{x}),$$

we come to a compact algebra M (in the sense of §11.2). If $B = -\Delta$, the generalized form (13.1) of the equation of the top (11.24) takes the form of the system of Navier-Stokes equations (13.5) (the details are given in §13.1). The spectral problem defining the principal moments and rotations of the commutative top under consideration (commutativity follows from the identities given at the end of §13.1) is written in the following way (cf. with (11.42)):

$$\operatorname{rot} \mathbf{u} = \lambda \mathbf{u}, \qquad \operatorname{div} \mathbf{u} = 0, \qquad \mathbf{x} \in V, \qquad \int_V \mathbf{u}\,dV = 0.$$

Considering the latter on M, we get an easily solvable system of equations for $\mathbf{u}_k$,

$$i\mathfrak{H}_k \times \mathbf{u}_k = \lambda \mathbf{u}_k, \qquad \mathfrak{H}_k \cdot \mathbf{u}_k = 0 \qquad (|\mathfrak{H}_k| \neq 0).$$

As a result, we come to the following set of solutions $\mathbf{u}_k = \overset{\pm p}{\mathbf{e}_k}$, $p = 1, 2$, and $\lambda = \lambda_k^\pm$ of the spectral problem under consideration which are described below in terms of the vector $\mathbf{a}_k$ orthogonal to $\mathfrak{H}_k$ and the vector $\mathbf{b}_k = |\mathfrak{H}_k|^{-1}\mathfrak{H}_k \times \mathbf{a}_k$, forming with $\mathfrak{H}_k$ and $\mathbf{a}_k$ a positively oriented frame

$$\left.\begin{aligned} \overset{+1}{\mathbf{e}_k} &= \mathbf{a}_k \cos(\mathfrak{H}_k \cdot \mathbf{x}) - \mathbf{b}_k \sin(\mathfrak{H}_k \cdot \mathbf{x}), \\ \overset{+2}{\mathbf{e}_k} &= \mathbf{a}_k \sin(\mathfrak{H}_k \cdot \mathbf{x}) + \mathbf{b}_k \cos(\mathfrak{H}_k \cdot \mathbf{x}), \end{aligned}\right\} \qquad \begin{aligned} \operatorname{rot} \overset{+p}{\mathbf{e}_k} &= \lambda_k^+ \overset{+p}{\mathbf{e}_k}, \qquad p = 1, 2, \\ \lambda_k^+ &= |\mathfrak{H}_k|, \end{aligned}$$

$$\left.\begin{aligned} \overset{-1}{\mathbf{e}_k} &= \mathbf{a}_k \cos(\mathfrak{H}_k \cdot \mathbf{x}) + \mathbf{b}_k \sin(\mathfrak{H}_k \cdot \mathbf{x}), \\ \overset{-2}{\mathbf{e}_k} &= \mathbf{a}_k \sin(\mathfrak{H}_k \cdot \mathbf{x}) - \mathbf{b}_k \cos(\mathfrak{H}_k \cdot \mathbf{x}), \end{aligned}\right\} \qquad \begin{aligned} \operatorname{rot} \overset{-p}{\mathbf{e}_k} &= \lambda_k^- \overset{-p}{\mathbf{e}_k}, \qquad p = 1, 2, \\ \lambda_k^- &= -|\mathfrak{H}_k|. \end{aligned}$$

As follows from these formulas, the inertia operator of the top under consideration is neither positive nor negative definite, and has neither the smallest nor the greatest eigenvalues.

14.4. An example of violation of quasicompactness. By acting on both sides of equality (13.7) by the operator rot and taking into account the identity

$$\begin{gathered} \operatorname{rot}(\mathbf{u} \times \mathbf{v}) = (\mathbf{v}, \nabla)\mathbf{u} - (\mathbf{u}, \nabla)\mathbf{v} \equiv \{\mathbf{v}, \mathbf{u}\} \\ (\operatorname{div} \mathbf{u} = \operatorname{div} \mathbf{v} = 0, \qquad (\mathbf{v}, \nabla)\mathbf{u} = (\mathbf{v} \cdot \nabla)\mathbf{u}), \end{gathered} \tag{5}$$

we come to the stationary Helmholtz equation for the curl ω,

$$-\varepsilon\Delta\omega + \{\mathbf{u}, \omega\} = \operatorname{rot} \mathbf{f}, \quad \omega = \operatorname{rot} \mathbf{u},$$

which can also be written in the form (11.24). The corresponding commutator $\{\mathbf{u}, \mathbf{v}\}$ preserves the solenoidal property of the fields $\mathbf{u}$, $\mathbf{v}$,

$$\operatorname{div}\{\mathbf{u}, \mathbf{v}\} = \operatorname{div}\operatorname{rot}(\mathbf{v} \times \mathbf{u}) = 0.$$

However, in this case the commutator in no longer consistent with the scalar product (4). More precisely, it can be easily established by straightforward verification, it violates quasicompactness of the corresponding algebra M. To analyze the corresponding alternative possibility, first we write out the corresponding nonstationary equation,

$$\frac{\partial}{\partial t}\omega - \varepsilon\Delta\omega + \{\mathbf{u}, \omega\} = \operatorname{rot}\mathbf{f}.$$

For the case of plane-parallel flows, this equation is known to reduce to relation (2). Considering the latter for the right-hand sides $f(x, y) = yh(x)$, or, for a more general approach, $f = f(t, x, y) = yh(t, x)$ with a given function $h(t, x)$, on the manifold of stream function ψ of the form $\psi = \psi(t, x, y) = y\varphi(t, x)$, we come to the following equations:

$$(6)\qquad \frac{\partial\theta}{\partial t} - \varepsilon\frac{\partial^2\theta}{\partial x^2} + [\varphi, \theta] = h(t, x), \qquad -\frac{\partial^2\varphi}{\partial x^2} = \theta,$$
$$[\varphi, \theta] \equiv \varphi\frac{\partial\theta}{\partial x} - \theta\frac{\partial\varphi}{\partial x} \qquad (\varepsilon = 1/\mathrm{Re}).$$

Considering system (6) on the interval $0 < x < 1$, we easily see that the commutator $[\theta, \varphi]$ is not consistent with the natural, for this case, scalar product

$$(\theta, \varphi) = \int_0^1 \theta\varphi\, dx.$$

At the same time equation (6) models completely the corresponding multidimensional system generated by the Helmholtz equation given above (or by the three-dimensional Navier-Stokes equations) which makes it possible to trace the appearance of turbulent solutions when the coefficient $\varepsilon > 0$ is sufficiently decreased. The corresponding specific problem is, in this case, the following problem on a plane jet of a viscous incompressible fluid impinging an immovable wall orthogonal to the direction of the flow:

$$u(0, y) = v(0, y) = 0, \qquad u(1, y) = -1, \qquad v(1, y) = 0,$$
$$(u(x, y) = \partial\psi/\partial y, \qquad v(x, y) = -\partial\psi/\partial x),$$
$$-\infty < y < \infty.$$

For a stationary regime of the flow, we come to the following nonlinear problem:

$$(7)\qquad \begin{gathered} \varepsilon\frac{d^4\varphi}{dx^4} - \varphi\frac{d^3\varphi}{dx^3} + \frac{d\varphi}{dx}\frac{d^2\varphi}{dx^2} = 0, \qquad 0 < x < 1, \\ \varphi(0) = \frac{d\varphi}{dx}(0) = 0, \qquad \varphi(1) = -1, \qquad \frac{d\varphi}{dx}(1) = 0. \end{gathered}$$

The main differential relation here is the well-known *Himmens equation* used in boundary layer theory **[119]**. In the limiting Stokes case, when $\varepsilon \to \infty$, the solution of problem (7) is

$$\varphi(x) = -3x^2 + 2x^3.$$

In the limiting Euler case, when $\varepsilon \to 0$, we come to the following equation and restrictions:

$$-\frac{d^2\varphi}{dx^2} = \lambda\varphi, \qquad 0 < x < 1, \quad \lambda = \text{const} > 0, \quad \varphi(0) = 0, \quad \varphi(1) = -1,$$

considered with a part of the boundary conditions of the original problem dropped (which is connected with the decrease of the order of the equation under consideration). The system of separating flows adjoining a solid wall, which are defined by the stream functions

$$\psi_\lambda(x, y) = y\varphi_\lambda(x)$$

corresponds to the set of solutions

$$\varphi_\lambda(x) = \begin{cases} \dfrac{\sinh(x\sqrt{-\lambda})}{\sinh(\sqrt{-\lambda})}, & \lambda < 0, \\ x, & \lambda = 0, \quad \lambda \neq \pi^2 n^2, \quad n = 1, 2, \ldots, \\ \dfrac{\sin(x\sqrt{\lambda})}{\sin(\sqrt{\lambda})}, & \lambda > 0, \end{cases}$$

of the obtained limiting problem. It is possible that in the "intermediate" regime of sufficiently small $\varepsilon > 0$, the topology of the flow defined by conditions (7) would be similar to that just described.

We shall not pursue this subject further.

CHAPTER 5

Specific Features of Turbulence Models

§15. Wave properties of turbulent flows

15.1. An "ideal" turbulent medium. A straightforward transfer of the dynamical Newton law to nonviscous and incompressible media leads to the Euler hydrodynamic equations usually written in the form

$$D_t u_j + \mathbf{u} \cdot \nabla u_j + D_j p = 0, \qquad \nabla \cdot \mathbf{u} = 0 \qquad (j = 1, 2, 3).$$

Here $D_t \equiv \partial/\partial t$ and $D_j \equiv \partial/\partial x_j$ are partial derivatives with respect to time t and the spatial variable x_j, respectively, $\nabla \equiv (D_j)$ is the gradient, $\mathbf{u} = (u_j)$, $j = 1, 2, 3$, is the velocity field, p is the pressure divided by a constant density (defined up to an additive function, the potential of external forces), $\mathbf{u} \cdot \nabla \equiv u_k D_k$ is the connective derivative $(k = 1, 2, 3)$, $\nabla \cdot \mathbf{u} \equiv D_k u_k \equiv \operatorname{div} \mathbf{u}$ is the divergence of the field $\mathbf{u}$. As usual, summation is assumed to be carried over repeated indices unless the contrary is not specified or follows explicitly from the context. Nonpotential mass forces are assumed to be equal to zero. The liquid is assumed to be turbulent. The substitution in the equations of the corresponding expansions $u = \overline{\mathbf{u}} + \mathbf{u}'$ and $p = \overline{p} + p'$ for the averaged, $\overline{\mathbf{u}}$ and $\overline{p}$, and pulsational, $\mathbf{u}'$ and p', components (such that $\overline{\mathbf{u}'} = 0$, $\overline{p'} = 0$) and subsequent averaging of the equalities obtained (denoted by bars and satisfying usual properties of linearity and homogeneity and not affecting nonrandom (in particular, average) values) lead to the following equations:

$$D_t \overline{u}_i + \overline{\mathbf{u}} \cdot \nabla \overline{u}_i + \overline{\mathbf{u}' \cdot \nabla u_i'} + D_i \overline{p} = 0, \qquad \nabla \cdot \overline{\mathbf{u}} = 0.$$

The subtraction of these equalities from the initial ones gives the equations for the pulsational components u_i'. By formally multiplying the indicated equations by the components u_s' $(s = 1, 2, 3)$, interchanging the indices i and s, summing the two equalities obtained, and averaging the sum, we come to the equations for the correlations $\overline{u_i' u_s'}$,

$$D_t \overline{u_i' u_s'} + \overline{\mathbf{u}} \cdot \nabla \overline{u_i' u_s'} + \overline{u_i' \mathbf{u}'} \cdot \nabla \overline{u}_s + \overline{u_s' \mathbf{u}'} \cdot \nabla \overline{u}_i + h_{is} = 0,$$
$$h_{is} \equiv \nabla \cdot (\overline{u_i' u_s' \mathbf{u}'} + \overline{p' u_i' \delta_s} + \overline{p' u_s' \delta_i}) - \overline{p'(D_i u_s' + D_s u_i')},$$

where $\delta_i = (\delta_{ki})$ is the unit vector such that $\delta_{ki} = 1$ $(\delta_{ki} = 0)$ for $k = i$ (for $k \neq i$, i, $k = 1, 2, 3$, respectively). In addition, the incompressibility condition $\nabla \cdot \mathbf{u} = 0$ and the identities

$$u_i' \mathbf{u}' \cdot \nabla u_s' + u_s' \mathbf{u}' \cdot \nabla u_i' = \mathbf{u}' \cdot \nabla u_i' u_s' = \nabla \cdot (u_i' u_s' \mathbf{u}'), \quad u_i' D_s p' = D_s(p' u_i') - p' D_s u_i'$$

are used. The term $\overline{\mathbf{u}' \cdot \nabla u_i'}$ in the equation for the averaged velocity reduces to the divergent form,

$$\overline{\mathbf{u}' \cdot \nabla u_i'} = \nabla \cdot \overline{u_i' \mathbf{u}'}.$$

This testifies that there exist additional stresses τ_{js} in the turbulent medium; they are called the *Reynolds stresses* and are defined by pairs of one-point correlations for components of the pulsational part of the velocity,

$$\tau_{js} \equiv \overline{u'_j(t,\mathbf{x})u'_s(t,\mathbf{x})} = \tau_{sj}, \qquad \mathbf{x} \equiv (x_1, x_2, x_3).$$

In the study of turbulent flow, the quantities τ_{js} are measured along with the functions $\bar{\mathbf{u}}$ and $\bar{p}$, and are systematically used in the discussion.

Another specific feature of turbulent flow is connected with the presence of the so-called generation terms

$$\overline{u'_i\mathbf{u}'} \cdot \nabla\bar{u}_s + \overline{u'_s\mathbf{u}'} \cdot \nabla\bar{u}_i$$

in the equations for the Reynolds stresses. These terms are usually considered as some sources of turbulence responsible for the energy exchange between the averaged macroscopic motion of a turbulent media and its micropulsations. The constructions carried out below are mainly devoted to the study of the mechanism of the indicated energy exchange. To this end, the "nonclosed" terms h_{js} are defined by approximate semiempirical dependencies which model the processes of turbulent diffusion (the divergent term on the right-hand side of the identity for h_{js}) and relaxation (the remaining term) **[102]**, and then, for a time, the analysis of diffusion is omitted (just as molecular viscosity was omitted before),

$$\nabla \cdot (\overline{u'_j u'_s \mathbf{u}'} + \overline{p'u'_j}\boldsymbol{\delta}_s + \overline{p'u'_s}\boldsymbol{\delta}_j) = 0.$$

As it was noted above, the Rotta approximation **[173]**

$$-\overline{p'(D_j u'_s + D_s u'_j)} = \frac{q}{L}\left(\tau_{js} - \frac{q^2}{3}\delta_{js}\right), \qquad q \equiv (\tau_{kk})^{1/2}$$

is used for the relaxation term, $L > 0$ is the relaxation scale which is assumed to be constant in what follows. From the above equations we obtain a certain closed but "truncated" (i.e., considered without diffusion terms) system of Reynolds equations, which in the terms of the columns

$$\boldsymbol{\tau}_s \equiv \overline{u'_s\mathbf{u}'}$$

of the matrix $\boldsymbol{\tau} = (\tau_{js})$ is written as

$$\begin{gathered}
D_t\bar{u}_j + \bar{\mathbf{u}} \cdot \nabla\bar{u}_j + \nabla \cdot \boldsymbol{\tau}_j + D_j\bar{p} = 0, \\
D_t\tau_{js} + \bar{\mathbf{u}} \cdot \nabla\tau_{js} + \boldsymbol{\tau}_j \cdot \nabla\bar{u}_s + \boldsymbol{\tau}_s \cdot \nabla\bar{u}_j + h_{js} = 0, \\
\nabla \cdot \mathbf{u} = 0, \\
h_{js} = \frac{q}{L}\left(\tau_{js} - \frac{q^2}{3}\delta_{js}\right), \qquad q = (\tau_{kk})^{1/2}, \qquad L = \text{const} > 0.
\end{gathered} \tag{1}$$

This system degenerates into the initial Euler equations for $\boldsymbol{\tau} \equiv 0$ (laminar liquid) and is interpreted as defining the state (or the regime) $\bar{\mathbf{u}}$, $\boldsymbol{\tau}$, $\bar{p}$ of some "ideal" turbulent medium for $\boldsymbol{\tau} \not\equiv 0$.

15.2. Field of small perturbations. Now let $\mathbf{u}^0$, $\boldsymbol{\tau}^0$, p^0 be a state of an ideal turbulent medium whose fields $\bar{\mathbf{u}}^0$, $\boldsymbol{\tau}^0$, $\bar{p}^0$ are constant. The substitution $\bar{\mathbf{u}} = \bar{\mathbf{u}}^0$, $\boldsymbol{\tau} = \boldsymbol{\tau}^0$, $\bar{p} = \bar{p}^0$ in (1) leads to the equation

$$h^0_{js} \equiv \frac{q_0}{L}\left(\tau^0_{js} - \frac{q_0^2}{3}\delta_{js}\right) = 0, \qquad q_0 \equiv (\tau^0_{kk})^{1/2};$$

the latter implies that for $q_0 > 0$ we have a homogeneous and isotropic distribution of the velocity pulsations, $\tau^0_{js} = (q_0^2/3)\delta_{js}$. Assuming the corresponding turbulent liquid to be at rest "on the average" ($\overline{\mathbf{u}}^0 = \mathbf{0}$), we have

$$\overline{\mathbf{u}}^0 = \mathbf{0}, \quad \tau^0_{js} = c^2\delta_{is}, \quad c = \frac{q_0}{3^{1/2}}, \quad q_0, \overline{p}^0 = \text{const} > 0.$$

The quantities q_0 (or c) and $\overline{p}^0$ are assumed to be given. Assuming also the new regime $\overline{\mathbf{u}}$, $\boldsymbol{\tau}$, $\overline{p}$ to be close to the given nonperturbed state in the sense that the differences

$$\xi_j = \overline{u}_j - \overline{u}^0_j, \quad \eta_{js} = \tau_{js} - \tau^0_{js}, \quad \zeta = \overline{p} - \overline{p}^0$$

are small, substituting $\overline{u}_j$, τ_{js} and $\overline{p}$ from the previous inequalities in (1), and ignoring the terms which are quadratic in the small perturbations ξ_j, $\eta_{js} = \eta_{sj}$, and ζ (including their derivatives), we come to the following linear system for the fields $\boldsymbol{\xi} = (\xi_j)$, $\boldsymbol{\eta} = (\eta_{js})$, and ζ,

$$\begin{gathered} D_t\xi_j + \nabla \cdot \boldsymbol{\eta}_j + D_j\zeta = 0, \quad (\boldsymbol{\eta}_j \equiv (\eta_{kj})), \\ D_t\eta_{js} + c^2(D_j\xi_s + D_s\xi_j) + f_{js} = 0, \\ \nabla \cdot \boldsymbol{\xi} = 0, \\ f_{js} = \frac{3^{1/2}c}{L}\left(\eta_{js} - \frac{\chi}{3}\delta_{js}\right), \qquad \chi = \eta_{kk}. \end{gathered} \tag{2}$$

As a consequence of the second relation of system (2),

$$D_t\chi = D_t\eta_{kk} + c^2(D_k\xi_k + D_k\xi_k) + f_{kk} = 0,$$

i.e., the function $\chi = \chi_0(\mathbf{x})$ is stationary. The latter is assumed to be given. Taking an arbitrary constant $\varkappa > 0$ and a smooth function $\mathcal{F}(t, x)$, we introduce new vector fields $\mathbf{E} = (E_s)$, $\mathbf{A}$, $\mathbf{H}$, and $\mathbf{j} = (j_s)$ and scalars φ and ρ by the identities

$$\begin{gathered} E_s = \varkappa\nabla \cdot \left(\boldsymbol{\eta}_s - \frac{\chi_0}{3}\boldsymbol{\delta}_s\right), \quad \mathbf{A} = \varkappa c(\boldsymbol{\xi} + \nabla\mathcal{F}), \quad \mathbf{H} = \varkappa c\nabla \times \boldsymbol{\xi}, \\ \supset_s = \frac{\varkappa}{4\pi}\nabla \cdot \mathbf{f}_s \quad (\mathbf{f}_s \equiv (f_{ks})), \\ \varphi = \varkappa\left(\zeta + \frac{\chi_0}{3} - D_t\mathcal{F}\right), \quad \rho = \frac{\varkappa}{4\pi}D_jD_k\left(\eta_{jk} - \frac{\chi_0}{3}\delta_{jk}\right). \end{gathered}$$

Multiplying the first inequality of system (2) by $\varkappa$, we come to the relation

$$\frac{1}{c}\frac{\partial \mathbf{A}}{\partial t} + \mathbf{E} + \operatorname{grad}\varphi = 0$$

connecting the electric field $\mathbf{E}$ with the scalar and vector potentials φ and $\mathbf{A}$. The constant c corresponds to the speed of light in the vacuum. The given relation implies the Faraday law of electromagnetic induction,

$$\frac{1}{c}\frac{\partial \mathbf{H}}{\partial t} + \operatorname{rot}\mathbf{E} = \operatorname{rot}\left(\frac{1}{c}\frac{\partial \mathbf{A}}{\partial t} + \mathbf{E} + \nabla\varphi\right) = 0 \qquad (\operatorname{rot} \equiv \nabla\times).$$

Acting on the second equality from (2) by the operator $\frac{\varkappa}{c}\operatorname{div}$ (differentiation with respect to x_j, summation with respect to j, and multiplication by $\varkappa/c$) and taking into account the third equality from (2) and the identity

$$D_jD_j\xi \equiv \Delta\xi = -\operatorname{rot}(\operatorname{rot}\xi) \qquad (\operatorname{div}\xi \equiv \nabla \cdot \xi = 0),$$

we obtain the Maxwell law connecting the conduction current $\mathbf{j}$ with the displacement current $(1/4\pi)\partial\mathbf{E}/\partial t$ and the magnetic field $\mathbf{H}$,

$$\partial\mathbf{E}/\partial t - c\,\mathrm{rot}\,\mathbf{H} + 4\pi\mathbf{j} = 0.$$

As a result, we come to the system of Maxwell equations,

$$\frac{1}{c}\frac{\partial\mathbf{H}}{\partial t} + \mathrm{rot}\,\mathbf{E} = 0, \qquad \mathrm{div}\,\mathbf{H} = 0,$$
$$\frac{1}{c}\frac{\partial\mathbf{E}}{\partial t} + \mathrm{rot}\,\mathbf{H} + \frac{4\pi}{c}\mathbf{j} = 0, \qquad \mathrm{div}\,\mathbf{E} = 4\pi\rho.$$

The last equation from (2) implies the Ohm law,

$$\mathbf{j} = \sigma\mathbf{E}, \quad \sigma = q_0/4\pi L \qquad (q_0 = 3^{1/2}c)$$

(after applying the operator $\frac{\varkappa}{4\pi}$ div to both sides of the equality), closing this system.

New potentials

$$\mathbf{A}' = \varkappa c(\boldsymbol{\xi} + \nabla\mathcal{F}), \quad \varphi' = \varkappa\left(\zeta + \frac{\chi_0}{3} - D_t\mathcal{F}'\right)$$

are related to the old ones by the equations

$$\mathbf{A}' = \mathbf{A} + \mathrm{grad}\,\mathcal{G}, \quad \varphi' = \varphi - \frac{1}{c}\frac{\partial\mathcal{G}}{\partial t},$$
$$\mathcal{G} = \varkappa(\mathcal{F}' - \mathcal{F})$$

(gauge invariance).

We can consider the given constructions as solving a problem stated long ago on obtaining the equations of electromagnetic field from the basic (dynamical) Newton law **[63, 182]**. The corresponding material carrier of electromagnetic waves is some ideal turbulent medium, the dynamical state of which $\bar{\mathbf{u}}$, $\boldsymbol{\tau}$, $\bar{p}$ is close to the regime $\bar{\mathbf{u}}^0$, $\boldsymbol{\tau}^0 = (\tau^0_{is})$, $\bar{p}^0$. Then the value of the speed of light is defined by the equality

$$c = \left((\overline{u'^2_1} + \overline{u'^2_2} + \overline{u'^2_3})/3\right)^{1/2}$$

(where $\overline{u'^2_i} = \tau^0_{ii}$) and depends only on the pulsational velocity component. Since adding to $\mathbf{u}$ a given (nonrandom) vector field $\mathbf{a} = \bar{\mathbf{a}}$ does not change the pulsational component $\mathbf{u}'$,

$$((\mathbf{u}+\mathbf{a})' = (\mathbf{u}+\mathbf{a}) - (\overline{\mathbf{u}+\mathbf{a}}) = \mathbf{u} - \bar{\mathbf{u}} = \mathbf{u}'),$$

the value of c turns out to be invariant (in particular, it does not depend on the choice of the inertial reference frame, thus substantiating the Einstein fundamental principle of special relativity). The procedure of obtaining the Maxwell equations, invariant under the Lorentz transformations, from the equations of classical mechanics which are invariant under the Galileo transformations reveals, it seems, a certain new "surprise" of statistical averaging complementing the well-known paradoxes **[60]**.

Further, when passing to dimensional physical fields, the constant $\varkappa$ acquires the dimension of the ratio of the mass m to the charge q. To complete the suggested "hydrodynamic" version of electrodynamics, we obtain the law of motion of a particle of mass m and charge q in the electrodynamic field $\mathbf{E}$, $\mathbf{H}$ generated by an ideal turbulent medium. The potential and kinetic energies of a particle moving in a fluid with relative velocity $\mathbf{v}$ will be expressed by the values

$$m(\zeta + \chi_0/3) \quad \text{and} \quad m|\boldsymbol{\xi} + \mathbf{v}|^2/2$$

($\zeta = \overline{p} - \overline{p}^0$, $|\mathbf{v}|$ is the length of the vector $\mathbf{v}$). We assume that the mean velocity of small perturbations of the fluid is much less than the relative velocity of the particle, and the mean velocity of the turbulent pulsations is much greater than $|\mathbf{v}|$, $|\boldsymbol{\xi}| \ll |\mathbf{v}| \ll c$. In this case, the quadratic term $m|\boldsymbol{\xi}|^2/2$ in the expression $m|\boldsymbol{\xi} + \mathbf{v}|^2/2$ can be ignored. The Lagrangian of the particle takes the following form:

$$L(t, \mathbf{x}, \mathbf{v}) = \frac{1}{2}m|\mathbf{v}|^2 + m\boldsymbol{\xi} \cdot \mathbf{v} - m(\zeta + \chi_0/3), \qquad \mathbf{x} = \mathbf{x}(t), \quad \frac{d\mathbf{x}}{dt} = \mathbf{v}.$$

Here $\boldsymbol{\xi} \cdot \mathbf{v}$ is the scalar product of the vectors $\boldsymbol{\xi}$ and $\mathbf{v}$, and $\boldsymbol{\xi} = \boldsymbol{\xi}(t, \mathbf{x})$ and $\zeta = \zeta(t, \mathbf{x})$ are considered as given functions. The motion of the particle is described by the equation

$$\frac{d}{dt}\left(\frac{\partial L}{\partial \mathbf{v}}\right) = \frac{\partial L}{\partial \mathbf{x}}, \qquad \frac{d}{dt} \equiv \frac{\partial}{\partial t} + \mathbf{v} \cdot \nabla,$$

or, since

$$\frac{\partial L}{\partial \mathbf{v}} = m\mathbf{v} + m\boldsymbol{\xi}$$

and

$$\frac{\partial L}{\partial \mathbf{x}} = m\nabla(\boldsymbol{\xi} \cdot \mathbf{v})|_{\mathbf{v}=\mathbf{v}(t)} - m\nabla(\zeta + \chi_0/3),$$

by the equation

$$m\frac{d\mathbf{v}}{dt} + m\frac{\partial \boldsymbol{\xi}}{\partial t} + m(\mathbf{v} \cdot \nabla)\boldsymbol{\xi} = m\nabla(\boldsymbol{\xi} \cdot \mathbf{v})|_{\mathbf{v}=\mathbf{v}(t)} - m\nabla)\zeta + \chi_0/3).$$

According to the first equation from (2),

$$\partial\boldsymbol{\xi}/\partial t = -\nabla \cdot \boldsymbol{\eta} - \nabla\zeta \qquad (\nabla \cdot \boldsymbol{\eta} \equiv (\nabla \cdot \boldsymbol{\eta}_i)).$$

Since

$$\mathbf{v} \times (\nabla \times \boldsymbol{\xi}) = \nabla(\boldsymbol{\xi} \cdot \mathbf{v})|_{\mathbf{v}=\mathbf{v}(t)} - (\mathbf{v} \cdot \nabla)\boldsymbol{\xi},$$

we find that

$$m\frac{d\mathbf{v}}{dt} = m(\nabla \cdot \boldsymbol{\eta} - \operatorname{grad} \chi_0/3) + m\mathbf{v} \times \operatorname{rot} \boldsymbol{\xi}.$$

Identifying the vector $m(\nabla \cdot \boldsymbol{\eta} - \operatorname{grad} \chi_0/3)$ with the electrostatic force $q\mathbf{E}$, or setting $\varkappa = m/q$, we come to the required equation

$$m\frac{d\mathbf{v}}{dt} = q\mathbf{E} + \frac{q}{c}\mathbf{v} \times \mathbf{H}.$$

We see that the Coriolis force $m\mathbf{v} \times \operatorname{rot} \boldsymbol{\xi}$ generates the Lorentz force $(q/c)\mathbf{v} \times \mathbf{H}$. The scheme of the arguments used is given in Figure 20.

15.3. Characteristics. Passing to the consideration of system (1) (or the corresponding group of quasilinear terms in the complete Reynolds system for the "second" moments, inclusive), we shall study its characteristics.

Recall that a *characteristic* of a system of quasilinear first-order partial differential equations in variables t, $\mathbf{x}$ is usually a smooth level surface

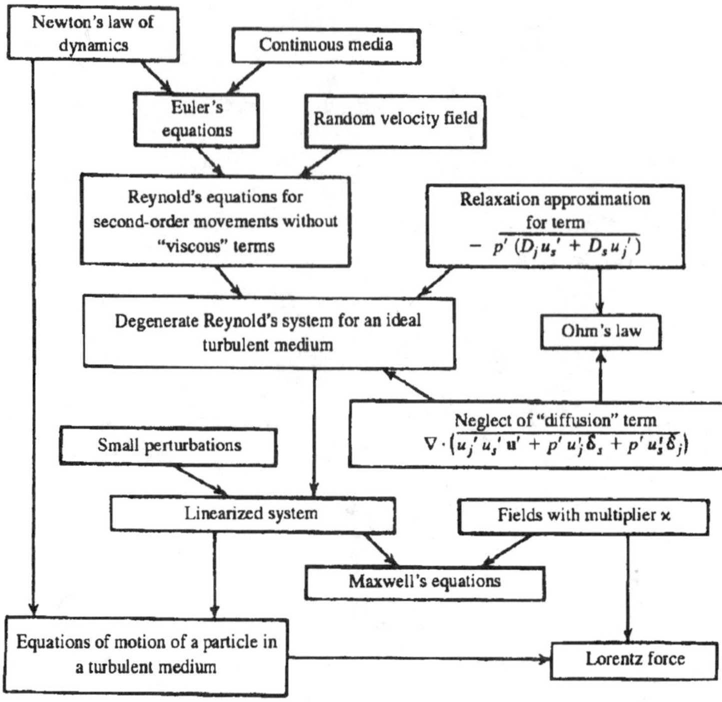

FIGURE 20. The scheme used for obtaining the equations of electrodynamics from the equations of classical mechanics.

$$\varphi(t, \mathbf{v}) = \text{const}, \qquad (D_t\varphi)^2 + |\nabla\varphi|^2 \neq 0 \qquad (|\nabla\varphi| \equiv (\nabla\varphi \cdot \nabla\varphi)^{1/2}),$$

such that the system under consideration cannot be uniquely solved with respect to the vector of the derivatives of the unknowns taken in the direction of the normal to this surface at an arbitrary point of the latter. The derivative along the normal is, obviously, proportional to the derivative with respect to $\varphi = \varphi(t, \mathbf{x})$. Consequently, this conditions is equivalent to the following restriction for $\varphi(t, \mathbf{x})$: there exists a nontrivial solution of the corresponding linear homogeneous system considered for the derivatives with respect to the variable φ and obtained by linearization of the initial quasilinear equations, and by subsequent removal of the derivatives with respect to the local coordinates complementing the variable φ, and also the terms not containing derivatives. This restriction means that the determinant of the linear homogeneous system under consideration is equal to 0. This determinant is a polynomial in $D_t\varphi$ and the components of the gradient $\nabla\varphi$. Therefore, the formulated restriction for the characteristic $\varphi(t, \mathbf{x}) = \text{const}$ means that one or several elementary algebraic equations for the partial derivatives of φ, generated by the roots of the given polynomial are satisfied. These equations are called *characteristic equations* of the initial system.

PROPOSITION 1. *The system of equations* (1) *has the following characteristic equations*:

$$D_t\varphi + \bar{\mathbf{u}} \cdot \nabla\varphi = 0 \qquad \textit{and} \qquad |\nabla\varphi| = 0,$$

exhausting the set of characteristic equations of system (1) *in the "laminar" case* (*when* $\boldsymbol{\tau} \equiv \mathbf{0}$), *and*

$$(3) \qquad (D_t\varphi + \bar{\mathbf{u}} \cdot \nabla\varphi)^2 = (\boldsymbol{\tau}\nabla\varphi, \nabla\varphi), \qquad (\boldsymbol{\tau}\nabla\varphi, \nabla\varphi) \equiv \tau_{is}\frac{\partial\varphi}{\partial x_i}\frac{\partial\varphi}{\partial x_s},$$

exhausting the set of additional characteristic equations of system (1) *in the "turbulent" case* (*when* $\boldsymbol{\tau} \not\equiv \mathbf{0}$) *under the additional assumption that the quadratic form of the matrix* $\boldsymbol{\tau} = (\tau_{js})$ *is strictly positive definite.*

The proof of this proposition for the case of an arbitrary dimension is given in the last subsection of this section. The condition that the quadratic form of the symmetric matrix $\boldsymbol{\tau} = (\tau_{js})$ is strictly positive definite corresponds to the assumption that the turbulent velocity pulsations are essentially three-dimensional. Equation (3) defines the phase φ of some wave of small perturbations of the turbulent medium, or of a "*turbulent wave*", and the corresponding equation $\varphi(t, \mathbf{x}) = \text{const}$, the front of the latter. Equation (3) immediately implies that for a given vector $\mathbf{y}$, $|\mathbf{y}| = 1$, the front of a plane turbulent wave moves with respect to the medium with some finite velocity $C_y > 0$ (or $C_y < 0$) in the direction of the vector $\mathbf{y}$ (in the opposite direction). The corresponding "main" direction and velocity are defined by the nontrivial solutions $\mathbf{y} = (y_j)$ and $C_y = \lambda^{1/2}$ of the equation $\boldsymbol{\tau}\mathbf{y} = \lambda\mathbf{y}$, or $\tau_{jk}y_k = \lambda y_j$. For example, for a turbulent jet injected into the flow in the direction $\mathbf{y} = (1, 0, 0)$ of the flow, the value of $C_y = \overline{(u_1'^2)}^{-1/2}$ is 0.65, 0.60 and 0.45 m/sec at the distance of 50, 100, and 150 diameters of the nozzle from the source, respectively, if we use the data of the paper **[149]**; the velocity of the jet is equal to 10.2 m/sec, and the velocity of the flow is 5–8 m/sec.

In the case of homogeneous and isotropic turbulence, when $\tau_{js} = c^2\delta_{js}$ with no dependence on time, equation (3) acquires the form

$$(\bar{\mathbf{u}} \cdot \nabla\varphi)^2 = c^2|\nabla\varphi|^2,$$

or, in the reference frame in which the vector $\bar{\mathbf{u}} \neq 0$ is parallel to the axis x_1,

$$|\bar{\mathbf{u}}|^2 \left(\frac{\partial\varphi}{\partial x_1}\right)^2 = c^2|\nabla\varphi|^2.$$

Introducing a *turbulent analog of the Mach number*, the number $M_c = |\bar{\mathbf{u}}|/c$, we find that the obtained equation has only the trivial solution $\varphi \equiv \text{const}$ if $M_c < 1$. For $M_c = \text{const} > 1$ this equation admits a characteristic cone,

$$\varphi = x_1 \mp (M_c^2 - 1)^{1/2}(|\mathbf{x}|^2 - x_1^2)^{1/2}.$$

Thus, equation (3) reveals an effect analogous to the "transonic" phenomenon in gas dynamics.

Using equation (3), we find the angles of inclination $\beta_+ > 0$ and $\beta_- < 0$ of the outer and, respectively, inner boundary lines of the mixing layer of the initial part of a plane stationary turbulent jet flowing from a channel into a fluid at rest. To this end, we associate the angles $\beta_\pm$ with the values characterizing the steady turbulent flow in the channel. These boundary lines are usually identified **[1]** with the lines of discontinuities of the derivative with respect to x_2 of the trapezoidal profile U of the longitudinal velocity component, which approximates the one observed near the efflux cross-section. The latter are defined by the "rays" generated by equation (3) considered for $\varphi = \varphi(x_1, x_2)$. Each ray passing through the point (x_1^0, x_2^0) is, by definition, the straight line

$$(x_2 - x_2^0)/(x_1 - x_1^0) = k(x_1^0, x_1^0), \qquad k \equiv -(\partial\varphi/\partial x_1)/(\partial\varphi/\partial x_2),$$

tangent to the characteristic $\varphi(x_1, x_2) = \varphi(x_1^0, x_2^0)$ at the point (x_1^0, x_2^0) and approximating the characteristic near this point. For $\bar{\mathbf{u}} = (U, 0, 0)$ and $\varphi = \varphi(x_1, x_2)$, equation (3) takes the form

$$(U^2 - \tau_{11})\left(\frac{\partial\varphi}{\partial x_1}\right)^2 - 2\tau_{12}\frac{\partial\varphi}{\partial x_1}\frac{\partial\varphi}{\partial x_2} - \tau_{22}\left(\frac{\partial\varphi}{\partial x_2}\right)^2 = 0$$

and has the solutions

$$k = k_\pm \equiv \frac{-\tau_{12} \pm (\tau_{12}^2 + (U^2 - \tau_{11})\tau_{22})^{1/2}}{U^2 - \tau_{11}}.$$

For a steady turbulent flow in a channel, the functions U, τ_{11}, τ_{12}, τ_{22} depend only on x_2. Then $k = k(x_2)$. Using dimensionless variables, we identify the outer (the inner) boundary of the mixing layer with the ray passing through the point $x_2 = 1$ on the wall of the channel (the point located on the outer boundary $x_2 = x_2^*$ of the boundary layer on the wall, respectively). Both points are located on the efflux cross-section of the channel $x_1 = 0$ (Figure 21). The location of the second point x_2^* is determined approximately by the locations of the maxima of the functions $\tau_{11}, \tau_{12}, \tau_{22}, 0 < x_2 < 1$, measured in the experiment **[46]**. According to the experiment, for $x_2 = x_2^*$ we have

$$\tau_{11}^{1/2}/U = 2.5^{1/2}/12, \qquad \tau_{22}^{1/2}/U = 1/12, \qquad \tau_{12}/\tau_{11}^{1/2}\tau_{22}^{1/2} = 0.4;$$

it follows from **[46]** that $\tau_{12}/U^2 \to 0$ and $\tau_{11}^{1/2}/U \to 0.3$ as $x \to 1$. The paper **[46]** does not contain any data concerning the value of

$$\gamma = \lim_{x_2 \to 1-0} (\tau_{22}^{1/2}/\tau_{11}^{1/2}).$$

However, as it follows from that paper, $\tau_{22} < \tau_{11}$. Assuming the existence of the limiting quantity γ, we find that $0 \leqslant \gamma \leqslant 1$. Substituting the obtained data in the formula for $k_\pm$ and taking into account that

$$\pm k_\pm = \tan\beta_\pm > 0,$$

we get

$$\beta_+ = -\arctan k_+(1) = \arctan(0.3\gamma/(0.91)^{1/2}), \qquad \beta_- = \arctan k_-(x_2^*) \cong -5^\circ.$$

For the greatest admissible value $\gamma = 1$, we have $\beta_+ \cong 17^\circ$. The experiment **[1]** gives $\beta_+ = 12^\circ \div 16^\circ$ and $\beta_- = -7^\circ$.

15.4. Effects of viscosity. To clarify the influence of molecular and turbulent diffusions, we include the "viscous" term $-\nu\Delta u_j$ $(\Delta \equiv \nabla \cdot \nabla)$ in the left-hand side of the initial evolution equation for u_j ($\nu = \text{const} > 0$ is the coefficient of kinematic (or molecular) viscosity) and pass to the Navier-Stokes equations. By averaging the

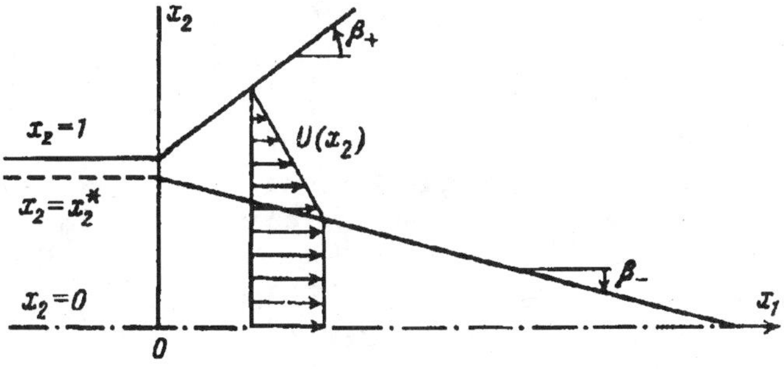

FIGURE 21. The initial part of the plane submerged turbulent jet.

equations, we come to the well-known *complete* (but nonclosed) system of Reynolds equations for the fields $\bar{\mathbf{u}}$, $\boldsymbol{\tau}$, $\bar{p}$:

$$D_t\bar{u}_j+\bar{\mathbf{u}}\cdot\nabla\bar{u}_j+\nabla\cdot\boldsymbol{\tau}_j-\nu\Delta\bar{u}_j+D_j\bar{p}=0,\qquad \nabla\cdot\bar{\mathbf{u}}=0,$$
$$D_t\tau_{js}+\bar{\mathbf{u}}\cdot\nabla\tau_{js}+\boldsymbol{\tau}_j\cdot\nabla\bar{u}_s+\boldsymbol{\tau}_s\cdot\nabla\bar{u}_j-\nu\Delta\tau_{js}+H_{js}=0,$$
$$H_{js}\equiv\nabla\cdot(\overline{u_j'u_s'\mathbf{u}'}+\overline{p'u_j'}\boldsymbol{\delta}_s+\overline{p'u_s'}\boldsymbol{\delta}_j)-\overline{p'(D_ju_s'+D_su_j')}+2\nu\overline{\nabla u_j'\cdot\nabla u_s'}.$$

Following the above-mentioned closure schemes, we approximate the first term in the formula for H_{js} by the terms of the turbulent diffusion of the correlations τ_{js}, see **[102, 134, 152, 158]**,

$$\begin{aligned}&\overline{u_j'u_s'\mathbf{u}'}+\overline{p'u_j'}\boldsymbol{\delta}_s+\overline{p'u_s'}\boldsymbol{\delta}_j\\&\qquad=-q\left\{L_1\nabla\tau_{js}+L_2(D_j\boldsymbol{\tau}_s+D_s\boldsymbol{\tau}_j)+L_3\left[(D_jq^2)\boldsymbol{\delta}_s+(D_sq^2)\boldsymbol{\delta}_j\right]\right\},\\&q\equiv(\tau_{11}+\tau_{22}+\tau_{33})^{1/2}.\end{aligned}$$

Here $L_1, L_2, L_3>0$ are given scales. As before, the pressure stresses term

$$-\overline{p'(D_ju_s'+D_su_j')}$$

is approximated by the Rotta formula. A similar algebraic approximation is also frequently used for the dissipation term

$$2\nu\overline{\nabla u_j'\cdot\nabla u_s'}.$$

Therefore, the indicated summands are modeled by the following terms:

$$-\overline{p'(D_ju_s'+D_su_j')}+2\nu\overline{\nabla u_j'\cdot\nabla u_s'}=\left(\frac{\nu}{L_4^2}+\frac{q}{L_5}\right)\tau_{js}-\frac{q^3}{L_6}\delta_{js},$$

where the scales $L_4, L_5, L_6>0$ are again assumed to be given. The approximate formula for H_{js} takes the following form:

$$\begin{aligned}H_{js}=&-q\left\{L_1\nabla\tau_{js}+L_2(D_j\boldsymbol{\tau}_s+D_s\boldsymbol{\tau}_j)+L_3\left[(D_jq^2)\boldsymbol{\delta}_s+(D_sq^2)\boldsymbol{\delta}_j\right]\right\}\\&+\left(\frac{\nu}{L_4^2}+\frac{q}{L_5}\right)\tau_{js}-\frac{q^3}{L_6}\delta_{js}.\end{aligned}$$

The domain of a turbulent flow is usually divided into three main parts: the *boundary layer* on the wall, where the mechanism of molecular viscosity prevails; the *mixing layer* within which the influence of the terms $\nabla\cdot\tau_j$ is essential; the *turbulent core*, where the turbulent waves under consideration seem to be least subject to the influence of molecular and turbulent diffusions. In this section we study to what extent the processes of molecular and turbulent diffusions damp these waves. Another goal of the constructions carried out below is to obtain the dispersion relation connecting the wave number k of a plane turbulent wave with the frequency ω of the transferred oscillations. Restricting ourselves to the turbulent core, we assume that the scales L_1, $\dots$, L_6 are constant. We also assume that the fields $\overline{\mathbf{u}}^0$, $\boldsymbol{\tau}^0$ and $\overline{p}^0$ of the unperturbed state to be constant. Substituting them for $\overline{\mathbf{u}}$, $\boldsymbol{\tau}$ and $\overline{p}$ in the Reynolds equations given above and considered under the indicated approximation for H_{js}, we come to the following necessary condition:

$$(\nu/L_4^2 + q_0/L_5)\tau_{js}^0 - q_0^3\delta_{js}/L_6 = 0 \qquad (q_0 = (\tau_{11}^0 + \tau_{22}^0 + \tau_{33}^0)^{1/2} = \text{const} > 0),$$

or

$$\tau_{js}^0 = c^2\delta_{js}, \qquad c = q_0(L_5/((1+\mu)L_6))^{1/2}, \qquad \mu = \nu L_5/(q_0 L_4^2).$$

The corresponding linearized system for small perturbations $\boldsymbol{\xi}, \boldsymbol{\eta}, \zeta$ of the fields $\overline{\mathbf{u}}, \boldsymbol{\tau}, p$, considered near the state $\overline{\mathbf{u}}^0$, $\boldsymbol{\tau}^0$, $\overline{p}^0$, after passing to the moving system of coordinates

$$t \to t \quad \text{and} \quad x_j \to x_j - t\overline{u}_j^0, \qquad \text{or} \quad D_j \to D_j \quad \text{and} \quad D_t + \overline{\mathbf{u}}^0 \cdot \nabla \to D_t,$$

takes the following form:

$$D_t\xi_j + \nabla\cdot\boldsymbol{\eta}_j - \nu\Delta\xi_j + D_j\zeta = 0, \quad \nabla\cdot\boldsymbol{\xi} = 0,$$
$$D_t\eta_{js} + c^2(D_j\xi_s + D_s\xi_j) + F_{js} = 0,$$
$$F_{js} = -(\nu + q_0L_1)\Delta\eta_{js} - q_0L_2\,(D_j\nabla\cdot\boldsymbol{\eta}_s + D_s\nabla\cdot\boldsymbol{\eta}_j) - 2q_0L_3D_jD_s\chi$$
$$+ \frac{q_0}{L_5}(1+\mu)\eta_{js} - \frac{q_0}{L_6}\frac{1+3\mu/2}{1+\mu}\chi\delta_{js},$$
$$\chi = \eta_{11} + \eta_{22} + \eta_{33}.$$

As a consequence,

$$D_t\chi - (\nu + q_0L_1 + 2q_0L_3)\Delta\chi + \frac{q_0}{L_5}(1+\mu)\left(1 - 3\frac{L_5}{L_6}\frac{1+3\mu/2}{(1+\mu)^2}\right)\chi = 2q_0L - 2D_jD_s\eta_{js}.$$

Choosing an arbitrary constant $\varkappa > 0$ and an auxiliary function $\psi = \psi(t, \mathbf{x})$ such that

$$D_t\psi - (\nu + q_0L_1 + 2q_0L_2)\Delta\psi + \frac{q_0}{L_5}(1+\mu)\psi = \frac{q_0}{L_6}\frac{1+3\mu/2}{1+\mu}\chi + 2q_0L_3\Delta\chi,$$

we introduce, as in the case of an ideal turbulent medium, the vector and scalar fields $\mathbf{E} = (E_s)$, $\mathbf{A}$, $\mathbf{H}$, $\mathbf{j} = (j_s)$, ρ, and φ by the equalities

$$E_s = \varkappa\nabla\cdot\boldsymbol{\eta}_s - \varkappa D_s\psi, \quad \mathbf{A} = \varkappa c(\boldsymbol{\xi} + \nabla\mathcal{F}), \quad \mathbf{H} = \varkappa c\nabla\times\boldsymbol{\xi},$$
$$j_s = \varkappa\frac{q_0}{4\pi L_5}(1+\mu)\,(\nabla\cdot\boldsymbol{\eta}_s - D_s\psi) - \varkappa\frac{q_0L_2}{4\pi}D_s\,(D_iD_k\eta_{ik} - \Delta\psi)$$
$$- \varkappa\frac{\nu + q_0L_1 + q_0L_2}{4\pi}\Delta\,(\nabla\cdot\boldsymbol{\eta}_s - D_s\psi),$$
$$\rho = \frac{\varkappa}{4\pi}\,(D_iD_k\eta_{ik} - \Delta\psi), \quad \varphi = \varkappa(\zeta + \psi - D_t\mathcal{F}).$$

Thus, we come to the following equations:

$$\frac{1}{c}D_t\mathbf{A} + \frac{\nu}{c}\operatorname{rot}\mathbf{H} + \mathbf{E} + \nabla\varphi = 0, \qquad \operatorname{rot}\mathbf{A} = \mathbf{H},$$

$$\frac{1}{c}D_t\mathbf{H} + \operatorname{rot}\mathbf{E} - \frac{\nu}{c}\Delta\mathbf{H} = 0, \qquad \operatorname{div}\mathbf{H} = 0,$$

$$\frac{1}{c}D_t\mathbf{E} - \operatorname{rot}\mathbf{H} + \frac{4\pi}{c}\mathbf{j} = 0, \qquad \operatorname{div}\mathbf{E} = 4\pi\rho,$$

$$\mathbf{j} = \frac{q_0}{4\pi L_5}(1+\mu)\mathbf{E} - q_0L_2\nabla\rho - \frac{\nu + q_0L_1 + q_0L_2}{4\pi}\Delta\mathbf{E},$$

modifying the Maxwell equations and the Ohm law and taking into account the presence of molecular viscosity and turbulent diffusion. The equations obtained are written in the moving system of coordinates. The substitution of $D_t + \overline{\mathbf{u}}^0 \cdot \nabla$ for D provides the passage to the corresponding system at rest.

Referring ρ to the density of the distribution of turbulent sources ("charges"), and introducing the corresponding current $\mathbf{j}$, defined as before, we come, as in the case of the Maxwell equations, to the equality

$$\frac{c}{4\pi}\operatorname{div}\left(\frac{1}{c}D_t\mathbf{E} - \operatorname{rot}\mathbf{H} + \frac{4\pi}{c}\mathbf{j}\right) = D_t\rho + \operatorname{div}\mathbf{j} = 0.$$

Substituting in the latter the above-given formula for $\mathbf{j}$, we obtain the effect of "smearing" of the charge due to diffusion,

$$D_t\rho - (\nu + q_0L_1 + 2q_0L_2)\Delta\rho + \frac{q_0}{L_5}(1+\mu)\rho = 0.$$

The equalities

$$-c\operatorname{rot}\mathbf{E} = D_t\mathbf{H} - \nu\Delta\mathbf{H}, \qquad -c^2\operatorname{rot}\left(\frac{1}{c}D_t\mathbf{E} - \operatorname{rot}\mathbf{H} + \frac{4\pi}{c}\mathbf{j}\right) = 0$$

lead to the following equation for the vorticity field $\mathbf{H}$:

$$\left(-D_t + (\nu + q_0L_1 + q_0L_2)\Delta - \frac{q_0}{L_5}(1+\mu)\right)(-D_t + \nu\Delta)\,\mathbf{H} - c^2\Delta\mathbf{H} = 0$$

(the equalities $\operatorname{div}\mathbf{H} = 0$ and $\operatorname{rot}(\operatorname{rot}\mathbf{H}) = -\Delta\mathbf{H}$ are used here). For the case of a plane monochromatic wave under the (usual) assumption that

$$\mathbf{H} = \widehat{\mathbf{H}}\exp(-\beta t - i\omega t + i\mathbf{k}\cdot\mathbf{x}), \qquad i^2 = -1,$$

with the real numbers β and ω, and real and complex vectors $\mathbf{k}$ and $\widehat{\mathbf{H}}$ such that $\mathbf{k}\cdot\widehat{\mathbf{H}} = 0$, we come to the following complex equality for $k = (\mathbf{k}\cdot\mathbf{k})^{1/2}$ and ω:

$$\left(i\omega + \beta - (\nu + q_0L_1 + q_0L_2)k^2 - \frac{q_0}{L_5}(1+\mu)\right)(i\omega + \beta - \nu k^2) + c^2k^2 = 0,$$

or to the following two real equalities:

$$\beta = \left(\nu + \frac{q_0L_1 + q_0L_2}{2}\right)k^2 + \frac{q_0}{2L_5}(1+\mu)$$

and

$$\omega^2 = q_0^2\frac{L_5}{(1+\mu)L_6}k^2 - q_0^2\left(\frac{L_1 + L_2}{2}k^2 + \frac{1+\mu}{2L_5}\right)^2$$

(we have used the identity $(1+\mu)L_6c^2 = L_5q_0^2$ in the latter case). The last equation expresses the required dispersion relation in the moving coordinate system. In the coordinate system at rest, ω is replaced by the sum $\omega + \mathbf{k}\cdot\overline{\mathbf{u}}^0$. In terms of the dimensionless wave number K and frequency Ω,

$$K = k\left(\frac{(L_1+L_2)L_5}{1+\mu}\right)^{1/2}, \qquad \Omega = \frac{\omega}{q_0}(L_1+L_2)^{1/2}L_6^{1/2},$$

$$\varepsilon = (1+\mu)\frac{(L_1+L_2)^{1/2}L_6^{1/2}}{L_5} \qquad (\varepsilon > 0, \quad \mu = \frac{\nu L_5}{q_0 L_4^2}),$$

the *dispersion relation* obtained is written in the following form:

$$\Omega^2 = K^2 - \frac{\varepsilon^2}{4}(K^2+1)^2.$$

When the parameter $\varepsilon > 0$ varies, the dispersion relation considered in the (K,Ω)-plane generates the ε-family of closed dispersion curves of the following form:

$$2(K^2-\Omega^2)^{1/2}/(K^2+1) = \varepsilon, \qquad |\Omega| \leqslant K$$

(see Figure 22). The case of an ideal turbulent fluid corresponds to the passage to the limit as $\varepsilon \to +0$, when $|\Omega|/K \to 1-0$. For a general $\varepsilon > 0$, the *condition of existence of turbulent waves*, or the inequality $\Omega \neq 0$ (or $\Omega^2 > 0$), is, obviously, equivalent to the inequality $K > \varepsilon(K^2+1)/2$ (since $K > 0$, and the dispersion relation is valid), or to the fact that the conditions

$$\varepsilon < 1, \qquad (1-(1-\varepsilon^2)^{1/2})/\varepsilon < K < (1+(1-\varepsilon^2)^{1/2})/\varepsilon$$

are simultaneously satisfied. The inversion of the dependence $\Omega(K)$ in the dispersion relation leads to the functions

$$K = K_\pm(\Omega) = \left(2-\varepsilon^2 \pm 2(1-\varepsilon^2-\varepsilon^2\Omega^2)^{1/2}\right)^{1/2}\varepsilon^{-1}, \quad 0 < |\Omega| < (1-\varepsilon^2)^{1/2}\varepsilon^{-1},$$

corresponding to the left ($K = K_-(\Omega)$) and right ($K = K_+(\Omega)$) parts of the dispersion curve, obtained as the curve is cut on the (K,Ω)-plane by the vertical line $K = (2-\varepsilon^2)^{1/2}/\varepsilon$.

Note that the necessary condition of existence of turbulent waves expressed by the inequality

(4) $$\varepsilon = (1+\mu)(L_1+L_2)^{1/2}L_6^{1/2}L_5^{-1} < 1, \qquad \mu = \nu L_5/q_0L_4^2$$

makes it much harder to detect them (e.g., in numerical experiments **[134]** the parameter ε exceeds 1). The inequalities

$$(1-(1-\varepsilon^2)^{1/2})/\varepsilon < K < (1+(1-\varepsilon^2)^{1/2})/\varepsilon$$

complementing this condition (to the corresponding sufficient one) narrow down the set of admissible wave numbers. For small values of ε, the lower bound of this set is approximated by the value $\varepsilon/2$, and the upper bound is approximately equal to $2/\varepsilon$.

The admissible values of the frequency are also bounded:

$$|\Omega| < (1-\varepsilon^2)^{1/2}\varepsilon^{-1}$$

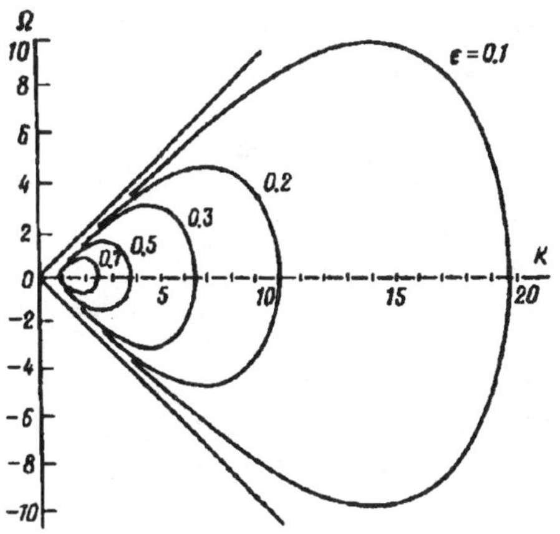

FIGURE 22. Dispersion curves of the linearized system of Reynolds equations.

(absence of high-frequency oscillations). The presence of two dispersion relations, $K = K_{\pm}(\Omega)$, testifies to the existence of two types of waves for an admissible frequency Ω, *long waves* with $K = K_{-}(\Omega)$, and *short waves* with $K = K_{+}(\Omega)$ $(K_{-}(\Omega) < K_{+}(\Omega))$.

Long (short) waves can be considered as corresponding to some normal (anomalous, respectively) dispersion, i.e., for $\Omega > 0$ the group velocity $d\Omega/dk$ of long (short) waves is greater (less) than 0. As already noted, the long waves asymptotically (as $\varepsilon \to 0$) coincide with the waves of an ideal turbulent medium, for which $K = |\Omega|$ (since

$$K_{-}(\Omega) = |\Omega| + O(\varepsilon^2), \qquad \varepsilon \to 0).$$

As for the short waves, they do not depend on the frequency Ω for sufficiently small $\varepsilon > 0$, i.e.,

$$K_{+}(\Omega) = 2/\varepsilon + O(\varepsilon), \qquad \varepsilon \to 0.$$

The latter are likely to correspond to some "background" waves generated by the mechanism of viscous friction present in a fluid.

Therefore, if the presence of the terms $\nabla \cdot \boldsymbol{\tau}_j$ in the equations for the components of the averaged velocity $\overline{u}_j$ testify to the existence of additional stresses τ_{js} in an incompressible medium, then the presence of "generation terms"

$$\boldsymbol{\tau}_j \cdot \nabla \overline{u}_s + \boldsymbol{\tau}_s \cdot \nabla \overline{u}_j$$

in the equations for the stresses τ_{js} provides this medium with the property of transferring bound oscillations of the fields $\overline{\mathbf{u}} = (\overline{u}_j)$ and $\boldsymbol{\tau} = (\tau_{js})$ in the form of elastic transverse waves formally subject to the laws of the electromagnetic field. Corresponding to the "magnetic" and "electric" components are small perturbations of the curl of the velocity field $\overline{\mathbf{u}}$ and the divergence of the Reynolds tensor $\boldsymbol{\tau}$. The corresponding "principal" directions and squares of the velocity of propagation of perturbations are defined by the eigenvectors and eigenvalues of the matrix $\boldsymbol{\tau}$. The characteristic surfaces are given by equation (3).

The mechanisms of molecular and turbulent diffusions present in a real medium "damp" the above-mentioned transversal waves. The condition under which the latter exist is inequality (4), where $L_1 + L_2$ is the characteristic scale of the turbulent diffusion, and L_4, L_5, L_6 are the characteristic scales of the corresponding relaxation and "generation" of the stresses τ_{js}.

Only indirect evidence allows us to speak about the practical existence of the waves under consideration. The wave character of the large-scale motion of a turbulized medium was observed in the well-known experiments of Mollo-Christensen, Brown, Roshko, and Dimotakis **[123, 130, 163]**, and in the spectral analysis of Landahl **[150]**, where, in fact, it was revealed (and stressed) that the wave mechanism prevails over the statistical properties of turbulent flows. Later the term *coherent structures* was used to denote large-scale motions of fluid which appear on the general background of the small-scale chaos of turbulent pulsations **[82, 103, 186]**. As mentioned in the introduction, in this respect these considerations can, indeed, provide an alternative approach to the problem of clarifying the nature of these motions based on the assumption of their weak dependence on the Reynolds number. It remains to note that the wave character of these motions (revealed theoretically) essentially distinguishes them from those large-scale flows, which admit the description by means of solutions of the Euler equations (also independent of the Reynolds number). It is desirable to separate (at least, formally) the assumed specific features of the turbulent medium from its well-known acoustic properties **[16, 74, 157]** exhibited by the corresponding longitudinal waves. The turbulent waves considered above do not depend on the compressibility of the fluid and are "transverse". Finally, it should be pointed out that the analysis given above is purely tentative. Further study of these properties confronts the problem of their experimental investigation.

15.5. A proposition on characteristics. We pass to the proof of the proposition on characteristics formulated in §15.3. According to the method of obtaining the equation for the function $\varphi = \varphi(t, \mathbf{x})$ generating the characteristics $\varphi(t, \mathbf{x}) = \text{const}$ of the quasilinear system (1) from §15.3, we substitute the dependencies $\xi_i(\varphi)$, $\eta_{is}(\varphi)$, and $\zeta(\varphi)$ for ξ_i, η_{is}, and ζ in the corresponding linearized system,

(4)
$$\begin{gathered}
D_t\xi_i + \bar{\mathbf{u}}^0 \cdot \nabla\xi_j + \boldsymbol{\xi} \cdot \nabla\bar{u}_i^0 + \nabla \cdot \eta_i + D_i\zeta = 0, \\
D_t\eta_{is} + \bar{\mathbf{u}}^0 \cdot \nabla\eta_{is} + \boldsymbol{\xi} \cdot \nabla\tau_{is}^0 + \boldsymbol{\tau}_i^0 \cdot \nabla\xi_s + \boldsymbol{\eta}_i \cdot \nabla\bar{u}_s^0 + \boldsymbol{\tau}_s^0 \cdot \nabla\xi_i + \boldsymbol{\eta}_s \cdot \nabla\bar{u}_i^0 + f_{is} = 0, \\
\nabla \cdot \boldsymbol{\xi} = 0, \\
f_{is} = \frac{q_0}{L}\left(\eta_{is} - \frac{\chi}{3}\delta_{is}\right), \qquad q_0 = \tau_{kk}^0, \qquad \chi = \eta_{kk}
\end{gathered}$$

(obtained by the linearization of the quasilinear system (1) in a neighborhood of an arbitrary regime $\bar{\mathbf{u}}^0 \equiv \bar{\mathbf{u}}$, $\boldsymbol{\tau}^0 \equiv \boldsymbol{\tau}$, $\bar{p}^0 \equiv \bar{p}$ satisfying (1)). Then, by removing the free terms, we come to the required linear homogeneous system for the derivatives

$$\dot{\xi}_i \equiv \frac{d\xi_i}{d\varphi}, \qquad \dot{\eta}_{is} \equiv \frac{d\eta_{is}}{d\varphi} = \dot{\eta}_{si}, \qquad \dot{\zeta} \equiv \frac{d\zeta}{d\varphi}.$$

With the use of the abbreviated notation

$$\varphi_t \equiv D_t\varphi, \qquad \varphi_i \equiv D_i\varphi, \qquad \varphi_u \equiv D_t\varphi + \bar{\mathbf{u}} \cdot \nabla\varphi,$$

the indicated system takes the form

$$\text{(A1)} \qquad \varphi_u \dot{\xi}_i + \nabla\varphi \cdot \dot{\boldsymbol{\eta}}_i + \varphi_i \dot{\zeta} = 0,$$

$$\text{(A2)} \qquad \varphi_u \dot{\eta}_{is} + (\boldsymbol{\tau}_i \cdot \nabla\varphi)\dot{\xi}_s + (\boldsymbol{\tau}_s \cdot \nabla\varphi)\dot{\xi}_i = 0,$$

$$\text{(A3)} \qquad \nabla\varphi \cdot \dot{\boldsymbol{\xi}} = 0.$$

The condition under which the system (A1)–(A3) admits a nontrivial solution $\dot{\xi}_i$, $\dot{\eta}_{is}$, $\dot{\zeta}$ at a fixed point $(t, \mathbf{x})$ determines the characteristic equation of system (1) at the given point. The corresponding solution $\dot{\xi}_i$, $\dot{\eta}_{is}$, $\dot{\zeta}$ gives the proper characteristic direction. The greatest number of linearly independent characteristic directions corresponding to a fixed characteristic equation gives, by definition, the *multiplicity* of the latter. To make the exposition more complete, we analyze the case of an arbitrary dimension $n \geqslant 2$: $i, s = 1, \dots, n$. The corresponding considerations are divided into three parts. The "turbulent flow" (the matrix $\boldsymbol{\tau}$ is nondegenerate) is considered in Part I and Part II. The "laminar" case is investigated in Part III ($\boldsymbol{\tau} \equiv 0$ and then (1) coincides with the Euler equations for an incompressible fluid). In Part I, we prove that each equation

$$\text{(B1)} \qquad \varphi_u = 0,$$

$$\text{(B2)} \qquad |\nabla\varphi| = 0,$$

$$\text{(B3)} \qquad \varphi_u^2 = (\boldsymbol{\tau}\nabla\varphi, \nabla\varphi)$$

is characteristic for (1), i.e., the validity of any of the equalities (B1)–(B3) ensures the existence of a nontrivial solution of system (A1)–(A3). Moreover, the multiplicities of the indicated characteristic equations are calculated here. In Part II it is proved that system (1) has no characteristic equations different from those given in (B1)–(B3), i.e., the existence of a nontrivial solution of the system (A1)–(A3) implies that equalities (B1), (B2), or (B3) are satisfied.

I. Passing to the first part of the proof, we note that the necessary condition

$$\varphi_t^2 + |\nabla\varphi|^2 \neq 0,$$

imposed on the characteristic surface $\varphi(t, \mathbf{x}) = \text{const}$, is equivalent to the restriction

$$\text{(C)} \qquad \varphi_u^2 + |\nabla\varphi|^2 \neq 0,$$

(which is obvious). We show that (B1) is a characteristic equation of system (1).

Indeed, (B1) and (C) imply $|\nabla\varphi| \neq 0$. Contracting (A2) with φ_i, that is, multiplying (A2) by φ_i and summing over i from 1 to n, we find, taking into account (A3), that

$$(\boldsymbol{\tau}\nabla\varphi, \nabla\varphi)\dot{\boldsymbol{\xi}} \equiv \tau_{ik}\varphi_i\varphi_k = 0.$$

Since $|\nabla\varphi| \neq 0$ and the matrix $\boldsymbol{\tau}$ is nondegenerate (namely, positive, $(\boldsymbol{\tau}\nabla\varphi, \nabla\varphi) > 0$ for $|\nabla\varphi| \neq 0$), we have $\dot{\boldsymbol{\xi}} = 0$. Then (A2) and (A3) are satisfied identically (since $\varphi_u = 0$), and (A1) reduces to the following system for $\dot{\eta}_{is}$ and $\dot{\zeta}$:

$$\dot{\eta}_{ik}\varphi_k + \dot{\zeta}\varphi_i = 0 \qquad (|\nabla\varphi|^2 \neq 0).$$

Contracting the latter with φ_i, we come to the equality $(\dot{\boldsymbol{\eta}}\nabla\varphi, \nabla\varphi) + \dot{\zeta}|\nabla\varphi|^2 = 0$, or

$$\dot{\zeta} = -\frac{(\dot{\boldsymbol{\eta}}\nabla\varphi, \nabla\varphi)}{|\nabla\varphi|^2}.$$

Substituting the orthogonal decomposition

$$\dot{\eta}_i = \mathbf{a}_i + \lambda \frac{\varphi_i}{|\nabla\varphi|^2}\nabla\varphi, \qquad \mathbf{a}_i \cdot \nabla\varphi = 0,$$

or

$$\dot{\eta}_{ik} = a_{ik} + \lambda \frac{\varphi_i \varphi_k}{|\nabla\varphi|^2}, \qquad a_{ik}\varphi_k = 0, \qquad \lambda, a_{ik} = \text{const},$$

in the equalities obtained, we come to $1 + \cdots + (n-1) + n$ easily solvable equations for a_{ik},

$$a_{ik}\varphi_k = 0, \qquad a_{ik} = a_{ki},$$

and the equality $\dot{\zeta} = -\lambda$. Hence, (B1) is a characteristic equation of system (1). Since the total number of constants a_{ik} and λ in this case is $n^2 + 1$, the multiplicity of (B1) is

$$n^2 + 1 - (1 + \cdots + n) = (n-1)n/2 + 1.$$

Now we show that (B2) is a characteristic equation of system (2).

Indeed, (B2) and (C) imply $\varphi_u \neq 0$. Then from equations (A1) and (A2) it follows that $\dot{\xi} = 0$ and $\dot{\eta} = 0$. Hence, (B2) is a characteristic equation of system (1) of multiplicity 1 ($\dot{\zeta} \neq 0$).

Finally, we show that (B3) is a characteristic equation of system (1).

Indeed, (B3) and (C) imply that $\varphi_u \neq 0$ and $|\nabla\varphi| \neq 0$ (since the matrix $\boldsymbol{\tau}$ is nondegenerate). Contracting (A2) with $\varphi_i\varphi_s$ and using (A3), we find that $(\dot{\eta}\nabla\varphi, \nabla\varphi)\varphi_u = 0$. Since $\varphi_u \neq 0$, we have

$$\text{(D)} \qquad (\dot{\eta}\nabla\varphi, \nabla\varphi) = 0.$$

Contracting (A1) with φ_i and again using (A3), we find that $(\dot{\eta}\nabla\varphi, \nabla\varphi) + |\nabla\varphi|^2\dot{\zeta} = 0$. Since $|\nabla\varphi| \neq 0$, taking into account (D), we come to $\dot{\zeta} = 0$. Then equality (A1) reduces to

$$\text{(E)} \qquad \dot{\xi}_i = -\frac{\nabla\varphi \cdot \dot{\eta}_i}{\varphi_u}, \qquad \dot{\zeta} = 0.$$

Substituting $\dot{\xi}_i$ from (E) in (A2), and using (B3), we obtain the following system for $\dot{\eta}_{is}$:

$$\text{(F)} \qquad (\boldsymbol{\tau}_i \cdot \nabla\varphi)(\nabla\varphi \cdot \dot{\eta}_s) + (\boldsymbol{\tau}_s \cdot \nabla\varphi)(\nabla\varphi \cdot \dot{\eta}_i) = (\boldsymbol{\tau}\nabla\varphi, \nabla\varphi)\dot{\eta}_{is}.$$

Now, (A3) follows from (F).

Indeed, contracting (F) with $\varphi_i\varphi_s$ and using the inequality $(\boldsymbol{\tau}\nabla\varphi, \nabla\varphi) > 0$, we come to (D). Then, contracting the first inequality from (E) with φ_i and using (D), we obtain (A3).

Thus, the values $\dot{\eta}_{is}$ defined by some nontrivial solution of system (F) with $\dot{\xi}_i$ and $\dot{\zeta}$ from (E) form a nontrivial solution of (A1)–(A3).

We pass to the solution of system (F) at a given point $t = t^{(0)}$, $\mathbf{x} = \mathbf{x}^{(0)}$ such that $|\nabla\varphi| \neq 0$. Without loss of generality, we assume that $\mathbf{x}^{(0)} = \mathbf{0}$ and $\frac{\partial\varphi}{\partial x_1}(t^{(0)}, \mathbf{0}) \neq 0$. We introduce a new system of the coordinates $x_1', \ldots, x_n'$ such that

$$x_1' = \varphi_1^{(0)}x_1 + \varphi_2^{(0)}x_2 + \cdots + \varphi_n^{(0)}x_n, \quad x_2' = x_2, \ldots, x_n' = x_n,$$
$$\varphi_i^{(0)} \equiv \frac{\partial\varphi}{\partial x_i}(t^{(0)}, \mathbf{0}),$$

or

$$x_1 = \frac{1}{\varphi_1^{(0)}} x_1' - \frac{\varphi_2^{(0)}}{\varphi_1^{(0)}} x_2' - \cdots - \frac{\varphi_n^{(0)}}{\varphi_1^{(0)}} x_n', \quad x_2 = x_2', \ldots, x_n = x_n'.$$

Then for any point $t, \mathbf{x}$, we have

$$\text{(G)} \qquad \frac{\partial(x_1', \ldots, x_n')}{\partial(x_1, \ldots, x_n)} = \begin{vmatrix} \varphi_1^{(0)} & \varphi_2^{(0)} & \ldots & \varphi_n^{(0)} \\ 0 & 1 & \ldots & 0 \\ \ldots & \ldots & \ldots & \ldots \\ 0 & 0 & \ldots & 1 \end{vmatrix} = \varphi_1^{(0)} \neq 0$$

and

$$\frac{\partial \varphi}{\partial x_1'} = \frac{\partial x_k}{\partial x_1'} \frac{\partial \varphi}{\partial x_k} = \frac{\varphi_1}{\varphi_1^{(0)}}, \quad \frac{\partial \varphi}{\partial x_2'} = -\frac{\varphi_2^{(0)}}{\varphi_1^{(0)}} \varphi_1 + \varphi_2, \ldots, \frac{\partial \varphi}{\partial x_n'} = -\frac{\varphi_n^{(0)}}{\varphi_1^{(0)}} \varphi_1 + \varphi_n.$$

For $t = t^{(0)}$ and $\mathbf{x} = \mathbf{0}$, when $\varphi_i = \varphi_i^{(0)}$, we find that

$$\frac{\partial \varphi}{\partial x_1'}(t^{(0)}, \mathbf{0}) = 1, \qquad \frac{\partial \varphi}{\partial x_2'}(t^{(0)}, \mathbf{0}) = \cdots = \frac{\partial \varphi}{\partial x_n'}(t^{(0)}, \mathbf{0}) = 0.$$

On the other hand, using the symmetry $\tau_{ks} = \tau_{sk}$, we can rewrite (F) as

$$\tau_{ik} \frac{\partial \varphi}{\partial x_k} \frac{\partial \varphi}{\partial x_m} \dot{\eta}_{ms} + \tau_{sk} \frac{\partial \varphi}{\partial x_k} \frac{\partial \varphi}{\partial x_m} \dot{\eta}_{mi} = \tau_{km} \frac{\partial \varphi}{\partial x_k} \frac{\partial \varphi}{\partial x_m} \dot{\eta}_{is},$$

or

$$\text{(H)} \qquad \tau_{ik} \frac{\partial x_p'}{\partial x_k} \frac{\partial \varphi}{\partial x_p'} \frac{\partial \varphi}{\partial x_q'} \frac{\partial x_q'}{\partial x_m} \dot{\eta}_{ms} + \tau_{sk} \frac{\partial x_p'}{\partial x_k} \frac{\partial \varphi}{\partial x_p'} \frac{\partial \varphi}{\partial x_q'} \frac{\partial x_q'}{\partial x_m} \dot{\eta}_{mi} = \tau_{km} \frac{\partial \varphi}{\partial x_k} \frac{\partial \varphi}{\partial x_m} \dot{\eta}_{is}.$$

Setting

$$\tau_{is} \frac{\partial x_p'}{\partial x_i} \frac{\partial x_q'}{\partial x_s} \equiv \tau_{pq}', \qquad \dot{\eta}_{is} \frac{\partial x_p'}{\partial x_i} \frac{\partial x_q'}{\partial x_s} \equiv \dot{\eta}_{pq}',$$

we find that

$$\tau_{km} \frac{\partial \varphi}{\partial x_k} \frac{\partial \varphi}{\partial x_m} = \tau_{pq}' \frac{\partial \varphi}{\partial x_p'} \frac{\partial \varphi}{\partial x_q'}.$$

Contracting (H) with

$$\frac{\partial x_\alpha'}{\partial x_i} \frac{\partial x_\beta'}{\partial x_s}$$

for fixed $\alpha, \beta = 1, \ldots, n$, we obtain the following system for $\dot{\eta}_{\alpha\beta}'$:

$$\text{(H}'\text{)} \qquad \tau_{\alpha p}' \frac{\partial \varphi}{\partial x_p'} \frac{\partial \varphi}{\partial x_q'} \dot{\eta}_{q\beta}' + \tau_{\beta p}' \frac{\partial \varphi}{\partial x_p'} \frac{\partial \varphi}{\partial x_q'} \dot{\eta}_{q\alpha}' = \tau_{pq}' \frac{\partial \varphi}{\partial x_p'} \frac{\partial \varphi}{\partial x_q'} \dot{\eta}_{\alpha\beta}'.$$

As a corollary of (G), the system (H$'$) is equivalent to the initial one (H). For $t = t^{(0)}$ and $\mathbf{x} = \mathbf{0}$, we have

$$\tau_{km} \frac{\partial \varphi}{\partial x_k} \frac{\partial \varphi}{\partial x_m}(t^{(0)}, \mathbf{0}) = \tau_{pq}' \frac{\partial \varphi}{\partial x_p'} \frac{\partial \varphi}{\partial x_q'}(t^{(0)}, \mathbf{0}) = \tau_{11}'(t^{(0)}, \mathbf{0}) \equiv \tau_{11}'^{0} > 0,$$

and (H$'$) takes the form

$$\text{(H}''\text{)} \qquad \tau_{\alpha 1}'^{0} \dot{\eta}_{1\beta}' + \tau_{\beta 1}'^{0} \dot{\eta}_{1\alpha}' = \tau_{11}'^{0} \dot{\eta}_{\alpha\beta}', \qquad \tau_{\alpha\beta}'^{0} \equiv \tau_{\alpha\beta}'(t^{(0)}, \mathbf{0}).$$

System (H$''$) is easily solvable with respect to $\dot{\eta}'_{\alpha\beta}$. It can be easily seen that the total set of its nontrivial solutions depends on $n-1$ arbitrary constants

$$\gamma_2, \dots, \gamma_n, \qquad \gamma_2^2 + \cdots + \gamma_n^2 \neq 0,$$

and is expressed by the dependencies

$$\dot{\eta}'_{11} = 0, \quad \dot{\eta}'_{1\beta} = \gamma_\beta, \quad \dot{\eta}'_{\alpha\beta} = \frac{\overset{0}{\tau}{}'_{\alpha 1}}{\overset{0}{\tau}{}'_{11}}\gamma_\beta + \frac{\overset{0}{\tau}{}'_{\beta 1}}{\overset{0}{\tau}{}'_{11}}\gamma_\alpha, \qquad 2 \leqslant \alpha \leqslant \beta \leqslant n,$$

$$\dot{\eta}'_{pq} = \dot{\eta}'_{qp}, \qquad p, q = 1, \dots, n.$$

The quantities

$$\dot{\eta}_{is} = \frac{\partial x_i}{\partial x'_p}\frac{\partial x_s}{\partial x'_q}\dot{\eta}'_{pq}$$

form a nontrivial solution of system (F). Thus, we have shown that (B3) is a characteristic equation of system (1) of multiplicity $n-1$.

II. Now we shall assume that at some point $(t, \mathbf{x})$ the system of equations (A1)–(A3) has a nontrivial solution $\dot{\xi}_i, \dot{\eta}_{is}, \dot{\zeta}$ and condition (C) is satisfied. The latter means that one of the following three mutually excluding possibilities is satisfied at the point $(t, \mathbf{x})$:

(a) $\varphi_u = 0$, $|\nabla\varphi| \neq 0$,
(b) $\varphi_u \neq 0$, $|\nabla\varphi| = 0$,
(c) $\varphi_u \neq 0$, $|\nabla\varphi| \neq 0$.

The cases (a) and (b) are reduced to the characteristic equations (B1) and (B2), respectively, and in the case (c), the vector $\dot{\boldsymbol{\xi}} \equiv (\dot{\xi}_i) \neq \mathbf{0}$.

Indeed, in this case the equality $\dot{\boldsymbol{\xi}} = \mathbf{0}$, together with (A2), implies $\dot{\boldsymbol{\eta}} \equiv (\dot{\eta}_{is}) = \mathbf{0}$ (since $\varphi_u \neq 0$). Then (A1) implies $\dot{\zeta} = 0$ (since $|\nabla\varphi| \neq 0$), and we come to the trivial solution $\dot{\xi}_i = 0$, $\dot{\eta}_{is} = 0$, and $\dot{\zeta} = 0$.

Further, contracting (A2) with $\varphi_i\varphi_s$ and taking into account (A3), we come to (D) (since $\varphi_u \neq 0$). Contracting (A1) with φ_i and taking into account (A3) and (D), we find $|\nabla\varphi|^2\dot{\zeta} = 0$, or $\dot{\zeta} = 0$ (since $|\nabla\varphi| \neq 0$). As a corollary, we come to (E). Then, contracting (A2) with φ_i and taking into account (A3), we find that

$$\varphi_u(\nabla\varphi \cdot \dot{\boldsymbol{\eta}}_s) + (\boldsymbol{\tau}\nabla\varphi, \nabla\varphi)\dot{\xi}_s = 0.$$

Comparing with (E), we conclude that

$$(\varphi_u^2 - (\boldsymbol{\tau}\nabla\varphi, \nabla\varphi))\dot{\boldsymbol{\xi}} = \mathbf{0}.$$

Since $\dot{\boldsymbol{\xi}} \neq \mathbf{0}$, we come to (B3). Consequently, there are no characteristic equations of system (1) different from (B1)–(B3).

III. Let us consider the case when $\boldsymbol{\tau} = \boldsymbol{\eta} = \dot{\boldsymbol{\eta}} = \mathbf{0}$. In this case, system (A1)–(A3) has the form

$$\text{(A}'\text{)} \qquad \varphi_u\dot{\xi}_i + \varphi_i\dot{\zeta} = 0, \qquad \nabla\varphi \cdot \dot{\boldsymbol{\xi}} = 0.$$

If the conditions (a) or (b), or, which is the same, (B1) or (B2) are satisfied, then system (A$'$) admits nontrivial solutions $\dot{\xi}_i, \dot{\zeta}$.

Indeed, in the case (a), $\dot{\zeta} = 0$ follows from equations (A$'$), and any vector $\dot{\boldsymbol{\xi}} \neq \mathbf{0}$

orthogonal to the gradient field $\nabla\varphi$ at the point $(t, \mathbf{x})$ under consideration gives rise in this case to a nontrivial solution of system (A′). In the case (b), $\dot{\boldsymbol{\xi}} = \mathbf{0}$ and $\dot{\zeta} \neq 0$ satisfy system (A′).

In the case (c), system (A′) has no nontrivial solutions. Indeed, contracting in this case the first equation from (A′) with $\dot{\xi}_i$ and using the second, we find that $\varphi_u|\dot{\boldsymbol{\xi}}|^2 = 0$, or $\dot{\boldsymbol{\xi}} = \mathbf{0}$ (since $\varphi_u \neq 0$). Since $|\nabla\varphi| \neq 0$, the first equation from (A′) implies $\dot{\zeta} = 0$.

Taking into account (C), we conclude that in the laminar case (when $\boldsymbol{\tau} \equiv \mathbf{0}$) there are no characteristic equations of system (1) different from (B1) and (B2). The proposition on the characteristics of the system of equations (1) is completely proved.

§16. Passage to turbulent regime and bifurcation

16.1. The closure scheme. Formal description of the passage to turbulent regime is usually connected with the analysis of the Orr-Sommerfeld equation (the linearized Navier-Stokes equations) **[43, 51, 61, 66, 78, 94, 105, 110–113, 129, 171]**. Its known solutions correspond to secondary hydrodynamic flows (e.g., Taylor vortices). The latter can be considered as defining individual realizations of a certain statistical ensemble describing a turbulent flow. However, each realization of this type cannot describe the flow as a whole. If we try to preserve the idea of the classical approach and apply it to the averaged equations (the Reynolds equations), then, in the framework of the closure schemes of second order, we obtain a new problem on bifurcation. This problem was considered in **[41]** for a channel. A specific feature of the passage in this case is the uniqueness of the bifurcating turbulent regime corresponding to the upper branch of the resistance curve given in Figure 23a (the graph of the known dependence of the resistance coefficient on the Reynolds number for a plane channel **[38, 46, 47, 62, 85, 105]**). The lower branch of the curve corresponds to the Poiseuille flow, while the upper one, to the statistically steady turbulent flow (see Figure 23b, c) bifurcating from the laminar one at an appropriate critical value of the Reynolds number.

The model proposed in **[41]** and investigated below describes this passage. The model is based on traditional second-order closure schemes (the Rotta scheme and the Nevzglyadov-Dryden relation **[71, 102, 135]** used before in **[122]**), but in terms of averaged variables admits a unique bifurcation (branching), as it is required by the experiment (Figure 23). Following **[41]**, we shall obtain the equations of this model. Under the assumption that the averaged flow is one-dimensional, we have $\bar{\mathbf{u}} = (U, 0, 0)$. Additionally, we assume the absence of the dependence on the variables $x_1 = x$ and x_3. We also assume that the constant component of the pressure gradient along the horizontal axis Ox is given,

$$-\frac{\partial p}{\partial x} = \sigma = \text{const} > 0.$$

After passing to dimensionless variables according to the formulas

$$t \to th/a, \quad y \to hy \quad (x_i \to hx_i),$$
$$U \to aU, \quad u_i' \to au_i' \quad (\tau \to a^2\tau), \quad p' \to a^2 p',$$
$$a = (\sigma h)^{1/2}, \quad h = \text{const} > 0, \quad -h < y = x_2 < h,$$

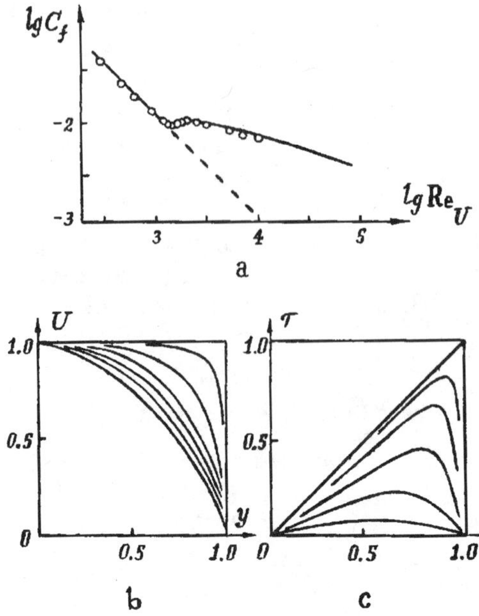

FIGURE 23. The resistance curve of a plane channel (a), profiles of the velocity $U(b)$ and correlation $\tau(c)$. ○ – experiment **[191, 192]**, — the model **[41**, or (16.1)–(16.4)]; $C_f = 2/\langle U\rangle^2$, $\mathrm{Re}_U = 2\langle U\rangle/\varepsilon$, $\langle U\rangle = \int_0^1 U(y)dy$, $U(y) = U_\varepsilon(y)$.

the first two equations of the complete Reynolds system (given at the beginning of §15.4) relating to $\overline{u}_1 = U(t,y)$ and $\tau_{12} = \tau(t,y)$, are written as

$$\frac{\partial U}{\partial t} + \frac{\partial \tau}{\partial y} - \varepsilon\frac{\partial^2 U}{\partial y^2} = 1, \qquad \frac{\partial \tau}{\partial t} + E\frac{\partial U}{\partial y} - \varepsilon\frac{\partial^2 \tau}{\partial y^2} + h_{12} = 0,$$

(1) $$\varepsilon = 1/\mathrm{Re} = \nu/ah, \quad E = \overline{v'^2}, \quad v' = u_2',$$

$$h_{12} = \frac{\partial}{\partial y}\overline{(u'v'v')} + 2\varepsilon\overline{\nabla u'\cdot\nabla v'} + \overline{u'\frac{\partial p'}{\partial y} + v'\frac{\partial p'}{\partial x}}, \qquad u' = u_1'$$

(we do not exclude the possibility of dependence on all spatial variables for the pulsational components). The corresponding stationary system takes the form

(2) $$\frac{d\tau}{dy} - \varepsilon\frac{d^2 U}{dy^2} = 1, \qquad E\frac{dU}{dy} - \varepsilon\frac{d^2\tau}{dy^2} + h_{12} = 0.$$

We shall consider the latter under the assumption that there is a symmetry expressed by the conditions

(3) $$U(-y) = U(y), \qquad \tau(-y) = -\tau(y).$$

Then, solving the first equation of system (2) with respect to the derivative

$$\frac{dU}{dy} = \frac{\tau(y) - y}{\varepsilon},$$

we come to the equation for $\tau = \tau(y)$,

$$-\varepsilon \frac{d^2\tau}{dy^2} + h_{12} = \frac{E}{\varepsilon}(y - \tau),$$

by means of the second equation. Using for h_{12} and E the standard diffusion-relaxation approximation and a dependence analogous to the Nevzglyadov-Dryden relation (modified with the geometry of the flow domain taken into account),

$$\begin{aligned} h_{12} &= -\frac{d}{dy}\left(\Gamma(y)\frac{d\tau}{dy}\right) + \varepsilon\alpha\tau + \frac{E}{\gamma(y)}\tau, \\ E &= \frac{c^2}{y}|\tau|, \qquad \Gamma(y) = \gamma(y) = \gamma_0(1 - y^2), \\ \alpha &= 2000, \quad c = 1, \quad \gamma_0 = 0.105 \end{aligned} \tag{4}$$

(the first term in the formula for h_{12} describes the turbulent diffusion of the correlation τ, the second, the process of relaxation to isotropy, the third, the anisotropic part of viscous relaxation; the corresponding scales of the diffusion $\Gamma(y)$ and relaxation $\gamma(y)$ are assumed to be equal and degenerating on the walls $y = \mp 1$, i.e., $\Gamma(\mp 1) = \gamma(\mp 1) = 0$), we obtain the equation for τ,

$$-\frac{d}{dy}\left((1 + k_\mu(y))\frac{d\tau}{dy}\right) + \alpha\tau = \mu \frac{|\tau|}{\varphi_\mu(y)}(\varphi_\mu(y) - \tau), \qquad 0 < y < 1, \tag{5}$$

where

$$\begin{aligned} &k_\mu(y) = \Gamma(y)/\varepsilon, \qquad \varphi_\mu(y) = y/(1 + \varepsilon/\gamma(y)), \qquad \mu = (c/\varepsilon)^2, \\ &\qquad\qquad\qquad \text{or} \\ &k_\mu(y) = \mu^{1/2}\Gamma(y)/c, \qquad \varphi_\mu(y) = y/(1 + c\mu^{-1/2}/\gamma(y)). \end{aligned} \tag{6}$$

Taking into account the imposed symmetry, we restrict ourselves to the domain $y > 0$. Passing to the problem on one-dimensional flow in the infinite channel $|y| < 1$, $-\infty < x < \infty$, we additionally assume that the boundary conditions

$$\tau(0) = \tau(1) = 0 \qquad (0 < y < 1, U(\mp 1) = 0) \tag{7}$$

are satisfied. The first of the given conditions is a corollary of the skew-symmetry of the correlation ($\tau(-y) = -\tau(y)$), the second is the requirement of absence of turbulent pulsations in the immediate proximity of the solid surface $y = 1$, where the no-slip condition $U(1) = 0$ is satisfied.

The profile of the averaged velocity $U(y)$,

$$U(y) = \frac{1}{\varepsilon}\int_y^1 (y' - \tau(y'))\, dy' \tag{8}$$

is calculated according to the solution $\tau(y)$ of the boundary problem (5)–(7) (the precise definition is given below) by means of the above-given formula for the derivative dU/dy and the boundary condition $U(1) = 0$. One of the possible solutions is the *trivial* one, $\tau \equiv 0$. In this case, formula (8) gives the well-known *Poiseuille profile*

$$U(y) = (1 - y^2)/2\varepsilon. \tag{9}$$

The goal of further analysis is to clarify the conditions for existence and uniqueness of some *nontrivial solution* $\tau(y) \not\equiv 0$ of problem (5)–(7) corresponding to the above-mentioned upper branch of the resistance curve shown in Figure 23.

Since this problem presents some interest by itself, we shall formulate it in a more general form than it is required in order to obtain the resistance curve. As it was agreed, equation (5) will be considered on the interval $0 < y < 1$ under the boundary conditions (7). We identify the given coefficients $k_\mu(y)$ and $\varphi_\mu(y)$ with nonnegative real functions of the set of variables $\mu > 0$ and $0 \leqslant y \leqslant 1$, analytic in y for any fixed $\mu > 0$. It is assumed that the function $\varphi_\mu(y)$ takes only positive values for $0 < y < 1$ and admits only simple zeros at the boundary points of this interval (i.e., the values of the derivative $d\varphi_\mu(y)/dy$ at $y = 0$ or $y = 1$ are different from zero if the function $\varphi_\mu(y)$ vanishes there). The constant $\alpha \geqslant 0$ is assumed to be fixed. The role of variable parameter is played by the constant $\mu > 0$.

By a (*classical*) *solution* of problem (5), (7) we mean a continuously differentiable real function $\tau(y)$ defined on the interval $0 \leqslant y \leqslant 1$, vanishing at its ends, twice continuously differentiable on the interval $0 < y < 1$, and having a square integrable second derivative on this interval. The set of such functions will be denoted by M in the sequel.

16.2. An illustrative example. Before passing to the main constructions, we analyze a very simple example. For

$$k_\mu(y) \equiv 0, \quad \varphi_\mu(y) \equiv \sin(\pi y), \quad \alpha = 0 \quad \text{and} \quad \tau(y) = A\sin(\pi y), \quad A = \text{const},$$

problem (5), (7) reduces to an algebraic equation for A:

$$\pi^2 A = \mu |A|(1 - A).$$

Along with the trivial solution $A = 0$, for $\mu > \pi^2$ the latter admits the unique nontrivial solution

$$A = A_\mu = 1 - \pi^2/\mu,$$

bifurcating from the trivial one $A = 0$ for the corresponding critical value $\mu_* = \pi^2$ of the variable parameter μ. The existence of the nontrivial solution can be established by reducing the equation under investigation to the corresponding problem of finding the minimum of the function (the *energy integral*),

$$E(A) = \pi^2 A^2/2 - \mu A|A|/2 + \mu |A|^3/3, \qquad -\infty < A < \infty,$$

whose derivative is the left-hand side of the equation for A. As it can be seen from Figure 24, for $\mu \leqslant \pi^2$ the function $E(A)$ has a unique critical point $A = 0$ (Figure 24a), and for $\mu > \pi^2$, one more critical point $A = A_\mu$ (Figure 24b). The latter is the required solution of the minimum problem.

Below, we reduce the analysis of problem (5), (7) to an analogous variational problem.

16.3. Analysis of the bifurcation problem. Let us formulate the main result (the space M is defined as at the end of §16.1, inf means the greatest lower bound).

THEOREM 1. *Let*

$$\begin{aligned} \mu_*(\mu) &= \alpha + \Lambda(\mu), \qquad \Lambda(\mu) = \inf\{R(\mu,\tau) : \tau \in M, \quad \tau \neq 0\}, \\ R(\mu,\tau) &\equiv \langle (1 + k_\mu)(\tau')^2 \rangle / \langle \tau^2 \rangle, \qquad \mu > 0, \qquad \tau \in M. \end{aligned} \tag{10}$$

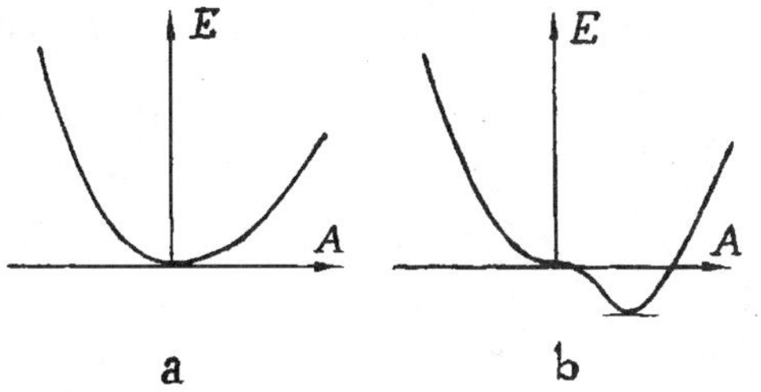

FIGURE 24. The energy integral for the equation $\pi^2 A = \mu|A|(1-A)$ for $\mu \leqslant \pi^2$ (a) and $\mu > \pi^2$ (b).

Then for $\mu \leqslant \mu_*(\mu)$, problem (5)–(7) has only the trivial solution $\tau \equiv 0$, and for every $\mu > \mu_*(\mu)$, a unique nontrivial solution $\tau = \tau_\mu(y) \geqslant 0$.

We pass to the proof of the theorem. Note that the existence of the greatest lower bound $\Lambda = \Lambda(\mu)$ of the Rayleigh function $R(\mu, \tau)$ given in the formulation and the inequality $\Lambda > 0$ immediately follow from Lemma 10.4 (§10.4) if we set in this lemma ($\psi' \equiv d\psi/dy$)

$$\{\psi, \chi\} \equiv \langle (1+k)\psi'\chi' \rangle, \qquad (\psi, \chi) \equiv \langle \psi\chi \rangle, \quad \langle \psi \rangle \equiv \int_0^1 \psi\, dy, \quad \psi, \chi \in M.$$

Another corollary of the indicated lemma is the fact that this bound is attained on some generalized solution ψ of the corresponding spectral problem

$$(11) \quad -\frac{d}{dy}\left((1+k_\mu(y))\frac{d\psi}{dy}\right) + \alpha\psi = \mu_*(\mu)\psi, \qquad 0 < y < 1, \quad \psi(0) = \psi(1) = 0$$

(considered for a fixed $\mu > 0$). It is not hard to prove that $\psi \in M$ indeed (the solution ψ has the required properties of smoothness). Omitting the details, we establish the validity of the following main properties A)–C):

A) *The solutions of problem* (5), (7) *take nonnegative values.*

Indeed, assuming that $\tau(y_0) < 0$ for some $0 < y_0 < 1$ (taking into account that $\tau(0) = \tau(1) = 0$), we obtain (due to the continuity of the function $\tau(y)$, $0 \leqslant y \leqslant 1$, and the indicated boundary conditions) that there exist the numbers

$$y_1, y_2, \qquad 0 \leqslant y_1 < y_0 < y_2 \leqslant 1,$$

such that $\tau(y) < 0$ for $y_1 < y < y_2$ and

$$\tau(y_1) = \tau(y_2) = 0.$$

Hence, $\tau'(y_1) \leqslant 0$ and $\tau'(y_2) \geqslant 0$; therefore,

$$J \equiv -\int_{y_1}^{y_2} ((1+k_\mu)\tau')'\, dy = (1+k_\mu)\tau'|_{y_1}^{y_2} \leqslant 0.$$

Simultaneously, according to (5), for $y_1 < y < y_2$ we have

$$-((1+k_\mu)\tau')' = -\alpha\tau + \mu\frac{|\tau|}{\varphi_\mu}(\varphi_\mu - \tau) = \alpha|\tau| + \mu\frac{|\tau|}{\varphi_\mu}(\varphi_\mu + |\tau|) > \mu|\tau| > 0,$$

which leads to the contradictory inequality $J > 0$ and completes the proof.

B) *The solutions of problem* (5), (7) *are analytic functions of the variable* $0 < y < 1$ *for any fixed* $\mu > 0$.

Indeed, property A) implies $|\tau| = \tau$. Then equation (5) reduces to the normal form

$$\tau'' = \mathfrak{F}(y, \tau, \tau'; \mu), \qquad 0 < y < 1,$$

with the right-hand side

$$\mathfrak{F}(y, u, v; \mu) \equiv \frac{\alpha u - \mu \frac{u}{\varphi_\mu(y)}(\varphi_\mu(y) - u) - \frac{dk_\mu}{dy}(y) v}{1 + k(y)},$$
$$0 < y < 1, \qquad -\infty < u, v < \infty,$$

which is an analytic function of the set of its variables (for every fixed $\mu > 0$). The analyticity of $\mathfrak{F}$ implies the analyticity of τ **[15]**.

C) *For every* $\mu > 0$, (5), (7) *admits at most one nontrivial solution.*

Indeed, let $\tau(y), \omega(y), 0 \leqslant y \leqslant 1$, be nontrivial solutions of problem (5), (7), and let there exist y_0, $0 < y_0 < 1$, such that $\tau(y_0) \neq \omega(y_0)$ (it is clear that $y_0 \neq 0, 1$ due to (7)). For the sake of definiteness we set $\tau(y_0) > \omega(y_0)$. Then there exist y_1, y_2, $0 \leqslant y_1 < y_0 < y_2 \leqslant 1$, such that $\tau(y) > \omega(y)$ for $y_1 < y < y_2$ and $\tau(y) = \omega(y)$ for $y = y_1, y_2$. Hence, $\tau'(y_1) \geqslant \omega'(y_1)$ and $\tau'(y_2) \leqslant \omega'(y_2)$, which implies the validity of the integral inequality

$$G \equiv \int_{y_1}^{y_2} \Delta(y)\, dy = (1 + k_\mu)\tau(\tau' - \omega')|_{y_1}^{y_2} \leqslant 0,$$
$$\Delta(y) \equiv \omega((1 + k_\mu)\tau')' - \tau((1 + k + \mu)\omega')'.$$

Simultaneously, from equation (5) and property A) it follows that ($|\tau| = \tau$ and $|\omega| = \omega$)

$$\Delta(y) = \mu \frac{\tau\omega}{\varphi_\mu}(\tau - \omega) \geqslant 0, \qquad y_1 < y < y_2.$$

Since the functions defining $\Delta(y)$ are analytic and are not identical zeros on the interval $y_1 < y < y_2$, it follows from the inequality obtained that $G > 0$ (the inequality $\tau(y) > \omega(y)$, $y_1 < y < y_2$, being additionally taken into account). The latter contradicts the previously obtained inequality $G \leqslant 0$ and completes the proof.

We show the *uniqueness of the trivial solution* $\tau \equiv 0$ *for* $\mu \leqslant \mu_*$ ($\mu_* \equiv \mu_*(\mu)$). To this end, we equip the linear manifold M defined at the end of §16.1 with the scalar product

$$\{\tau, \omega\} \equiv \int_0^1 \tau'\omega'\, dy, \qquad \tau, \omega \in M,$$

and the norm

$$\|\tau\| \equiv \{\tau, \tau\}^{1/2}.$$

Introducing on M the functional

$$(12) \quad E(\tau) \equiv \frac{1}{2}\langle(1 + k_\mu)\tau'^2\rangle + \frac{\mu}{3}\langle|\tau|^3/\varphi_\mu\rangle - \frac{\mu}{2}\langle\tau|\tau|\rangle + \frac{\alpha}{2}\langle\tau^2\rangle, \quad \langle\tau^2\rangle \equiv \int_0^1 \tau^2\, dy$$

(the existence of the integral $\langle|\tau|^3/\varphi_\mu\rangle$ follows from the boundedness of the function $|\tau(y)|^3/\varphi_\mu(y)$ in a neighborhood of the boundary points of the interval $0 < y < 1$, which, in turn, is guaranteed by the above-mentioned conditions imposed on $\varphi_\mu(y)$), we establish by means of a straightforward verification that this functional on M (with the norm introduced) is continuously differentiable and that the identity

$$E'(\tau)w = \langle(1+k_\mu)\tau'w'\rangle + \mu\langle w\tau|\tau|/\varphi_\mu\rangle - \mu\langle|\tau|w\rangle + \alpha\langle\tau w\rangle \tag{13}$$

is valid for its (Frechét) derivative. A *generalized solution* of problem (5), (7) is defined as an element τ of the Hilbert space $\mathbb{W}$ (obtained by completion of M with respect to the above norm) such that for any $w \in M$ (or $w \in \mathbb{W}$) it satisfies the identity $E'(\tau)w = 0$. Here the functional E is identified with its continuous extension to $\mathbb{W}$ and Lemma 10.1 (§10.1) is used. As a result of using the identity

$$\langle(1+k_\mu)\tau'\omega'\rangle = -\langle((1+k_\mu)\tau')'\omega\rangle, \qquad \tau, \omega \in M,$$

we come to the statement that any classical solution of problem (5), (7) is its generalized solution. The corresponding converse statement is also valid. In order to prove the latter it is sufficient to verify that the generalized solution $\tau \in \mathbb{W}$ has the required derivatives, and use again the above-mentioned identity considering it now for the equality $E'(\tau)w = 0$. Omitting the details, we conclude that the *classical and generalized solutions of the boundary problem coincide in* $\mathbb{W}$, *and that they correspond to critical points of the functional* $E(\tau)$ *extended (by continuity) to* $\mathbb{W}$.

Now, setting $w = \tau$ in (13) and using A) ($\tau = |\tau|$), for $\mu_* = \mu_*(\mu)$ from (10) we have

$$\begin{aligned} E'(\tau)\tau &= \langle(1+k_\mu)\tau'^2\rangle + \mu\langle|\tau|^3/\varphi_\mu\rangle - \mu\langle\tau^2\rangle + \alpha\langle\tau^2\rangle \\ &\geqslant (\mu_* - \alpha)\langle\tau^2\rangle + \mu\langle|\tau|^3/\varphi_\mu\rangle - \mu\langle\tau^2\rangle + \alpha\langle\tau^2\rangle \\ &= (\mu_* - \mu)\langle\tau^2\rangle + \mu\langle|\tau|^3/\varphi_\mu\rangle. \end{aligned}$$

The required uniqueness of the trivial solution immediately follows from the obtained inequality, so that for any $0 \leqslant \mu \leqslant \mu_*$ the critical point of the functional $E = E(\tau)$ defined on $\mathbb{W}$ coincides with the zero element of the space $\mathbb{W}$, which provides the uniqueness of the trivial solution of problem (5), (7) on the indicated set of values of the varying parameter μ.

We pass to the proof of the *existence of a nontrivial solution of problem* (5), (7) *for* $\mu > \mu_*$. Let us prove that the functional $E(\tau)$ is bounded from below on $\mathbb{W}$. To this end, fixing arbitrary $0 < y < 1$ and $\mu > 0$, we set

$$F(t) \equiv |t|^3/(3\varphi_\mu(y)) - t|t|/2, \qquad -\infty < t < \infty.$$

We have

$$dF/dt = |t|(t/\varphi_\mu(y) - 1) \geqslant 0 \quad (\leqslant 0) \quad \text{for } t \geqslant \varphi_\mu(y) \quad (t \leqslant \varphi_\mu(y)).$$

Hence,

$$F(t) \geqslant F(\varphi_\mu(y)) = -\varphi_\mu^2(y)/6;$$

thus,

$$\begin{aligned} E(\tau) &= \frac{1}{2}\langle(1+k_\mu)\tau'^2\rangle + \mu\langle F(\tau)\rangle + \frac{\alpha}{2}\langle\tau^2\rangle \\ &\geqslant \frac{1}{2}\langle\tau'^2\rangle + \mu\langle F(\varphi_\mu)\rangle = \frac{1}{2}\|\tau\|^2 - \frac{1}{6}\mu\langle\varphi_\mu^2\rangle \end{aligned} \tag{14}$$

($\alpha \geqslant 0$ and $k_\mu(y) \geqslant 0$ are used here). This inequality completes the proof of the required property for $E(\tau)$ bounded from below.

The boundedness from below of the functional $E(\tau)$ guarantees the existence of its greatest lower bound

$$d = \inf\{E(\tau) : \tau \in M\} = \inf\{E(\tau), \tau \in \mathbb{W}\}$$

(the latter equality is the corollary of M being dense in $\mathbb{W}$), attained on some sequence of functions $\tau_i \in M$, $i = 1, 2, \ldots$. In the last subsection of this section it is proved that among the minimizing sequences there exists a fundamental sequence such that

$$\|\tau_i - \tau_j\| \to 0 \quad \text{as } i, j \to \infty.$$

Due to the continuity of the functional $E(\tau)$ (extended to a smooth one) on $\mathbb{W}$, the required minimum of values of $E(\tau)$,

$$E(\tau_*) = d, \qquad \tau_* \in \mathbb{W},$$

is attained on the limit element $\tau_* \in \mathbb{W}$ of the fundamental (hence, converging) in $\mathbb{W}$ sequence τ_i, $i = 1, 2, \ldots$; then τ_* is a critical point of $E(\tau)$ (which is obvious) corresponding to some generalized solution of the problem under consideration. Due to the equivalence of this solution to the classical one introduced above (among the representatives of the equivalence class of the generalized solution there exists a classical one with the required smoothness properties), some classical solution of problem (5), (7) can be considered to play the role of τ_*.

The latter coincides with the trivial one for any $0 \leqslant \mu \leqslant \mu_*$. Indeed, as a corollary of (10), we have

$$\begin{aligned} E(\tau) &= \langle (1 + k_\mu)\tau'^2\rangle/2 + \mu\langle|\tau|^3/\varphi_\mu\rangle/3 - \mu\langle\tau|\tau|\rangle/2 + \alpha\langle\tau^2\rangle/2 \\ &\geqslant (\mu_* - \alpha)\langle\tau^2\rangle/2 + \mu\langle|\tau|^3/\varphi_\mu\rangle/3 - \mu\langle\tau^2\rangle/2 + \alpha\langle\tau^2\rangle/2 \\ &= (\mu_* - \mu)\langle\tau^2\rangle/2 + \mu\langle|\tau|^2/\varphi_\mu\rangle/3 \geqslant 0 \end{aligned}$$

for $0 \leqslant \mu \leqslant \mu_*$, and $E(\tau) > 0$ for $\tau \not\equiv 0$ (the inequality $\mu_* \geqslant \pi > 0$, which follows from (10), is additionally taken into account). Since $E(0) = 0$, we conclude that $\tau_* = 0$ for $0 \leqslant \mu \leqslant \mu_*$. Taking into account the uniqueness of the critical point of the functional $E(\tau)$ for $0 \leqslant \mu \leqslant \mu_*$, we have shown that for the indicated values of μ the minimum of the given functional is attained on the unique element $\tau = 0$ (corresponding to the trivial solution).

For $\mu > \mu_*$, the equality $E(0) = 0$ still holds; however, in this case there exists a function $\tau^- = \tau^-(y)$ such that

$$E(\tau^-) < 0, \qquad \tau^- \in \mathbb{W},$$

and hence (since, simultaneously, $E(\tau_*) < E(\tau^-)$), $\tau_* \not\equiv 0$. The required $\tau^-(y)$ is defined by the formula

$$(15) \quad \tau^-(y) = A\psi(y), \qquad A = \text{const} > 0, \quad \langle\psi^2\rangle = 1, \quad \psi(y) \geqslant 0, \quad 0 \leqslant y \leqslant 1.$$

Simultaneously, it is assumed that $\psi = \psi(y)$ is a fixed eigenfunction of the spectral problem (11) corresponding to the smallest eigenvalue μ_* (the existence of ψ with the required properties easily follows from Lemma 10.4). Using the identity

$$\langle(1 + k_\mu)\psi'^2\rangle/\langle\psi^2\rangle = \langle(1 + k_\mu)\psi'^2\rangle = \mu_* - \alpha \qquad (\langle\psi^2\rangle = 1),$$

we find, after substituting $\tau = \tau^-$ in formula (12), that

$$E(\tau^-) = \frac{1}{2}A^2\langle(1+k_\mu)\psi'^2\rangle + \frac{\mu}{3}A^3\langle|\psi|^2/\varphi_\mu\rangle - \frac{\mu}{2}A^2 + \frac{\alpha}{2}A^2$$
$$= \frac{1}{2}(\mu_* - \mu)A^2 + \frac{\mu}{3}\langle|\psi|^3/\varphi_\mu\rangle A^3$$

(the existence of the integral $\langle|\psi|^3/\varphi_\mu\rangle$ again follows from the above restrictions on the behaviour of the function $\varphi_\mu(y)$ near the boundary points of the interval $0 < y < 1$). Since the function ψ corresponds to the smallest eigenvalue of problem (11), we have $\psi(y) > 0$ for all $0 < y < 1$ (the Jentsch theorem). The latter guarantees that the strict inequality

$$\langle|\psi|^3/\varphi_\mu\rangle > 0$$

holds. Setting

$$A = \frac{3(\mu - \mu_*)}{4\langle|\psi|^3/\varphi_\mu\rangle\mu},$$

we obtain

$$E(\tau^-) = A^2(\mu_* - \mu)/4.$$

Hence, $E(\tau^-) < 0$ for $\mu > \mu_*$.

Due to the uniqueness of the nontrivial solution for a fixed μ proved above, we find that $\tau_* = \tau_\mu$ is a function of the parameter $\mu > 0$ different from the identical zero for $\mu > \mu_*$ (and coinciding with it for $\mu \leqslant \mu_*$).

Taking into account the existence of the required minimizing element established in §16.4, we see that Theorem 1 is proved.

As follows from Theorem 1, the critical values of the parameter μ (such that a nontrivial solution bifurcates from the trivial one) are defined by the solutions of the equation

$$\mu = \mu_*(\mu), \qquad \mu > 0, \tag{16}$$

where the function $\mu_*(\mu)$ is defined as in (10). We show that *for the case* (4), (6) (§16.1), *the solution of equation* (16) *is unique*. Thus, we shall establish that the bifurcation of the trivial solution is "global", i.e., that the critical value of the parameter $\mu = (c/\varepsilon)^2$ (or $\varepsilon = c/\mu^{1/2}$) corresponding to the turbulent (upper) branch on the graph of the resistance curve (Figure 23a) is unique.

In this case, for every fixed $\mu > 0$ we solve the minimum problem

$$R(\tau; \mu) \to \min, \qquad \tau \in M,$$

for the Rayleigh function

$$R(\tau; \mu) = \langle(1+k_\mu)(\tau')^2\rangle/\langle\tau^2\rangle = \frac{\langle\tau'^2\rangle}{\langle\tau^2\rangle} + \frac{1}{c}\frac{\langle\gamma\tau'^2\rangle}{\langle\tau^2\rangle}\mu^{1/2}$$

with $k_\mu = k_\mu(y)$ and $\gamma = \gamma(y)$ defined by equalities (4), (6). Then

$$\Lambda(\mu) = \inf\{R(\tau; \mu) : \tau \in M\} = R(\psi; \mu),$$

where

$$\psi = \psi_\mu \geqslant 0, \qquad \langle\psi'^2\rangle = 1,$$

is the solution of the indicated variational problem in $\mathbb{W}$ with the smoothness properties that guarantee $\psi \in M$ (the corresponding details are omitted). Simultaneously, ψ corresponds to the critical point of the smooth functional $R(\tau;\mu)$ considered in $\mathbb{W}$,

$$R'(\psi;\mu)w = 0, \qquad w \in \mathbb{W},$$

(identified with its extension by continuity to $\mathbb{W}$). Using the Jentsch theorem, we can easily show that the dependence of the solution $\psi = \psi_\mu$ on the parameter $\mu > 0$ is single valued. We shall additionally use the theorem on the differentiability of the corresponding function ψ_μ (taking, generally speaking, the values in $\mathbb{W}$) implicitly defined by the indicated functional equation (and proved in the general form in **[44]**). To this end, it is sufficient to show that the functional $R(\tau;\mu)$ is twice continuously differentiable with respect to the variables $\tau \in \mathbb{W}$, $\tau \neq 0$, and $\mu > 0$ and that its Hessian $R''(\tau;\mu)(w,w)$ (obtained by restricting the corresponding second Frechét derivative with respect to τ to the diagonal of the direct product $\mathbb{W} \times \mathbb{W}$) is nondegenerate. It is well known that the Rayleigh functions of the standard form $\langle \tau'^2\rangle/\langle\tau^2\rangle$ and $\langle\gamma\tau'^2\rangle/\langle\tau^2\rangle$ are twice continuously differentiable in $\tau \in \mathbb{W}$, $\tau \neq 0$. The function $R(\tau;\mu)$ is a linear combination of the two indicated standard numbers, the second with the coefficient $\mu^{1/2}/c$ ($c = \text{const} > 0$) analytically depending on $\mu > 0$. As can be easily verified, this guarantees that the required properties of the Rayleigh function $R(\tau;\mu)$ under investigation are satisfied. Differentiating the composite function $R(\psi_\mu;\mu) = \Lambda(\mu)$ and using the identity $R'(\psi_\mu;\mu)w = 0$, $w \in \mathbb{W}$, we find

$$\frac{d\Lambda}{d\mu} = R'(\psi_\mu;\mu)\frac{d\psi_\mu}{d\mu} + \frac{\partial R}{\partial\mu}(\psi_\mu;\mu) = \frac{\partial R}{\partial\mu}(\psi_\mu;\mu) = \frac{1}{c}\frac{\langle\gamma\psi_\mu'^2\rangle}{2\langle\psi_\mu^2\rangle}\frac{1}{\mu^{1/2}} = \frac{\langle\gamma\psi_\mu'^2\rangle}{2c}\frac{1}{\mu^{1/2}}.$$

Using the obtained identity, we establish the validity of the inequality

$$\frac{d}{d\mu}(\mu - \mu_*(\mu)) > 0 \quad \text{for } \mu - \mu_*(\mu) = 0, \quad \mu > 0,$$

obviously providing the required uniqueness of the positive solution of the equation

$$\mu - \mu_*(\mu) = 0.$$

For $\gamma = \gamma(y) = \gamma_0(1-y^2)$, from (4) we have ($k_\mu = \mu^{1/2}\gamma/c$)

$$\frac{d\Lambda}{d\mu} = \frac{\langle\gamma\psi_\mu'^2\rangle}{2c}\frac{1}{\mu^{1/2}} = \frac{\langle k_\mu\psi_\mu'^2\rangle}{2\mu} \leqslant \frac{\langle(1+k_\mu)\psi_\mu'^2\rangle}{2\mu} = \frac{\Lambda}{2\mu}.$$

Simultaneously,

$$\Lambda = \mu_* - \alpha = \mu - \alpha \leqslant \mu.$$

Hence,

$$\frac{d}{d\mu}(\mu - \mu_*(\mu)) = 1 - \frac{d\Lambda}{d\mu} \geqslant 1 - \frac{1}{2} > 0.$$

The uniqueness of a positive solution of equation (16) for the case (4), (6) is established.

16.4. The existence of a minimizing element. We pass to the proof of the fact that the sequence $\tau_i \in M$, $i = 1, 2, \ldots$, (on which the greatest lower bound of the functional $E(\tau)$, extended by continuity to $\mathbb{W}$ and initially defined on the elements of the set M (dense in $\mathbb{W}$) by equation (12), is attained) is fundamental. We recall that this fact is equivalent to the existence of the required element $\tau_* \in \mathbb{W}$ (limit for the indicated sequence in $\mathbb{W}$) such that $E(\tau_*) = d$. Note that the existence of the sequence $\tau_i \in M$,

$i = 1, 2, \ldots$, can be established by means of the theorems given in **[22,** Theorem 8.1]. To make the presentation more complete, we give an alternative proof.

Note that for any fixed $\mu > 0$ the functional $E(\tau)$ (as it follows from (12)) can be represented as the sum of two functions on $\mathbb{W}$,

$$E(\tau) = f(\tau) + g(\tau),$$

$$f(\tau) = \frac{1}{2}\langle(1 + k_\mu)\tau'^2\rangle + \frac{\mu}{3}\langle|\tau|^3/\varphi_\mu\rangle, \qquad g(\tau) = -\frac{\mu}{2}\langle\tau|\tau|\rangle + \frac{\alpha}{2}\langle\tau^2\rangle,$$

the first of which, $f(\tau)$, is convex downwards (in the usual sense, as in §10.2), and the second, $g(\tau)$, is strong continuous in the sense of §13.3. The former is obvious, and the latter immediately follows from the explicit formula for $g(\tau)$ and the complete continuity of the Poincaré-Steklov embedding $\langle\tau'^2\rangle/\langle\tau^2\rangle \geqslant \pi^2$ (here we use the corollary from Lemma 3 from §13.3). In what follows, we additionally use the reflexivity criterion (mentioned in §10.3) applied here to the Hilbert space $\mathbb{W}$ and the Mazur lemma **[31, 73]** according to which from the elements of any sequence $\{\tau_j\} \subset \mathbb{W}$ weakly converging to some element $\tau_* \in \mathbb{W}$, we can form a sequence of convex combinations of the form

$$\widetilde{\tau}_k = \sum_{j=1}^{k} \lambda_{kj}\tau_j, \qquad \lambda_{kj} \geqslant 0, \quad \sum_{j=1}^{k} \lambda_{kj} = 1, \quad k = 1, 2, \ldots,$$

converging to τ_* in $\mathbb{W}$ (i.e., converging "in the norm", $\|\tau_* - \widetilde{\tau}_k\| \to 0$, $k \to \infty$). An elementary proof of the formulated lemma in the case of a Hilbert space is given in **[84]**. The existence of the minimizing element follows from the following lemma.

LEMMA 1. *Let $\mathbb{B}$ be a reflexive Banach space over the field of real numbers and let $f = f(\tau)$ and $g = g(\tau)$ be continuous real functions defined on $\mathbb{B}$ such that f is convex downwards and g is strong continuous. If their sum $E = f + g$ is coercive (in the sense that for any sequence $\{x_i\} \subset \mathbb{B}$ $(i = 1, 2, \ldots)$ such that $\|x_i\| \to \infty$ the sequence of numbers $E(x_i) \to +\infty$) and is bounded from below on $\mathbb{B}$, then E attains its minimal value on some element of the space $\mathbb{B}$.*

PROOF. As a corollary of the assumed boundedness from below of the sum E, the latter has a minimizing sequence $\{x_i\} \subset \mathbb{B}$ such that

$$E(x_i) \to d = \inf\{E(x) : x \in \mathbb{B}\}.$$

As an immediate corollary of the fact that the sum under consideration is coercive, the minimizing sequence is bounded. Then it follows from the reflexivity of $\mathbb{B}$ that there exists a subsequence of the sequence $\{x_i\}$ weakly converging in $\mathbb{B}$ to some element $x_* \subset \mathbb{B}$. Moreover, the strong continuity of the function g implies that some subsequence of the sequence of the images $\{g(x_k)\}$ converges to the value $g(x_*)$. Therefore, this sequence $\{x_i\}$ is identified without loss of generality with its subsequence satisfying the following conditions:

$$E(x_k) \leqslant d + 1/k, \qquad d = \inf\{E(x) : x \in \mathbb{B}\},$$
$$|g(x_*) - g(x_k)| \leqslant 1/k, \qquad k = 1, 2, \ldots.$$

Further, for any given $m = 1, 2, \ldots$, the sequence $\{x_{m+n}\}$ weakly converges to x_*. Then, according to the Mazur lemma, a sequence of convex combinations of the form

$$\widetilde{x}_s^{(m)} = \sum_{p=1}^{s} \lambda_{sp}^{(m)} x_{m+p}, \qquad \lambda_{sp}^{(m)} \geqslant 0, \quad \sum_{p=1}^{s} \lambda_{sp}^{(m)} = 1, \quad s = 1, 2, \ldots,$$

corresponding to a fixed m converges to x_* in $\mathbb{B}$ as $s \to \infty$. It is obvious that in this case for every $m = 1, 2, \ldots$ there is a number $s = s_m \in \{1, 2, \ldots\}$ such that the sequence $\widetilde{x}_{s_m}^{(m)}$, $m = 1, 2, \ldots$, converges to x_* in $\mathbb{B}$ as $m \to \infty$. Then the convexity of f implies that

$$f(\widetilde{x}_{s_m}^{(m)}) \leqslant \sum_{p=1}^{s_m} \lambda_{s_m p}^{(m)} f(x_{m+p}), \qquad m = 1, 2, \ldots.$$

Hence

$$\begin{aligned} E(\widetilde{x}_{s_m}^{(m)}) &\leqslant \sum_{p=1}^{s_m} \lambda_{s_m p}^{(m)} f(x_{m+p}) + g(\widetilde{x}_{s_m}^{(m)}) \\ &= \sum_{p=1}^{s_m} \lambda_{s_m p}^{(m)} E(x_{m+p}) + \sum_{p=1}^{s_m} \lambda_{s_m p}^{(m)} (g(x_*) - g(x_{m+p})) + g(\widetilde{x}_{s_m}^{(m)}) - g(x_*) \\ &\leqslant \sum_{p=1}^{s_m} \lambda_{s_m p}^{(m)} \left(d + \frac{1}{m+p}\right) + \sum_{p=1}^{s_m} \lambda_{s_m p}^{(m)} \frac{1}{m+p} + |g(x_*) - g(\widetilde{x}_{s_m}^{(m)})| \\ &\leqslant d + 1/m + 1/m + \varepsilon_m, \end{aligned}$$

where

$$\varepsilon_m = |g(x_*) - g(\widetilde{x}_{s_m}^{(m)})|.$$

Setting

$$\delta_m = |E(x_*) - E(\widetilde{x}_{s_m}^{(m)})|,$$

and using the obtained inequality, we find that

$$E(x_*) = E(x_*) - E(\widetilde{x}_{s_m}^{(m)}) + E(\widetilde{x}_{s_m}^{(m)}) \leqslant d + 2/m + \varepsilon_m + \delta_m.$$

The convergence of $\{\widetilde{x}_{s_m}^{(m)}\}$ to x_* and the continuity of E and g imply that $\varepsilon_m, \delta_m \to 0$ as $m \to \infty$. Passing to the limit as $m \to \infty$ in the last inequality, we find $E(x_*) \leqslant d$; this immediately implies $E(x_*) = d$ (because simultaneously $E(x_*) \geqslant d$), and completes the proof of Lemma 1.

To complete the proof of Theorem 1, it remains to use the lemma and to note that the coercitivity of the functional E from (12) straightforwardly follows from the inequalities (14).

APPENDIX

Formal Constructions Connected with Euler Equations

§17. Continuous models of some Lagrangian systems

17.1. Notation. The constructions given below are mainly of local character and pertain to a connected neighborhood of a fixed interior point of some compact smooth manifold V of (finite) dimension $n \geqslant 2$, equipped with a symmetric covariant tensor field $g_{\alpha\beta}$, $\alpha, \beta = 1, \dots, n$, of rank 2, generating a nondegenerate Riemannian metric on V. The system of local coordinates $x^1, \dots, x^n$ on V is considered as defining a varying point x of the manifold V. The components of the metric tensor and other functions under consideration are assumed to be sufficiently smooth (e.g., of class C^2). The differentials $dx^1, \dots, dx^n$ are identified with the elements of a linear basis of the cotangent space **[91]**. The latter is assumed to be equipped with the structure of a Grassmann algebra over the field of real numbers, defined by a bilinear associative operation of *exterior multiplication* $\wedge$ such that

$$dx^\alpha \wedge dx^\beta = -dx^\beta \wedge dx^\alpha \quad \text{and} \quad \varphi \wedge dx^\alpha = dx^\alpha \wedge \varphi = \varphi\, dx^\alpha,$$

if φ is a scalar. The abbreviated notation

$$dx^{\alpha_1 \dots \alpha_p} \equiv dx^{\alpha_1} \wedge \dots \wedge dx^{\alpha_p}$$

is used for the elements of an arbitrary geometric basis of the cotangent space which have the form

$$dx^{\alpha_1} \wedge \dots \wedge dx^{\alpha_p}, \qquad 1 \leqslant \alpha_1 < \dots < \alpha_p \leqslant n.$$

Summation (from 1 to n) is assumed with respect to the repeated index if it is not stipulated otherwise or is not obvious from the context. The overlined neighboring indices

$$\alpha_1 \dots \overline{\alpha_{p_1} \dots \alpha_{p_2} \dots \alpha_{p_s}} \dots \alpha_p$$

are taken in the increasing order

$$\alpha_{p_1} < \dots < \alpha_{p_2} < \dots < \alpha_{p_s}.$$

For example, each component $\omega_{\alpha_1 \dots \alpha_p}$ of a covariant skew-symmetric tensor field of rank p given on V can be expressed via one of its $\frac{n!}{p!(n-p)!}$ basis components $\omega_{\overline{\alpha_1 \dots \alpha_p}}$ by the formula

$$\omega_{\alpha_1 \dots \alpha_p} = \varepsilon_{\alpha_1 \dots \alpha_p} \omega_{\overline{\alpha_1 \dots \alpha_p}},$$

where

$$\varepsilon_{\alpha_1\dots\alpha_p} \equiv \delta^{\overline{\alpha_1\dots\alpha_p}}_{\alpha_1\dots\alpha_p}$$

$$\equiv \begin{cases} 1, & \text{if the transposition } \alpha_1\dots\alpha_p \to \overline{\alpha_1\dots\alpha_p} \text{ is even,} \\ -1, & \text{if the transposition } \alpha_1\dots\alpha_p \to \overline{\alpha_1\dots\alpha_p} \text{ is odd,} \\ 0, & \text{if some of the numbers } \alpha_1,\dots,\alpha_p \text{ coincide,} \end{cases}$$

is the Kronecker symbol.

A homogeneous linear combination $\overset{p}{\omega}$ of the basis elements $dx^{\overline{\alpha_1\dots\alpha_p}}$ with the coefficients $\omega_{\overline{\alpha_1\dots\alpha_p}}$ having the form

$$\overset{p}{\omega} \equiv \omega_{\overline{\alpha_1\dots\alpha_p}}\, dx^{\overline{\alpha_1\dots\alpha_p}} = \frac{1}{p!}\omega_{\alpha_1\dots\alpha_p}\, dx^{\alpha_1\dots\alpha_p},$$

is invariant with respect to the change of the system of local coordinates (from the one without the prime to the one with it),

$$\begin{aligned} \omega_{\alpha_1\dots\alpha_p}\, dx^{\alpha_1\dots\alpha_p} &= \omega_{\alpha'_1\dots\alpha'_p}\frac{\partial x^{\alpha'_1}}{\partial x^{\alpha_1}}\cdots\frac{\partial x^{\alpha'_p}}{\partial x^{\alpha_p}}\frac{\partial x^{\alpha_1}}{\partial x^{\beta'_1}}\cdots\frac{\partial x^{\alpha_p}}{\partial x^{\beta'_p}}\, dx^{\beta'_1\dots\beta'_p} \\ &= \omega_{\alpha'_1\dots\alpha'_p}\frac{\partial x^{\alpha'_1}}{\partial x^{\alpha_1}}\frac{\partial x^{\alpha_1}}{\partial x^{\beta'_1}}\cdots\frac{\partial x^{\alpha'_p}}{\partial x^{\alpha_p}}\frac{\partial x^{\alpha_p}}{\partial x^{\beta'_p}}\, dx^{\beta'_1\dots\beta'_p} \\ &= \omega_{\alpha'_1\dots\alpha'_p}\delta^{\alpha'_1}_{\beta'_1}\dots\delta^{\alpha'_p}_{\beta'_p}\, dx^{\beta'_1\dots\beta'_p} \\ &= \omega_{\alpha'_1\dots\alpha'_p}\, dx^{\alpha'_1\dots\alpha'_p}; \end{aligned}$$

(here we use the identity

$$dx^{\alpha_1\dots\alpha_p} = \frac{\partial x^{\alpha_1}}{\partial x^{\beta'_1}}\cdots\frac{\partial x^{\alpha_p}}{\partial x^{\beta'_p}}\, dx^{\beta'_1\dots\beta'_p}).$$

After passing to arbitrary local coordinates, it defines some *p-form* on V (an exterior homogeneous even differential form of order p **[32, 37, 83]**). The set $\wedge^p = \wedge^p(V)$ of such forms $\overset{p}{\omega}$ is, obviously, linear. By adjoining to the sets $\wedge^1,\dots,\wedge^n$ the linear manifold of scalars $\wedge^0$, and by completing the obtained set of *scalars* $\wedge^0$, *covector fields* $\wedge^1,\dots,$ "*volume elements*" $\wedge^n$ by the trivial linear spaces

$$\dots,\wedge^{-2},\wedge^{-1},\wedge^{n+1},\wedge^{n+2},\dots = \{0\}$$

(coinciding with the zero element of the cotangent space), we can extend the operation $\wedge$ to all possible pairs of forms $\overset{p}{\omega}$, $\overset{q}{\chi}$ as acting from the tensor product $\wedge^p \otimes \wedge^q$ of the spaces $\wedge^p$ and $\wedge^q$ to $\wedge^{p+q}$ according to the rule

$$\begin{aligned} \overset{p}{\omega} \wedge \overset{q}{\chi} &\equiv \omega_{\overline{\alpha_1\dots\alpha_p}}\chi_{\overline{\beta_1\dots\beta_q}}\, dx^{\overline{\alpha_1\dots\alpha_p}} \wedge dx^{\overline{\beta_1\dots\beta_q}} \\ &= \frac{1}{p!q!}\omega_{\alpha_1\dots\alpha_p}\chi_{\beta_1\dots\beta_q}\, dx^{\alpha_1\dots\alpha_p\beta_1\dots\beta_q}. \end{aligned}$$

Along with $\wedge$, we shall use the operation of *exterior differentiation* d acting from $\wedge^p$ to $\wedge^{p+1}$ as

$$d\left(\omega_{\alpha_1\dots\alpha_p}\, dx^{\alpha_1\dots\alpha_p}\right) \equiv \left(D_\beta\omega_{\alpha_1\dots\alpha_p}\right) dx^{\beta\alpha_1\dots\alpha_p}, \qquad D_\beta \equiv \frac{\partial}{\partial x^\beta},$$

and the operation of *metric conjugation* $*$ acting from $\wedge^p$ to $\wedge^{n-p}$ as

$$* \left(\omega_{\overline{\alpha_1 \dots \alpha_p}}\, dx^{\overline{\alpha_1 \dots \alpha_p}}\right) \equiv h\varepsilon_{\overline{\alpha_1 \dots \alpha_p \beta_1 \dots \beta_{n-p}}}\, \omega^{\overline{\alpha_1 \dots \alpha_p}}\, dx^{\overline{\beta_1 \dots \beta_{n-p}}},$$
$$h = |g|^{1/2}, \qquad g = \det(g_{\alpha\beta})$$

(g is the determinant of the metric tensor). Here

$$\omega^{\alpha_1 \dots \alpha_p} \equiv g^{\alpha_1 \sigma_1} \dots g^{\alpha_p \sigma_p} \omega_{\sigma_1 \dots \sigma_p}$$

is the contravariant component of the form $\overset{p}{\omega}$.

Repeated action of the operation d results in zero,

$$\begin{aligned} dd\overset{p}{\omega} &= \left(D_\beta D_\sigma \omega_{\overline{\alpha_1 \dots \alpha_p}}\right) dx^{\beta\sigma\overline{\alpha_1 \dots \alpha_p}} \\ &= \left(D_\beta D_\sigma \omega_{\overline{\alpha_1 \dots \alpha_p}} - D_\sigma D_\beta \omega_{\overline{\alpha_1 \dots \alpha_p}}\right) dx^{\overline{\beta\sigma}\,\overline{\alpha_1 \dots \alpha_p}} = 0. \end{aligned}$$

Repeated action of $*$ reduces to the multiplication by a constant,

$$** \overset{p}{\omega} = \operatorname{sgn}(g)(-1)^{p(n-p)} \overset{p}{\omega}.$$

Indeed,

$$\begin{aligned} ** \overset{p}{\omega} &= *(*\overset{p}{\omega}) = h\varepsilon_{\overline{\sigma_1 \dots \sigma_{n-p} \tau_1 \dots \tau_p}} (*\overset{p}{\omega})^{\overline{\alpha_1 \dots \alpha_{n-p}}}\, dx^{\overline{\tau_1 \dots \tau_p}} \\ &= \frac{h\varepsilon_{\sigma_1 \dots \sigma_{n-p} \overline{\tau_1 \dots \tau_p}}}{(n-p)!} g^{\sigma_1 \beta_1} \dots g^{\sigma_{n-p} \beta_{n-p}} (*\overset{p}{\omega})_{\beta_1 \dots \beta_{n-p}}\, dx^{\overline{\tau_1 \dots \tau_p}} \\ &= \frac{h^2 \varepsilon_{\sigma_1 \dots \sigma_{n-p} \overline{\tau_1 \dots \tau_p}}}{p!(n-p)!} g^{\sigma_1 \beta_1} \dots g^{\sigma_{n-p} \beta_{n-p}} \varepsilon_{\alpha_1 \dots \alpha_p \beta_1 \dots \beta_{n-p}} \\ &\qquad \times g^{\alpha_1 \gamma_1} \dots g^{\alpha_p \gamma_p} \omega_{\gamma_1 \dots \gamma_p}\, dx^{\overline{\tau_1 \dots \tau_p}}. \end{aligned}$$

Since

$$\varepsilon_{\sigma_1 \dots \sigma_{n-p} \tau_1 \dots \tau_p} = (-1)^{n-p} \varepsilon_{\tau_1 \sigma_1 \dots \sigma_{n-p} \tau_2 \dots \tau_p} = \dots = (-1)^{p(n-p)} \varepsilon_{\tau_1 \dots \tau_p \sigma_1 \dots \sigma_{n-p}},$$

we find (additionally using the symmetry of the metric tensor) that

$$\begin{aligned} ** \overset{p}{\omega} &= \frac{h^2 (-1)^{p(n-p)}}{p!(n-p)!} g^{\alpha_1 \gamma_1} \dots g^{\alpha_p \gamma_p} g^{\beta_1 \sigma_1} \dots g^{\beta_{n-p} \sigma_{n-p}} \varepsilon_{\alpha_1 \dots \alpha_p \beta_1 \dots \beta_{n-p}} \\ &\quad \times \varepsilon_{\overline{\tau_1 \dots \tau_p} \sigma_1 \dots \sigma_{n-p}} \omega_{\gamma_1 \dots \gamma_p}\, dx^{\overline{\tau_1 \dots \tau_p}}. \end{aligned}$$

Since (obviously)

$$g^{\nu_1 \lambda_1} \dots g^{\nu_n \lambda_n} \varepsilon_{\nu_1 \dots \nu_n} = \det(g^{\alpha\beta}) \varepsilon_{\lambda_1 \dots \lambda_n}, \qquad \det(g^{\alpha\beta}) = 1/g,$$

taking into account $h^2 = |g|$, we obtain that

$$** \overset{p}{\omega} = \frac{|g|(-1)^{p(n-p)}}{g\,p!(n-p)!} \varepsilon_{\gamma_1 \dots \gamma_p \sigma_1 \dots \sigma_{n-p}} \varepsilon_{\overline{\tau_1 \dots \tau_p} \sigma_1 \dots \sigma_{n-p}} \omega_{\gamma_1 \dots \gamma_p}\, dx^{\overline{\tau_1 \dots \tau_p}}.$$

Finally, using the identity

$$\begin{aligned} \frac{1}{(n-p)!} \varepsilon_{\overline{\gamma_1 \dots \gamma_p} \sigma_1 \dots \sigma_{n-p}} \varepsilon_{\overline{\tau_1 \dots \tau_p} \sigma_1 \dots \sigma_{n-p}} &= \delta^{\overline{\gamma_1 \dots \gamma_p}}_{\tau_1 \dots \tau_p} \\ &= \begin{cases} 1, & \text{if } (\gamma_1, \dots, \gamma_p) = (\tau_1, \dots, \tau_p), \\ 0, & \text{if } (\gamma_1, \dots, \gamma_p) \neq (\tau_1, \dots, \tau_p), \end{cases} \end{aligned}$$

we conclude that

$$**\overset{p}{\omega} = \operatorname{sgn}(g)(-1)^{p(n-p)}\omega_{\overline{\tau_1\dots\tau_p}}\,dx^{\tau_1\dots\tau_p},$$

q.e.d.

Moreover, the operations $\wedge$, d, $*$ have the following properties:

1) $\overset{p}{\omega}\wedge\overset{q}{\chi} = (-1)^{pq}\overset{q}{\chi}\wedge\overset{p}{\omega}$. Indeed,

$$\begin{aligned}\overset{p}{\omega}\wedge\overset{q}{\chi} &= \omega_{\overline{\alpha_1\dots\alpha_p}}\chi_{\overline{\beta_1\dots\beta_q}}\,dx^{\overline{\alpha_1\dots\alpha_p}\overline{\beta_1\dots\beta_q}}\\ &= (-1)^p\omega_{\overline{\alpha_1\dots\alpha_p}}\chi_{\overline{\beta_1\dots\beta_q}}\,dx^{\beta_1\overline{\alpha_1\dots\alpha_p}\beta_2\dots\beta_q} = \dots\\ &= (-1)^{pq}\chi_{\overline{\beta_1\dots\beta_q}}\omega_{\overline{\alpha_1\dots\alpha_p}}\,dx^{\overline{\beta_1\dots\beta_q}\overline{\alpha_1\dots\alpha_p}} = (-1)^{pq}\overset{q}{\chi}\wedge\overset{p}{\omega}.\end{aligned}$$

2) $d(\overset{p}{\omega}\wedge\overset{q}{\chi}) = d\overset{p}{\omega}\wedge\overset{q}{\chi} + (-1)^p\overset{p}{\omega}\wedge d\overset{q}{\chi}$. Indeed,

$$\begin{aligned}d(\overset{p}{\omega}\wedge\overset{q}{\chi}) &= \left(D_\gamma\left(\omega_{\overline{\alpha_1\dots\alpha_p}}\chi_{\overline{\beta_1\dots\beta_q}}\right)\right)dx^{\gamma\overline{\alpha_1\dots\alpha_p}\overline{\beta_1\dots\beta_q}}\\ &= \left(D_\gamma\omega_{\overline{\alpha_1\dots\alpha_p}}\right)\chi_{\overline{\beta_1\dots\beta_q}}\,dx^{\gamma\overline{\alpha_1\dots\alpha_p}}\wedge dx^{\overline{\beta_1\dots\beta_p}}\\ &\quad+\omega_{\overline{\alpha_1\dots\alpha_p}}D_\gamma\left(\chi_{\overline{\beta_1\dots\beta_q}}\right)(-1)^p\,dx^{\overline{\alpha_1\dots\alpha_p}}\wedge dx^{\gamma\overline{\beta_1\dots\beta_q}}\\ &= d\overset{p}{\omega}\wedge\overset{q}{\chi} + (-1)^p\overset{p}{\omega}\wedge d\overset{q}{\chi}.\end{aligned}$$

3) $\overset{p}{\omega}\wedge *\overset{p}{\chi} = \overset{p}{\chi}\wedge *\overset{p}{\omega}$. Indeed,

$$\begin{aligned}\overset{p}{\omega}\wedge *\overset{p}{\chi} &= \omega_{\overline{\alpha_1\dots\alpha_p}}\,dx^{\overline{\alpha_1\dots\alpha_p}}\wedge h\varepsilon_{\overline{\beta_1\dots\beta_p}\overline{\gamma_1\dots\gamma_{n-p}}}\chi^{\overline{\beta_1\dots\beta_p}}\,dx^{\overline{\gamma_1\dots\gamma_{n-p}}}\\ &= \omega_{\overline{\alpha_1\dots\alpha_p}}\chi^{\overline{\beta_1\dots\beta_p}}\frac{1}{(n-p)!}\varepsilon_{\overline{\beta_1\dots\beta_p}\gamma_1\dots\gamma_{n-p}}\varepsilon_{\overline{\alpha_1\dots\alpha_p}\gamma_1\dots\gamma_{n-p}}h\,dx^{1\dots n}\\ &= \omega_{\overline{\alpha_1\dots\alpha_p}}\chi^{\overline{\alpha_1\dots\alpha_p}}*1 = \chi_{\overline{\alpha_1\dots\alpha_p}}\omega^{\overline{\alpha_1\dots\alpha_p}}*1 = \overset{p}{\chi}\wedge *\overset{p}{\omega}.\end{aligned}$$

Here $*1 = h dx^{1\dots n} = dV$ is the *volume element* on V.

Identity 3) allows us to introduce in $\wedge^p$ the structure of a Hilbert (if the metric is definite, or if the metric tensor is positive definite) or (otherwise) a pseudo-Hilbert space defined by the scalar (pseudoscalar, respectively) product of the following form:

$$(\overset{p}{\omega},\overset{p}{\chi}) \equiv \int_V \overset{p}{\omega}\wedge *\overset{p}{\chi} = \int_V \overset{p}{\chi}\wedge *\overset{p}{\omega}.$$

In this case, introducing the operation δ formally adjoint to $-d$ as acting from $\wedge^p$ to $\wedge^{p-1}$ and satisfying the identity

$$(-d\overset{p-1}{\omega},\overset{p}{\chi}) = (\overset{p-1}{\omega},\delta\overset{p}{\chi}),$$

for any form $\overset{p-1}{\omega}$ vanishing on the boundary ∂V of the manifold V (if the latter is not closed) and for any form $\overset{p}{\chi}$, we find an explicit formula for δ,

$$\delta\overset{p}{\chi} = \operatorname{sgn}(g)(-1)^{(p-1)(n-p)}*d*\overset{p}{\chi}.$$

Indeed, according to 2), we have

$$d(\overset{p-1}{\omega}\wedge *\overset{p}{\chi}) = d\overset{p-1}{\omega}\wedge *\overset{p}{\chi} + (-1)^{p-1}\overset{p-1}{\omega}\wedge d*\overset{p}{\chi}$$

and (since $d * \overset{p}{\chi} \in \wedge^{n-p+1}$)

$$* * d * \overset{p}{\chi} = \operatorname{sgn}(g)(-1)^{(p-1)(n-p+1)} d * \overset{p}{\chi},$$

which, together with the property 3) above and the identity

$$\int_V d(\overset{p-1}{\omega} \wedge * \overset{p}{\chi}) = 0$$

(the Stokes theorem and the condition introduced above that the form $\overset{p-1}{\omega}$ vanishes on the boundary of V if V is nonclosed), leads to the indicated formula for δ.

Note the connection of these operations with the classical operators of vector analysis. The operation δ on 1-forms $u = u_1 dx^1 + u_2 dx^2 + u_3 dx^3$ in three-dimensional Euclidean space coincides with the divergence

$$\begin{aligned} \delta u &= *d * (u_1 dx^1 + u_2 dx^2 + u_3 dx^3) \\ &= *d(u^1 dx^{23} + u^2 dx^{31} + u^3 dx^{12}) \\ &= D_1 u^1 + D_2 u^2 + D_3 u^3 \equiv \operatorname{div} \mathbf{u}. \end{aligned}$$

Here $\mathbf{u} = (u^1, u^2, u^3)$ is the contravariant vector field with the components $u^\alpha = u_\alpha$ (the fact that the metric is Euclidean is used here) associated to u. The identity defining the operation δ takes the form of the well-known relation

$$(-\nabla\varphi, \mathbf{u}) = (\varphi, \operatorname{div} \mathbf{u})$$

"conjugating " the gradient $\nabla\varphi = \operatorname{grad}\varphi$ of a scalar function φ (taking zero values on the boundary of the domain of definition) with the divergence $\operatorname{div}\mathbf{u}$ of the vector field $\mathbf{u}$. The identity

$$\Delta \overset{p}{\omega} = d\delta \overset{p}{\omega} + \delta d \overset{p}{\omega}$$

corresponds to and unites the other two equalities

$$\begin{gathered} \operatorname{div}\operatorname{grad}\varphi = (D_1^2 + D_2^2 + D_3^2)\varphi, \\ \operatorname{grad}\operatorname{div}\mathbf{u} - \operatorname{rot}\operatorname{rot}\mathbf{u} = (D_1^2 + D_2^2 + D_3^2)\mathbf{u}; \end{gathered}$$

this identity defines a multidimensional analog Δ (from $\wedge^p$ to $\wedge^p$) of the Laplace operator $D_1^2 + D_2^2 + D_3^2$. The symmetric bilinear form

$$\langle \overset{p}{\omega}, \overset{p}{\chi} \rangle \equiv \operatorname{sgn}(g) * (\overset{p}{\omega} \wedge * \overset{p}{\chi}) = \langle \overset{p}{\chi}, \overset{p}{\omega} \rangle$$

corresponds to the scalar product of vectors of Euclidean space.

17.2. Natural motions and the Euler equations. According to **[36]**, a *flow* on the manifold V is identified with a certain covector field u on V (belonging to the space $\wedge^1$). A *path* on V is defined as a smooth image of an interval I of the real line R; a *motion of a volume* (a domain) Q of the manifold V is a smooth mapping of the direct product of the interval I and the volume Q to V. The motion is defined as consisting of the paths

$$z(x) = \{z_t(x) : \text{ time } t \text{ runs over the interval } I\}$$

defined according to $u(t, x)$ (generally speaking, nonstationary) by the trajectories (integral curves) of the problem

$$\text{(1)} \qquad \frac{d}{dt} z_t(x) = \mathbf{u}(t, z_t(x)) \quad \text{in } I, \qquad z_{t_0}(x) = x \quad \text{everywhere in } Q.$$

Here t_0 is a fixed (initial) moment of time from the interval I,

$$\mathbf{u} = (u^1, \dots, u^n), \qquad u^\alpha = g^{\alpha\beta} u_\beta,$$

is the contravariant field of the flow $u = u_1\, dx^1 + \dots + u_n\, dx^n$.

We identify Q with a fixed coordinate neighborhood Q_x of the variable $x = (x^1, \dots, x^n)$. Under the assumption of smoothness of $\mathbf{u}(t, x)$, for a sufficiently small I and any t from I the motion of the volume Q in V modeled by the mapping $x \to z_t(x)$ is a diffeomorphism of Q onto the new coordinate neighborhood

$$Q_{z_t(x)} = \{z_t(x) : \ x \text{ runs over } Q\}$$

with the variable $z_t(x) = (z_t^1(x), \dots, z_t^n(x))$ **[12]**. Introducing a scalar field $\rho(t, x) > 0$ from $\wedge^0$ (possibly, nonstationary), we require that the *element of mass* $*\rho$ (from $\wedge^n$) distributed with the density $\rho = \rho(t, x)$ be preserved on the trajectories from (1),

$$(2) \qquad (*\rho)_{z_t(x)} = (*\rho)_x \quad \text{for all } t \text{ from } I \qquad (z_{t_0}(x) = x).$$

Introducing another scalar function $w(t, x, \rho)$ (depending, generally speaking, on the first), we require that for any sufficiently small $\tau > 0$ (such that $t_0 + \tau$ lies in I) the trajectories of the motion $z = z(x) = z_t(x)$ defined by (1) realize the extremals of the action

$$S(z) = \int_{t_0}^{t_0+\tau} L(t, Q_{z_t})\, dt$$

on the interval $t_0 < t < t_0 + \tau$, generated on the moving volume $Q_z = Q_{z_t(x)}$ by a direct continual analog $L(t, Q_z)$ of the *Lagrangian of a natural mechanical system* of the form

$$L(t, Q) \equiv \frac{1}{2} \int_Q \rho u \wedge *u - \int_Q *\rho w,$$

i.e., defined by the difference of the kinetic and potential energies of the moving volume. The scalar function w defines the distribution density of the potential energy of the medium generated, in particular, by the interaction of the constituent particles (dependence of w on ρ). The concretization of the above requirement (the stationary action principle) results in the following condition. For any smooth motion $\zeta = \zeta_t(x)$, $t_0 \leqslant t \leqslant t_0 + \tau$, of the volume Q with a fixed point x_0 and a prescribed direction at this point defined by a nonzero vector $\mathbf{a}$ such that

$$(2') \qquad \left.\frac{d}{dt}\zeta_t(x_0)\right|_{t=t_0} = \mathbf{a}, \qquad \zeta_{t_0}(x_0) = \zeta_{t_0+\tau}(x_0) = 0,$$

the variation of the action S on the motion $z = z_t(x)$,

$$\delta_\zeta S(z) \equiv \lim_{\varepsilon \to 0} \frac{S(z + \varepsilon\zeta) - S(z)}{\varepsilon}$$

exists for any (arbitrarily small) volume Q containing the fixed point x_0 and any (arbitrarily small) $\tau > 0$ and vanishes on the indicated motion z. Moreover, it is additionally required that the prescribed motion $z + \varepsilon\zeta$ obey requirement (2) uniformly in t, x, up to $o(\varepsilon)$, i.e.,

$$(2'') \qquad (*\rho)_{z_t(x)+\varepsilon\zeta_t(x)} - (*\rho)_{z_t(x)} = (o(\varepsilon))(*1)_{z_t(x)}.$$

The motions $z_t(x)$ satisfying identity (2) for some smooth mass distribution density $\rho(t, x) > 0$ and realizing, for a given distribution density of the potential energy

$w(t, x, \rho)$, the extremals of the action S of the natural mechanical system are called (according to the common usage) *natural*. Simultaneously, we shall consider the *Euler flows* which, again following **[36]**, we define as smooth covector fields $u(t, x)$ satisfying, for some smooth scalars $\rho(t, x) > 0$ and $\gamma(t, x)$, the following invariant relations (*Dezin's relations* **[36]**):

$$\frac{\partial}{\partial t} * \rho + d * \rho u = 0, \tag{3}$$

$$\frac{\partial}{\partial t} u + *(\omega \wedge u) + d\gamma = 0, \tag{4}$$

where

$$\omega = \operatorname{sgn}(g) * du \tag{5}$$

is the *curl* (or *vorticity*) of the field u. The scalars ρ and γ are called the *density* and the *Bernoulli integral corresponding to* u, respectively. As it will be seen below, flows u with Bernoulli integral of the form

$$\gamma = \gamma_w = \frac{1}{2}\langle u, u\rangle + w \qquad (\langle u, u\rangle \equiv \operatorname{sgn}(g) * (u \wedge *u)) \tag{6}$$

are covector fields of natural flows. It will be shown that the corresponding converse statement is also valid. First, we clarify the connection of these relations with the well-known Euler equations.

Equalities (3), (4), and (5) are invariant forms of the following relations used in hydrodynamics: the *continuity equation*

$$\frac{\partial \rho}{\partial t} + \operatorname{div}(\rho \mathbf{u}) = \mathbf{0},$$

the *Gromeka-Lamb form* of the dynamical Euler equations

$$\frac{\partial \mathbf{u}}{\partial t} + \boldsymbol{\omega} \times \mathbf{u} + \nabla \gamma = 0$$

(the symbol $\times$ is the vector multiplication), and the standard definition of curl $\boldsymbol{\omega}$

$$\boldsymbol{\omega} = \operatorname{rot} \mathbf{u} = \nabla \times \mathbf{u};$$

in the case of the Euclidean metric, they lead to the indicated relations. Identity (6) generalizes the well-known formula for the classical Bernoulli integral

$$\gamma = |\mathbf{u}|^2/2 + \varphi(t, x) + \int \frac{dp}{\rho}, \qquad |\mathbf{u}| = (\mathbf{u} \cdot \mathbf{u})^{1/2}$$

(the dot $\cdot$ is the scalar product) used in the case when the *pressure* in a fluid, $p = p(t, x)$, is connected with the density ρ by a given invertible dependence of the form $\rho = \rho(p)$ (*barotropic fluid or gas*). Using the identity

$$(\mathbf{u} \cdot \nabla)\mathbf{u} = \operatorname{rot} \mathbf{u} \times \mathbf{u} + \nabla(|\mathbf{u}|^2/2),$$

we come to the dynamical Euler equations,

$$\frac{\partial \mathbf{u}}{\partial t} + (\mathbf{u} \cdot \nabla)\mathbf{u} + \frac{1}{\rho}\nabla p + \nabla \varphi = 0.$$

The scalar $\varphi = \varphi(t, x)$ is the *specific* (i.e., divided by the density ρ) density of a given potential field of external mass forces. The density w of the potential energy of the corresponding natural motion is given by the dependence

$$w = \varphi(t, x) + \int \frac{dp}{\rho}.$$

Nonpotential mass forces are absent.

17.3. The stationary action principle. Let us formulate the statement concerning relations (3)–(5) which is of primary interest to us.

THEOREM 1. *The covector field $\mathbf{u}(t, x)$ of a natural motion with the density of the potential energy $w(t, x, \rho)$ is an Euler flow (satisfies (3)–(5)) with the same mass density $\rho(t, x)$ as for the given natural motion, and the Bernoulli integral γ defined via $w = w(t, x, \rho)$ by means of (6). Conversely, the motion defined (by means of (1)) by a given Euler flow $\mathbf{u}(t, x)$ is natural with the same density $\rho(t, x)$ as for $\mathbf{u}(t, x)$ and a new density $w = w(t, x)$ defined according to the given $\gamma = \gamma(t, x)$ again by means of (6).*

We pass to the proof of Theorem 1. Assuming that the given motion

$$z \equiv z_t \equiv z_t(x)$$

is natural, we establish that equality (3) is valid for the covector field $\boldsymbol{u}$ generated by it. To this end, by setting $\lambda \equiv \rho h$ ($h = |g|^{1/2}$) and passing to the corresponding moving local coordinates $z_t^\alpha(x)$, we find that

$$\lambda(t_0 + \tau, z_{t_0+\tau}(x)) = \lambda(t_0, x) + \tau \frac{d}{dt}\lambda(t, z_t(x))\Big|_{t=t_0} + o(\tau),$$

$$\frac{d}{dt}\lambda(t, z_t(x))\Big|_{t=t_0} = \frac{\partial \lambda}{\partial t}(t_0, x) + u^\alpha D_\alpha \lambda,$$

$$\lambda = \lambda(t_0, x), \qquad u^\alpha = u^\alpha(t_0, x) = \frac{d}{dt} z_t^\alpha(x)\Big|_{t=t_0}, \qquad D_\alpha = \frac{\partial}{\partial x^\alpha}.$$

Simultaneously,

$$\begin{aligned} dz_{t+\tau}^{1\dots n} &= (dx^1 + \tau du^1) \wedge \dots \wedge (dx^n + \tau du^n) + (o(\tau))\, dx^{1\dots n} \\ &= (1 + \tau D_\alpha u^\alpha + o(\tau))\, dx^{1\dots n}. \end{aligned}$$

Hence,

$$\begin{aligned} (*\rho)_{z_{t_0+\tau}} &= \lambda(t_0 + \tau, z_{t_0+\tau}(x))\, dz_{t_0+\tau}^{1\dots n} \\ &= \lambda(t_0, x)\, dx^{1\dots n} + \tau \left(\frac{\partial \lambda}{\partial t}(t_0, x) + D_\alpha(\lambda u^\alpha) \right) dx^{1\dots n} + (o(\tau))\, dx^{1\dots n}. \end{aligned}$$

Since, simultaneously,

$$\begin{aligned} \frac{\partial \lambda}{\partial t}\, dx^{1\dots n} &= \frac{\partial}{\partial t}(\rho h\, dx^{1\dots n}) = \frac{\partial}{\partial t}(*\rho)_x \quad (\lambda = \rho h), \\ D_\alpha(\lambda u^\alpha)\, dx^{1\dots n} &= \left(D_\sigma (h\rho u^\alpha \varepsilon_{\overline{\alpha\beta_1\dots\beta_{n-1}}}) \right) dx^{\overline{\sigma\beta_1\dots\beta_{n-1}}} \\ &= d(h\rho \varepsilon_{\overline{\alpha\beta_1\dots\beta_{n-1}}} u^\alpha\, dx^{\overline{\beta_1\dots\beta_{n-1}}}) \\ &= (d * \rho \boldsymbol{u})_x, \end{aligned}$$

we conclude that

$$\frac{d}{dt}(*\rho)_{z_t(x)}\bigg|_{t=t_0} = \left(\frac{\partial}{\partial t} * \rho + d * \rho \boldsymbol{u}\right)_x$$

(the index x means that the invariant expression supplied by it is considered in the corresponding coordinate neighborhood). Substituting in the relation obtained $z_{t_0}(x)$ for x and taking into account the arbitrariness of the choice of the initial point t_0 of the interval I, we find that for any t', t'' from this interval the identity

$$(*\rho)_{z_{t'}(x)} - (*\rho)_{z_{t''}(x)} = \int_{t''}^{t'} \left(\frac{\partial}{\partial t} * \rho + d * \rho \boldsymbol{u}\right)_{z_t(x)} dt,$$

ensuring the required equivalence of (2) and (3), is valid.

Now we obtain relation (4). Using the abbreviated notation

$$f[z_t] \equiv f_{z_t} \equiv f(t, z_t), \qquad z_t \equiv z_t(x), \qquad w[z_t] \equiv w(t, z_t, \rho[z_t]),$$
$$L[z_t] \equiv L(t, Q_{z_t}), \qquad l \equiv \frac{1}{2} u_\alpha u^\alpha - w,$$

by means of (2) we find

$$L[z_t] = \int_{Q_{z_t}} l[z_t]\rho_{z_t}\, dV_{z_t} = \int_{Q_x} l[z_t(x)]\rho_x\, dV_x \qquad (x = z_{t_0}(x)),$$

$$\Delta L \equiv L[z_t + \varepsilon\zeta_t] - L[z_t] = \int_{Q_x} (\Delta l)\rho_x\, dV_x,$$

$$\begin{aligned}
\Delta l &\equiv l[z_t + \varepsilon\zeta_t] - l[z_t] \\
&= \frac{1}{2} u_\alpha[z_t + \varepsilon\zeta_t] u^\alpha[z_t + \varepsilon\zeta_t] - \frac{1}{2} u_\alpha[z_t] u^\alpha[z_t] - w[z_t + \varepsilon\zeta_t] + w[z_t] \\
&= \frac{1}{2}(u_\alpha[z_t + \varepsilon\zeta_t] - u_\alpha[z_t]) u^\alpha[z_t + \varepsilon\zeta_t] + \frac{1}{2} u_\alpha[z_t](u^\alpha[z_t + \varepsilon\zeta_t] - u^\alpha[z_t]) \\
&\quad - w[z_t + \varepsilon\zeta_t] + w[z_t].
\end{aligned}$$

According to (1),

$$u^\alpha[z_t] = \frac{d}{dt} z_t^\alpha.$$

Simultaneously, $u_\alpha = g_{\alpha\beta} u^\beta$. Hence,

$$u^\alpha[z_t + \varepsilon\zeta_t] - u^\alpha[z_t] = \varepsilon \frac{d}{dt} \zeta_t^\alpha,$$

$$\begin{aligned}
u_\alpha[z_t + \varepsilon\zeta_t] - u_\alpha[z_t] &= g_{\alpha\beta}[z_t + \varepsilon\zeta_t] \frac{d}{dt}(z_t^\beta + \varepsilon\zeta_t^\beta) - g_{\alpha\beta}[z_t] \frac{d}{dt} z_t^\beta \\
&= (g_{\alpha\beta}[z_t + \varepsilon\zeta_t] - g_{\alpha\beta}[z_t]) \frac{d}{dt} z_t^\beta + \varepsilon g_{\alpha\beta}[z_t + \varepsilon\zeta_t] \frac{d}{dt} \zeta_t^\beta \\
&= \varepsilon \left(u^\beta[z_t] \zeta_t^\sigma D_\sigma g_{\alpha\beta}[z_t] + g_{\alpha\beta}[z_t] \frac{d}{dt} \zeta_t^\beta \right) + o(\varepsilon),
\end{aligned}$$

$$\begin{aligned}
\Delta l &= \frac{1}{2}\left[\varepsilon \left(u^\beta[z_t] \zeta_t^\sigma D_\sigma g_{\alpha\beta}[z_t] + g_{\alpha\beta}[z_t] \frac{d}{dt} \zeta_t^\beta \right) + o(\varepsilon) \right] u^\alpha[z_t + \varepsilon\zeta_t] \\
&\quad + \frac{\varepsilon}{2} u_\alpha[z_t] \frac{d}{dt} \zeta_t^\alpha - w[z_t + \varepsilon\zeta_t] + w[z_t] \\
&= \varepsilon \delta_{\zeta_t} l[z_t] + o(\varepsilon),
\end{aligned}$$

where

$$\delta_{\zeta_t} l[z_t] \equiv \frac{1}{2}\left(u^\beta \zeta_t^\sigma D_\sigma g_{\alpha\beta} + g_{\alpha\beta}\frac{d}{dt}\zeta_t^\beta\right) u^\alpha + \frac{1}{2} u_\alpha \frac{d}{dt}\zeta_t^\alpha - \zeta_t^\sigma D_\sigma w.$$

Taking into account the equalities

$$g_{\alpha\beta} u^\alpha \frac{d}{dt}\zeta_t^\beta = u_\beta \frac{d}{dt}\zeta_t^\beta = u_\alpha \frac{d}{dt}\zeta_t^\alpha = \frac{d}{dt}(u_\alpha \zeta_t^\alpha) - \zeta_t^\alpha \frac{d}{dt} u_\alpha,$$
$$\frac{d}{dt} u_\alpha[z_t] = \frac{\partial}{\partial t} u_\alpha + u^\beta D_\beta u_\alpha,$$

we find that

$$\delta_{\zeta_t} l[z_t] = \frac{d}{dt}(u_\alpha \zeta_t^\alpha) - \zeta_t^\alpha \left(\frac{\partial}{\partial t} u_\alpha + u^\beta D_\beta u_\alpha - \frac{1}{2} u^\beta u^\sigma D_\alpha g_{\beta\sigma} + D_\alpha w\right).$$

Taking into account (2″), we obtain

$$\begin{aligned} S(z+\varepsilon\zeta) - S(z) &= \int_{t_0}^{t_0+\tau}\left[\int_{Q_{z_t}+\varepsilon\zeta_t} l[z_t+\varepsilon\zeta_t](*\rho)_{z_t+\varepsilon\zeta_t} - \int_{Q_{z_t}} l[z_t](*\rho)_{z_t}\right] dt \\ &= \int_{t_0}^{t_0+\tau}\left[\int_{Q_x} l[z_t+\varepsilon z_t](*\rho)_x - \int_{Q_x} l[z_t](*\rho)_x\right] dt + o(\varepsilon) \\ &= \int_{t_0}^{t_0+\tau}\left[\int_{Q_x} (\Delta l)(*\rho)_x\right] dt + o(\varepsilon) \\ &= \int_{Q_x}\left[\int_{t_0}^{t_0+\tau} (\Delta l)\, dt\right](*\rho)_x + o(\varepsilon) \\ &= \varepsilon \int_{Q_x}\left[\int_{t_0}^{t_0+\tau} \delta_{\zeta_t} l[z_t]\, dt\right](*\rho)_x + o(\varepsilon). \end{aligned}$$

Hence,

$$\begin{aligned} \delta_\zeta S(z) &= \int_{Q_x}\left[\int_{t_0}^{t_0+\tau} \delta_{\zeta_t} l[z_t]\, dt\right](*\rho)_x \\ &= \int_{Q_x}\left[u_\alpha \zeta_t^\alpha \Big|_{t_0}^{t_0+\tau}\right](*\rho)_x - \int_{t_0}^{t_0+\tau}\left[\int_{Q_x} f_\alpha \zeta_t^\alpha (*\rho)_x\right] dt, \end{aligned}$$

where

$$f_\alpha = \frac{\partial}{\partial t} u_\alpha + u^\beta D_\beta u_\alpha - \frac{1}{2} u^\beta u^\sigma D_\alpha g_{\beta\sigma} + D_\alpha w.$$

With (2′) taken into account, we see that the validity, for any arbitrarily small $\tau > 0$ and any volume Q_x containing the fixed point x_0, of the equality $\delta_\zeta S(z) = 0$ (the *stationary action principle*) implies $f_\alpha(t_0, x_0)a^\alpha = 0$ (the expansion

$$\zeta_t(x_0^\alpha) = (t - t_0)a^\alpha + o(|t - t_0|)$$

is used here). The latter equation for an arbitrary nonzero vector $\mathbf{a} = (a^1, \dots, a^n)$ is, obviously, equivalent to the equality

$$f_\alpha(t_0, x_0) = 0 \qquad (\alpha = 1, \dots, n).$$

Taking into account the arbitrariness of the choice of the initial moment t_0 in the interval I and the point x_0 on the manifold V, we come to the equations

$$\text{(7)}\qquad \begin{gathered}\frac{d}{dt}u_\alpha - \frac{1}{2}u^\beta u^\sigma D_\alpha g_{\beta\sigma} + D_\alpha w = 0, \qquad \frac{d}{dt} \equiv \frac{\partial}{\partial t} + u^\beta D_\beta \\ \left(D_\alpha \equiv \frac{\partial}{\partial x^\alpha}, \qquad \alpha = 1, \dots, n\right).\end{gathered}$$

Conversely, (7) implies the equality $f_\alpha(t_0, x_0) = 0$, and the latter ensures the validity of the stationary action principle $\delta_\zeta S(z) = 0$.

Relations (7) represent the dynamical Euler equations in an arbitrary system of local (curvilinear) coordinates $x^1, \dots, x^n$. Our goal with the subsequent constructions is to rewrite equations (7) in the required invariant form (4). We have

$$u^\beta u^\sigma D_\alpha g_{\beta\sigma} = D_\alpha(g_{\beta\sigma}u^\beta u^\sigma) - g_{\beta\sigma}u^\beta D_\alpha u^\sigma - g_{\beta\sigma}u^\sigma D_\alpha u^\beta,$$

or, since

$$g_{\beta\sigma}u^\beta u^\sigma = u_\beta u^\beta = \operatorname{sgn}(g) * (\boldsymbol{u} \wedge *\boldsymbol{u}) \equiv \langle \boldsymbol{u}, \boldsymbol{u} \rangle \qquad \text{and} \qquad g_{\beta\sigma} = g_{\sigma\beta},$$

we find that

$$\begin{aligned}\frac{1}{2}u^\beta u^\sigma D_\alpha g_{\beta\sigma} &= \frac{1}{2}D_\alpha(u_\beta u^\beta) - u_\beta D_\alpha u^\beta \\ &= D_\alpha\left(\frac{1}{2}u_\beta u^\beta - u_\beta u^\beta\right) + u^\beta D_\alpha u_\beta = -D_\alpha \frac{\langle \boldsymbol{u}, \boldsymbol{u}\rangle}{2} + u^\beta D_\alpha u_\beta.\end{aligned}$$

This allows us to rewrite (7) in the following form:

$$\text{(8)}\qquad \frac{\partial}{\partial t}u_\alpha + u^\beta(D_\beta u_\alpha - D_\alpha u_\beta) + D_\alpha \gamma = 0,$$

where the scalar $\gamma = \gamma_w$ is defined as in (6). Now we show that (8) is the coordinate form of the invariant relation (4).

In the terms of covariants

$$\begin{gathered}\mathcal{F}_\alpha \equiv \frac{\partial}{\partial t}u_\alpha + D_\alpha \gamma \quad \text{and} \quad (d\boldsymbol{u})_{\alpha\beta} = D_\alpha u_\beta - D_\beta u_\alpha \\ (d\boldsymbol{u} = du_\beta \wedge dx^\beta = (D_\alpha u_\beta)\, dx^{\alpha\beta} = (D_\alpha u_\beta - D_\beta u_\alpha)\, dx^{\overline{\alpha\beta}}),\end{gathered}$$

we rewrite (8) in the form

$$u^\beta (d\boldsymbol{u})_{\alpha\beta} + \mathcal{F}_\alpha = 0.$$

By lifting in the obtained (tensor) equality the indices (by means of contraction with $g^{\sigma\alpha}$, i.e., by multiplying by $g^{\sigma\alpha}$ and summing over α),

$$g^{\sigma\alpha}\left(u^\beta(d\boldsymbol{u})_{\alpha\beta} + \mathcal{F}_\alpha\right) = u_\varkappa g^{\varkappa\beta} g^{\sigma\alpha}(d\boldsymbol{u})_{\alpha\beta} + g^{\sigma\alpha}\mathcal{F}_\alpha = u_\varkappa (d\boldsymbol{u})^{\varkappa\sigma} + \mathcal{F}^\sigma,$$

we come to the equations

$$\text{(9)}\qquad u_\beta (d\boldsymbol{u})^{\beta\alpha} + \mathcal{F}^\alpha = 0$$

equivalent to (8). We show that (9) is the coordinate form of the invariant relation

$$\text{(10)}\qquad (-1)^{n-1}(*d\boldsymbol{u}) \wedge \boldsymbol{u} + *\mathfrak{F} = 0,$$

where $\mathfrak{F} = \mathcal{F}_\alpha\, dx^\alpha$. To this end, by rewriting the first summand of equality (10) in the coordinate form, we find

$$
\begin{aligned}
(-1)^{n-1}(*d\boldsymbol{u}) \wedge \boldsymbol{u} &= (-1)^{n-1}(*d(u_\lambda\, dx^\lambda)) \wedge u_\nu\, dx^\nu \\
&= (-1)^{n-1}(*((d\boldsymbol{u})_{\overline{\alpha\lambda}}\, dx^{\overline{\alpha\lambda}})) \wedge u_\nu\, dx^\nu \\
&= (-1)^{n-1}(h\varepsilon_{\overline{\alpha\lambda\gamma_1\cdots\gamma_{n-2}}}(d\boldsymbol{u})^{\overline{\alpha\lambda}}\, dx^{\overline{\gamma_1\cdots\gamma_{n-2}}}) \wedge u_\nu\, dx^\nu \\
&= \frac{1}{2}(-1)^{n-1}h\varepsilon_{\alpha\lambda\overline{\gamma_1\cdots\gamma_{n-2}}}u_\nu(d\boldsymbol{u})^{\alpha\lambda}\, dx^{\overline{\gamma_1\cdots\gamma_{n-2}\nu}} \\
&= \frac{h}{2}u_\nu(d\boldsymbol{u})^{\alpha\lambda}\varepsilon_{\lambda\overline{\gamma_1\cdots\gamma_{n-2}}\alpha}\, dx^{\overline{\gamma_1\cdots\gamma_{n-2}\nu}} \\
&= \frac{h}{2}u_\nu(d\boldsymbol{u})^{\lambda\alpha}\varepsilon_{\alpha\overline{\gamma_1\cdots\gamma_{n-2}}\lambda}\, dx^{\overline{\gamma_1\cdots\gamma_{n-2}\nu}}.
\end{aligned}
$$

Summing over ν for fixed remaining indices, we come to the equality

$$
\begin{aligned}
&\frac{h}{2}\sum_{\nu=1}^{n} u_\nu(d\boldsymbol{u})^{\lambda\alpha}\varepsilon_{\alpha\overline{\gamma_1\cdots\gamma_{n-2}}\lambda}\, dx^{\overline{\gamma_1\cdots\gamma_{n-2}\nu}} \\
&\quad = \frac{h}{2}u_\alpha(d\boldsymbol{u})^{\lambda\alpha}\varepsilon_{\alpha\overline{\gamma_1\cdots\gamma_{n-2}}\lambda}\, dx^{\overline{\gamma_1\cdots\gamma_{n-2}\alpha}} + \frac{h}{2}u_\lambda(d\boldsymbol{u})^{\lambda\alpha}\varepsilon_{\alpha\overline{\gamma_1\cdots\gamma_{n-2}}\lambda} dx^{\overline{\gamma_1\cdots\gamma_{n-2}\lambda}},
\end{aligned}
\tag{11}
$$

i.e., only the summands with $\nu = \alpha, \lambda$ are left of the whole sum, the others vanish.

Let us consider the latter summands in more detail. For $n = 2$, we have

$$\varepsilon_{\alpha\overline{\gamma_1\cdots\gamma_{n-2}}\lambda} = \varepsilon_{\alpha\lambda} \qquad (n = 2)$$

and, since in this case the indices α, λ can take only two values (1 or 2), equality (11) in this situation is an identity. For $n \geqslant 3$ and $\nu \neq \alpha, \lambda$, the coefficient

$$\varepsilon_{\alpha\overline{\gamma_1\cdots\gamma_{n-2}}\lambda}$$

(with n pairwise distinct indices) is different from 0 only if the group of indices $\overline{\gamma_1 \dots \gamma_{n-2}}$ contains the index ν. However, then we have

$$dx^{\overline{\gamma_1\cdots\gamma_{n-2}\nu}} = 0,$$

which again implies that the corresponding summand on the left-hand side of equality (11) vanishes; this proves the validity of (11) in the general case.

Further, by summing the right-hand side of equality (11) over the indices α, λ,

$$
\begin{aligned}
&\sum_{\alpha,\lambda=1}^{n}\left(\frac{h}{2}u_\alpha(d\boldsymbol{u})^{\lambda\alpha}\varepsilon_{\alpha\overline{\gamma_1\cdots\gamma_{n-2}}\lambda}dx^{\overline{\gamma_1\cdots\gamma_{n-2}\alpha}} + \frac{h}{2}u_\lambda(d\boldsymbol{u})^{\lambda\alpha}\varepsilon_{\alpha\overline{\gamma_1\cdots\gamma_{n-2}}\lambda}dx^{\overline{\gamma_1\cdots\gamma_{n-2}\lambda}}\right) \\
&\qquad = h\sum_{\alpha,\lambda=1}^{n}\left(u_\lambda(d\boldsymbol{u})^{\lambda\alpha}\varepsilon_{\alpha\overline{\gamma_1\cdots\gamma_{n-2}}\lambda}dx^{\overline{\gamma_1\cdots\gamma_{n-2}\lambda}}\right)
\end{aligned}
$$

(the skew-symmetry of the contravariant $(d\boldsymbol{u})^{\lambda\alpha}$, $(d\boldsymbol{u})^{\lambda\alpha} = -(d\boldsymbol{u})^{\alpha\lambda}$, and of the Kronecker symbol $\varepsilon_{\alpha\overline{\gamma_1\cdots\gamma_{n-2}}\lambda}$,

$$\varepsilon_{\alpha\overline{\gamma_1\cdots\gamma_{n-2}}\lambda} = -\varepsilon_{\lambda\overline{\gamma_1\cdots\gamma_{n-2}}\alpha},$$

is used here), we find that

$$(-1)^{n-1}(*d\boldsymbol{u}) \wedge \boldsymbol{u} = hu_\lambda(d\boldsymbol{u})^{\lambda\alpha}\varepsilon_{\alpha\overline{\gamma_1\cdots\gamma_{n-2}}\lambda}dx^{\overline{\gamma_1\cdots\gamma_{n-2}\lambda}};$$

the coordinate form of equality (1), after dividing by $h > 0$, takes the form

$$(12) \qquad u_\lambda (d\boldsymbol{u})^{\lambda\alpha} \varepsilon_{\alpha\overline{\gamma_1 \ldots \gamma_{n-2}\lambda}} dx^{\overline{\gamma_1 \ldots \gamma_{n-2}\lambda}} + \mathcal{F}^\alpha \varepsilon_{\overline{\alpha\sigma_1 \ldots \sigma_{n-2}\beta}} dx^{\overline{\sigma_1 \ldots \sigma_{n-2}\beta}} = 0$$

(summation is carried over the repeated indices). With respect to the set of $n!/(1!(n-1)!) = n$ basis elements $dx^{\overline{\sigma_1 \ldots \sigma_{n-2}\beta}}$ in (12), this relation decomposes into n scalar equations. To distinguish the latter from (12) we fix the set of indices $\sigma_1 < \cdots < \sigma_{n-2} < \beta$ in the second summand in (12) and collect all the coefficients at the corresponding basis element

$$dx^{\overline{\sigma_1 \ldots \sigma_{n-2}\beta}}$$

in the first. We show that the total coefficient obtained at the indicated basis element in the first summand on the left-hand side of equality (12) has the following form:

$$u_\lambda (d\boldsymbol{u})^{\lambda\alpha} \varepsilon_{\overline{\alpha\sigma_1 \ldots \sigma_{n-2}\beta}}.$$

Indeed, this is obvious for $n = 2$. For $n \geqslant 3$, the components of the sum

$$u_\lambda (d\boldsymbol{u})^{\lambda\alpha} \varepsilon_{\alpha\overline{\gamma_1 \ldots \gamma_{n-2}\lambda}} dx^{\overline{\gamma_1 \ldots \gamma_{n-2}\lambda}}$$

defining the first summand from (12) and taken over $\gamma_1, \ldots, \gamma_{n-2}, \lambda$ for a fixed α, containing the set of indices $\sigma_1 < \cdots < \sigma_{n-2} < \beta$ excluding α are, obviously, the terms

$$u_{\sigma_1} (d\boldsymbol{u})^{\sigma_1\alpha} \varepsilon_{\overline{\alpha\sigma_1 \ldots \sigma_{n-2}\beta}} dx^{\overline{\sigma_1 \ldots \sigma_{n-2}\beta}},$$

$$\cdots\cdots\cdots\cdots\cdots$$

$$u_{\sigma_{n-2}} (d\boldsymbol{u})^{\sigma_{n-2}\alpha} \varepsilon_{\overline{\alpha\sigma_1 \ldots \sigma_{n-2}\beta}} dx^{\overline{\sigma_1 \ldots \sigma_{n-2}\beta}},$$

$$u_\beta (d\boldsymbol{u})^{\beta\alpha} \varepsilon_{\overline{\alpha\sigma_1 \ldots \sigma_{n-2}\beta}} dx^{\overline{\sigma_1 \ldots \sigma_{n-2}\beta}}$$

(or, reduced by an appropriate transposition of the indices to the required form,

$$\varepsilon_{\alpha\beta\sigma_2 \ldots \sigma_{n-2}\sigma_1} dx^{\beta\sigma_2 \ldots \sigma_{n-2}\sigma_1} = \varepsilon_{\overline{\alpha\sigma_1 \ldots \sigma_{n-2}\beta}} dx^{\overline{\sigma_1 \ldots \sigma_{n-2}\beta}},$$

$$\cdots\cdots\cdots\cdots\cdots\cdots$$

$$\varepsilon_{\alpha\sigma_1 \ldots \sigma_{n-3}\beta\sigma_{n-2}} dx^{\beta\sigma_2 \ldots \sigma_{n-3}\beta\sigma_{n-2}} = \varepsilon_{\overline{\alpha\sigma_1 \ldots \sigma_{n-2}\beta}} dx^{\overline{\sigma_1 \ldots \sigma_{n-2}\beta}}).$$

Completing the sum of the indicated terms by the zero summand

$$u_\alpha (d\boldsymbol{u})^{\alpha\alpha} \varepsilon_{\alpha\overline{\sigma_1 \ldots \sigma_{n-2}\alpha}} dx^{\overline{\sigma_1 \ldots \sigma_{n-2}\beta}} = 0,$$

we find (taking into account that a set of indices $\alpha, \sigma_1, \ldots, \sigma_{n-2}, \beta$ such that

$$\varepsilon_{\alpha\sigma_1 \ldots \sigma_{n-2}\beta} \neq 0$$

can only be a transposition of the numbers $1, \ldots, n$) that the required total coefficient at the basis element

$$dx^{\overline{\sigma_1 \ldots \sigma_{n-2}\beta}}$$

in the first summand from (12) is

$$\left(u_1 (d\boldsymbol{u})^{1\alpha} + \cdots + u_n (d\boldsymbol{u})^{n\alpha}\right) \varepsilon_{\overline{\alpha\sigma_1 \ldots \sigma_{n-2}\beta}}.$$

Here the index α takes only one value, uniquely defined by the given set of indices

$$\sigma_1, \ldots, \sigma_{n-2}, \beta.$$

Equation (12), which is equivalent to (10), is rewritten as

$$\left(u_\lambda (d\mathbf{u})^{\lambda\alpha} + \mathcal{F}^\alpha\right) \varepsilon_{\overline{\alpha\sigma_1\dots\sigma_{n-2}\beta}} dx^{\overline{\sigma_1\dots\sigma_{n-2}\beta}} = 0$$

and can be easily reduced to the required form (9).

To complete the proof, it remains to act on both sides of equality (10) by the operator $(-1)^{n-1}\operatorname{sgn}(g)*$ and obtain, taking into account the identities

$$\operatorname{sgn}(g)(-1)^{n-1} * *\mathcal{F} = \mathcal{F} = \frac{\partial}{\partial t}\mathbf{u} + d\gamma,$$

relation (4).

17.4. Translational movements and motions along geodesics. A motion $z_t(x)$ defined by a solution of problem (1) is called the *translational movement* (or the *translation*) if its contravariant field $\mathbf{u}(t, z_t(x))$ does not depend on the initial value of x, i.e.,

(14) $\quad \mathbf{u}(t, z_t(x')) = \mathbf{u}(t, z_t(x''))$ for all x', x'' from Q and t from I.

Since in this case

$$\frac{\partial z_t^\sigma(x)}{\partial x^\beta} \frac{\partial u^\alpha}{\partial z_t^\sigma(x)}(t, z_t(x)) = \frac{\partial u^\alpha}{\partial x^\beta}(t, z_t(x)) = 0$$

and

$$\det\left(\frac{\partial z_t^\sigma(x)}{\partial x^\beta}\right) \neq 0$$

(the change of coordinates $x \to z_t(x)$, $z_{t_0}(x) = x$ is a diffeomorphism), the contravariant field $\mathbf{u}(t, z_t(x))$ of a translational motion does not depend on $z_t(x)$ and, hence, does not depend on arbitrary local coordinates on $Q_{z_t(x)}$, i.e., at each moment t of motion all the points of the indicated volume have the same velocity $\mathbf{u} = \mathbf{u}(t)$ (not depending on the spatial variable),

$$D_\beta u^\alpha = 0 \qquad (\alpha, \beta = 1, \dots, n).$$

Then relation (3) defines the integral of motion $\lambda = \rho h$,

$$\frac{\partial\lambda}{\partial t} + D_\alpha(\lambda u^\alpha) = \frac{\partial\lambda}{\partial t} + u^\alpha D_\alpha \lambda = 0,$$

and the density ρ is uniquely defined by its fixed initial value $\rho(t_0, x)$,

(15) $$\rho(t, z_t(x)) = \frac{h(t_0, z)}{h(t, z_t(x))}\rho(t_0, x).$$

Moreover, relation (4) rewritten in the form (7) leads in this case to the equations

$$\frac{d}{dt}u_\alpha - \frac{1}{2}u^\beta u^\sigma D_\alpha g_{\beta\sigma} + D_\alpha w = 0 \qquad (\alpha = 1, \dots, n).$$

For $w = w(t)$ and $\partial g_{\alpha\beta}/\partial t = 0$, the latter expression gives an equivalent form of the equations for geodesics **[40]**. Namely, we have

THEOREM 2. *In the case of a translational movement, relation* (4) *provides an invariant form of equations* (16) *coinciding, in the absence of external force field* ($w = w(t)$) *and explicit dependence of the metric on time* ($\partial g_{\alpha\beta}/\partial t = 0$), *with the equations for geodesics. Formula* (15) *follows from relation* (3).

Indeed, by setting in the indicated equation $u_\alpha = g_{\alpha\varkappa}u^\varkappa$ and contracting the obtained equality with $g^{\theta\alpha}$, we obtain

$$\begin{aligned} g^{\theta\alpha}\Big(\frac{d}{dt}(g_{\alpha\varkappa}u^\varkappa) - \frac{1}{2}u^\varkappa u^\sigma D_\alpha g_{\beta\sigma} + D_\alpha w\Big) \\ = g^{\theta\alpha}g_{\alpha\varkappa}\frac{d}{dt}u^\beta + \left(g^{\theta\alpha}\frac{\partial}{\partial t}g_{\alpha\varkappa}\right)u^\varkappa + g^{\theta\alpha}u^\varkappa u^\lambda D_\lambda g_{\alpha\varkappa} \\ - \frac{1}{2}g^{\theta\alpha}(D_\alpha g_{\beta\sigma})u^\beta u^\sigma + g^{\theta\alpha}D_\alpha w. \end{aligned}$$

Taking into account the equalities

$$g^{\theta\alpha}g_{\alpha\varkappa} = \delta^\theta_\varkappa,$$

we have

$$\begin{aligned} u^\varkappa u^\lambda D_\lambda g_{\alpha\varkappa} = u^\lambda u^\varkappa D_\varkappa g_{\alpha\lambda} = \frac{1}{2}u^\varkappa u^\lambda D_\lambda g_{\alpha\varkappa} + \frac{1}{2}u^\lambda u^\varkappa D_\varkappa g_{\alpha\lambda} \\ = \frac{1}{2}u^\beta u^\sigma D_\beta g_{\alpha\sigma} + \frac{1}{2}u^\beta u^\sigma D_\sigma g_{\alpha\beta}, \end{aligned}$$

and the equations of motion take the standard form

$$\frac{du^\theta}{dt} + \Gamma^\theta_{\beta\sigma}u^\beta u^\sigma + \left(g^{\theta\alpha}\frac{\partial}{\partial t}g_{\alpha\varkappa}\right)u^\varkappa + g^{\theta\alpha}D_\alpha w = 0 \qquad (\theta = 1, \dots, n), \tag{16}$$

where

$$\Gamma^\theta_{\beta\sigma} \equiv \frac{1}{2}g^{\theta\alpha}\left(D_\beta g_{\alpha\sigma} + D_\sigma g_{\alpha\beta} - D_\alpha g_{\beta\sigma}\right)$$

is the *Christoffel symbol*. For $w = w(t)$ and $g_{\alpha\beta} = g_{\alpha\beta}(x)$ equations (16) coincide with the classical equations for the geodesics.

Equality (15) was obtained above. The theorem is completely proved.

Therefore, *the motions generated by equations* (3), (4) *form a certain physically admissible extension of the class of motions along the geodesics such that the mass conservation law* (*relation* (3)) *and the stationary action principle* (*relation* (4)), *valid for natural mechanical systems, still hold*. It is natural to consider the motions along geodesics as providing some special (nontrivial **[5, 12]**) class of solutions of Euler equations (4).

17.5. Equations of relativistic hydrodynamics. It can be easily proved by a straightforward verification (returning to previous arguments) that relations (3), (4) do not exclude the presence of an explicit dependence of the metric on time. Taking this dependence into account in the framework of relativistic mechanics, we shall come to new equations of motion (relations (19), (21)) considered on a Riemannian manifold M, with the point $x = (x^0, x^1, \dots, x^n)$, whose dimension exceeds the dimension n of the initial manifold V by 1. The varying action (20) corresponds to the "relativistic" action in the theory of relativity. The "physical" interpretation of the equations under consideration consists in identifying the covector field $\boldsymbol{v} = \boldsymbol{v}(x) = v_k\, dx^k$ (belonging to the space $\wedge^1 = \wedge^1(M)$) defined by these equations, with a flow of some "relativistic" ideal fluid in the gravitational field defined by the metric of the manifold M. It is additionally assumed that

$$\langle \boldsymbol{v}, \boldsymbol{v} \rangle \equiv v_k v^k > 0.$$

As it will be shown below, in the class of translational motions satisfying the normalization condition $\langle \boldsymbol{v}, \boldsymbol{v} \rangle = 1$, the corresponding relation (21) provides an invariant form of the equations for the geodesics on M. For the case of slow motions in weak gravitational fields, this relation passes into equation (4) and relation (19) takes the form of equation (3).

Among the well-known investigations concerning the variational approach to the construction of models of relativistic mechanics of continuous media, we mention **[39, 76, 87, 104]**. The scheme proposed below is close to that given in **[39, §27]**. As in that work, the varying functional is a straightforward continuous analog of the well-known relativistic action. However, an additional constraint $\langle \boldsymbol{v}, \boldsymbol{v} \rangle = 1$ borrowed from mechanics of a material point is imposed in **[39]** on the motion of the continuous medium. The use of this constraint in the continuous case under consideration is not yet justified from the physical point of view because of the following reasons.

As it is known **[40]**, the variation, under the condition $\langle \boldsymbol{v}, \boldsymbol{v} \rangle = 1$, of the relativistic action of a mass point (defined by the integral of the length element over the parameter of motion t) leads to equations of geodesics,

$$(17) \qquad \frac{dv_i}{dt} = \frac{1}{2}\left(D_i g_{jk}\right) v^j v^k \quad \text{or} \quad \frac{dv^i}{dt} + \Gamma^i_{jk} v^j v^k = 0$$

(the Latin indices used here run over the values $1, \dots, n$). Conversely, by contracting the first equation from (17) with v^i and the second (equivalent to the first) with v_i, and summing the obtained equalities we come to the identity $\langle \boldsymbol{v}, \boldsymbol{v} \rangle = \text{const}$ which under a suitable normalization provides the relation $\langle \boldsymbol{v}, \boldsymbol{v} \rangle = 1$. Due to this, the condition $\langle \boldsymbol{v}, \boldsymbol{v} \rangle = 1$ can be considered as characterizing only the geodesic motions on the manifold V. On the other hand, as we have already seen, the motions along geodesics form only a subclass of motions of an ideal fluid. The latter was demonstrated using natural mechanical systems as examples, and will be proved below for relativistic motions of an ideal gas. There are no obvious reasons for carrying the property $\langle \boldsymbol{v}, \boldsymbol{v} \rangle = 1$, inherent to translational motions, over to general motions of a continuous medium. Instead, in the general case it is natural to use the continuous analog of the mass conservation law expressed by relation (19) below (and analogous to (15)). This relation was also given in **[39]**. However, the condition $\langle \boldsymbol{v}, \boldsymbol{v} \rangle = 1$ was also used; this condition, as already noted, is not necessarily satisfied for the motion of a continuous medium. As it follows from the work mentioned above, the additional use of this condition in the general situation under consideration leads to a very "severe" constraint on the choice of the corresponding density $\varkappa$ in equality (19), which not justified from the physical point of view. The condition $\langle \boldsymbol{v}, \boldsymbol{v} \rangle = 1$ is used below only in the class of translations.

Now we turn to the main arguments. Let $z = z_t = z_t(x)$ be a motion on a compact smooth connected $(n+1)$-dimensional Riemannian manifold M, locally defined by a solution of the autonomous system

$$(18) \qquad \frac{dz_t}{dt} = \mathbf{v}(z_t) \quad \text{for } t \text{ from } I \text{ and } z_{t_0} = x \text{ from } \Omega.$$

Here Ω is a fixed "volume" (subdomain) of the manifold M,

$$\mathbf{v} = \mathbf{v}(x) = (v^0, v^1, \dots, v^n)$$

is the contravariant velocity of the stationary (not depending on the parameter of motion) covector field v (belonging to $\wedge^1(M)$) defined on M. The metric of the manifold M and other functions on M are also assumed to be stationary. We require that for v there exists a scalar $\varkappa$ (belonging to $\wedge^0$) such that the equality

$$d * \varkappa v = 0 \tag{19}$$

is satisfied everywhere in M. The varying action will be defined on the tube of motions

$$\{\Omega_z\} \equiv \{z_t(\Omega) : t_0 < t < t_0 + \tau\}$$

by the integral

$$S(z) \equiv -\int_{\{\Omega_z\}} *\varkappa|v|, \qquad |v| \equiv \langle v, v\rangle^{1/2}. \tag{20}$$

The class of motions under consideration is constrained by the requirement that the inequality

$$\langle v, v\rangle \equiv \operatorname{sgn}(g) * (v \wedge *v) = v_k v^k > 0$$

be satisfied everywhere in M. The motion z is called *relativistic* if, in addition to (19), the action (20) is stationary on z, i.e., the equality $\delta_\zeta S(z) = 0$ holds for any motion $\zeta = \zeta_t$ satisfying the requirements formulated above while considering the stationary action principle for natural mechanical systems (the variation of the action is defined as before).

THEOREM 3. *The motion on M generated by a covector field v with $\langle v, v\rangle > 0$, satisfying equation* (19) *for some scalar $\varkappa$, is relativistic if and only if*

$$v \wedge *d\frac{v}{|v|} = 0 \tag{21}$$

everywhere in M.

Indeed, as in the case of a natural motion, relation (19) is equivalent to the identity

$$(*\varkappa)_{z_t(x)} = (*\varkappa)_x \qquad (x = z_{t_0}(x)). \tag{22}$$

As a result of (22), on the moving volume $\Omega_z = z(\Omega)$ (i.e., the image of the volume Ω under the mapping $x \to z = z_t(x)$)

$$-S(z) = \int_{\{\Omega_z\}} |v|(*\varkappa)_z = \int_{\{\Omega_{z(x)}\}} |v(z(x))|(*\varkappa)_x$$

and (due to the arguments used above for natural motions)

$$\delta_\zeta S(z) = -\int_{\{\Omega_{z(x)}\}} (\delta_\zeta |v(z(x))|)\,(*\varkappa)_x.$$

Here (as above)

$$\delta_\zeta |v(z(x))| \equiv \frac{d}{d\varepsilon}|v(z(x) + \varepsilon\zeta(x))|\Big|_{\varepsilon=0}.$$

Further,

$$\begin{aligned}\frac{d}{d\varepsilon}|\boldsymbol{v}(z+\varepsilon\zeta)|\Big|_{\varepsilon=0} &= \frac{d}{d\varepsilon}\langle \boldsymbol{v},\boldsymbol{v}\rangle^{1/2}(z+\varepsilon\zeta)\Big|_{\varepsilon=0}\\ &= \frac{1}{2|\boldsymbol{v}(z)|}\,\frac{d}{d\varepsilon}(g_{ij}v^i v^j)(z+\varepsilon\zeta)\Big|_{\varepsilon=0}\\ &= \frac{1}{2|\boldsymbol{v}(z)|}\left((D_k g_{ij}(z))\,v^i(z)v^j(z)\zeta^k + 2v_k(z)\frac{d\zeta^k}{dt}\right)\end{aligned}$$

(here $D_k \equiv \partial/\partial z_k$). Since

$$\frac{v_k}{|\boldsymbol{v}|}\frac{d\zeta^k}{dt} = \frac{d}{dt}\frac{v_k\zeta^k}{|\boldsymbol{v}|} - \zeta^k\frac{d}{dt}\frac{v_k}{|\boldsymbol{v}|} \qquad\text{and}\qquad \frac{d}{dt}f(z_t) = v^i D_i f,$$

we find, omitting the argument $z = z(x)$ in the integrand,

(23)
$$\begin{aligned}\delta_\zeta S(z) &= -\int_{\{\Omega_z\}}\left(\frac{v^i v^j\zeta^k}{2|\boldsymbol{v}|}D_k g_{ij} + \frac{v_k}{|\boldsymbol{v}|}\frac{d\zeta^k}{dt}\right)(*\varkappa)_x\\ &= \int_{\{\Omega_z\}}\left(v^i D_i\frac{v_k}{|\boldsymbol{v}|} - \frac{v^i v^j}{2|\boldsymbol{v}|}D_k g_{ij}\right)\zeta^k(*\varkappa)_x - \int_{\{\Omega_z\}}\left(v^i D_i\frac{v_k\zeta^k}{|\boldsymbol{v}|}\right)(*\varkappa)_x.\end{aligned}$$

Rewriting relation (19) in the local coordinates $z^0, z^1, \dots, z^n$, we have

(24)
$$D_i(\varkappa H v^i) = 0, \qquad H = |g|^{1/2},$$

(here again $D_i \equiv \partial/\partial z_i$); the mass element takes the form

$$(*\varkappa)_z = \varkappa(z)H(z)\,dz^{01\dots n}, \qquad z = (z^0, z^1, \dots, z^n).$$

Then, using (22) and (24), we reduce the integrand of the second summand from (23) to the form

$$\begin{aligned}\left(v^i D_i\frac{v_k\zeta^k}{|\boldsymbol{v}|}\right)(z(x))(*\varkappa)_x &= \left(v^i D_i\frac{v_k\zeta^k}{|\boldsymbol{v}|}\right)(z)(*\varkappa)_z\\ &= v^i\left(D_i\frac{v_k\zeta^k}{|\boldsymbol{v}|}\right)(z)\varkappa(z)H(z)\,dz^{01\dots n}\\ &= \left(D_i\left(\varkappa H v^i\frac{v_k\zeta^k}{|\boldsymbol{v}|}\right)(z)\right)dz^{01\dots n},\end{aligned}$$

or, in the invariant notation,

$$D_i\left(\varkappa H v^i\frac{v_k\zeta^k}{|\boldsymbol{v}|}\right)(z)\,dz^{01\dots n} = \left(d*\frac{\varkappa\langle\boldsymbol{v},\zeta\rangle}{|\boldsymbol{v}|}\boldsymbol{v}\right)_z.$$

Using the Stokes theorem connected with the passage to integration over the boundary $\partial\{\Omega_z\}$ of the domain $\{\Omega_z\}$, we obtain

$$\int_{\{\Omega_z\}}\left(v^i D_i\frac{v_k\zeta^k}{|\boldsymbol{v}|}\right)(*\varkappa)_x = \int_{\{\Omega_z\}} d*\frac{\varkappa\langle\boldsymbol{v},\zeta\rangle}{|\boldsymbol{v}|}\boldsymbol{v} = \int_{\partial\{\Omega_z\}}\frac{\varkappa\langle\boldsymbol{v},\zeta\rangle}{|\boldsymbol{v}|}*\boldsymbol{v}.$$

The domain under consideration $\{\Omega_z\}$ is the tube of motions or a curvilinear cylinder in a multidimensional space with the boundary $\partial\{\Omega_z\}$ consisting of the "ends" of the cylinder $z_{t_0}(\Omega)$ and $z_{t_0+\tau}(\Omega)$ and the "lateral surface" σ_z complementing them. The equalities $\zeta_{t_0}(x) = \zeta_{t_0+\tau}(x) = 0$ (for all x from Ω) imply

$$\int_{\partial\{\Omega_z\}} \frac{\varkappa\langle v, \zeta\rangle}{|v|} = \int_{\sigma_z} \frac{\varkappa\langle v, \zeta\rangle}{|v|} * v.$$

Parametrizing the surface σ_z by the variables $\varphi^1, \dots, \varphi^n$ such that on σ_z the coordinates $z^i = z^i(\varphi^1, \dots, \varphi^n)$, $i = 0, 1, \dots, n$, we find that $(\alpha_1, \dots, \alpha_n = 1, \dots, n)$

$$\begin{aligned}(*v)|_{\sigma_z} &= H\varepsilon_{ij_1\dots j_n} v^i\, dz^{j_1} \wedge \dots \wedge dz^{j_n} \\ &= \varepsilon_{ij_1\dots j_n} v^i \frac{\partial z^{j_1}}{\partial \varphi^{\alpha_1}} \cdots \frac{\partial z^{j_n}}{\partial \varphi^{\alpha_n}} H\, d\varphi^{\alpha_1} \wedge \dots \wedge d\varphi^{\alpha_n} \\ &= \begin{vmatrix} v^0 & v^1 & \dots & v^n \\ \frac{\partial z^0}{\partial \varphi^{\alpha_1}} & \frac{\partial z^1}{\partial \varphi^{\alpha_1}} & \dots & \frac{\partial z^n}{\partial \varphi^{\alpha_1}} \\ \dots & \dots & \dots & \dots \\ \frac{\partial z^0}{\partial \varphi^{\alpha_n}} & \frac{\partial z^1}{\partial \varphi^{\alpha_n}} & \dots & \frac{\partial z^n}{\partial \varphi^{\alpha_n}} \end{vmatrix} H\, d\varphi^{\alpha_1} \wedge \dots \wedge d\varphi^{\alpha_n}.\end{aligned}$$

Since on σ_z the first row of the determinant is a linear combination of the others,

$$v^i = \frac{dz^i}{dt} = \frac{d\varphi^1}{dt}\frac{\partial z^i}{\partial \varphi^1} + \dots + \frac{d\varphi^n}{dt}\frac{\partial z^i}{\partial \varphi^n}, \qquad i = 0, 1, \dots, n,$$

we conclude that

$$(*v)|_{\sigma_z} = 0,$$

and the second summand on the right-hand side of the last equality in (23) vanishes. Then the equality $\delta_\zeta S(z) = 0$ considered for arbitrary ζ, Ω is equivalent (by the same arguments, used in an analogous situation above) to the relations

$$v^i D_i \frac{v_k}{|v|} - \frac{v^i v^j}{2|v|} D_k g_{ij} = 0, \qquad k = 0, 1, \dots, n. \tag{25}$$

Noting now that

$$\begin{aligned}\frac{v^i v^j}{2|v|} D_k g_{ij} &= \frac{1}{2} D_k \frac{v^i v^j}{2|v|} g_{ij} - \frac{1}{2|v|} g_{ij} v^i D_k v^j - \frac{1}{2|v|} g_{ij} v^j D_k v^i - \frac{1}{2} g_{ij} v^i v^j D_k \frac{1}{|v|} \\ &= \frac{1}{2} D_k \frac{|v|^2}{|v|} - \frac{1}{2|v|} v_j D_k v^j - \frac{1}{2|v|} v_i D_k v^i - \frac{|v|^2}{2} D_k \frac{1}{|v|} \\ &= \frac{1}{2} D_k |v| - \frac{1}{|v|} v_i D_k v^i + \frac{|v|^2}{2} \frac{1}{|v|^2} D_k |v| \\ &= D_k |v| - D_k \frac{1}{|v|} v_i v^i + v^i D_k \frac{1}{|v|} v_i = v^i D_k \frac{1}{|v|} v_i,\end{aligned}$$

we rewrite (25) in the form

$$v^i D_i \frac{v_k}{|v|} - v^i D_k \frac{v_i}{|v|} = 0 \qquad (k = 0, 1, \dots, n), \tag{26}$$

or, in terms of the 2-form

$$d\frac{v}{|v|} = d\left(\frac{v_k}{|v|} dx^k\right) = \left(D_i \frac{v_k}{|v|}\right) dx^{ik} = \left(D_i \frac{v_k}{|v|} - D_k \frac{v_i}{|v|}\right) dx^{\overline{ik}},$$

in the form

$$v^i \left(d \frac{\boldsymbol{v}}{|\boldsymbol{v}|} \right)_{ik} = 0,$$

which is equivalent to

(27) $$v_i \left(d \frac{\boldsymbol{v}}{|\boldsymbol{v}|} \right)^{ik} = 0 \qquad (k = 0, 1, \ldots, n).$$

Using arguments analogous to those following equality (9), we obtain that (27) is the coordinate form of relation (21).

Theorem 3 is established.

As before, a motion $z_t(x)$ (now stationary) with covariant field $\boldsymbol{v} = v_i\, dx^i$ defined on M is called *translational* if the contravariant field $\mathbf{v}(z_t(x))$ of the motion does not depend on the initial position on M, i.e.,

$$\mathbf{v}(z_t(x')) = \mathbf{v}(z_t(x'')) \quad \text{for all } x', x'' \text{ from } M \text{ and } t \text{ from } I.$$

Further, the parameter of motion t is called *proper* if

$$|\boldsymbol{v}(z_t(x))| \equiv \langle \boldsymbol{v}, \boldsymbol{v} \rangle^{1/2} = (\mathbf{v} \cdot \mathbf{v})^{1/2} = v_i v^i = 1.$$

THEOREM 4. *A relativistic translational movement with a proper parameter t is a motion along the geodesics of the manifold M, i.e., relation* (21) *for this motion, written in coordinate form, takes the form of one of the equivalent equations in* (17). *Relation* (19) *defines the integral of the motion $\Lambda \equiv \varkappa H$*

$$\Lambda(z_{t'}(x)) = \Lambda(z_{t''}(x)) \quad \textit{for all } t', t'' \textit{ from } I \textit{ and } x \textit{ from } M.$$

This theorem immediately follows from the coordinate form (25) of relation (21), and the identity

$$v^i D_i v_k(z) = \frac{dv_k}{dt} \qquad (z = z_t(x))$$

valid for any stationary flow on M. The equivalence of identity (19) to the conservation of the function $\varkappa H$ on the trajectories of translational motion is established by the arguments above concerning natural mechanical systems.

Now we establish the formal connection between relations (4) and (21).

Rewriting (21) in the coordinate form (26), we note that *for $v_0 \neq 0$ the first equation of system* (26) *is a corollary of the other n equations in* (26) *having the following form*:

(28) $$v^0 \left(D_0 \frac{v_\alpha}{|\boldsymbol{v}|} - D_\alpha \frac{v_0}{|\boldsymbol{v}|} \right) + v^\beta \left(D_\beta \frac{v_\alpha}{|\boldsymbol{v}|} - D_\alpha \frac{v_\beta}{|\boldsymbol{v}|} \right) = 0 \qquad (\alpha, \beta = 1, \ldots, n).$$

Indeed, contracting (28) with v^α, we obtain, using the obvious identity

$$v^\alpha v^\beta \left(D_\beta \frac{v_\alpha}{|\boldsymbol{v}|} - D_\alpha \frac{v_\beta}{|\boldsymbol{v}|} \right) = 0,$$

that

$$v^\alpha v^0 \left(D_0 \frac{v_\alpha}{|\boldsymbol{v}|} - D_\alpha \frac{v_0}{|\boldsymbol{v}|} \right) = 0.$$

After dividing by the nonzero factor v^0 and adding the zero term

$$v^0\left(D_0\frac{v_0}{|\boldsymbol{v}|}-D_0\frac{v_0}{|\boldsymbol{v}|}\right)=0,$$

we find that the equality obtained coincides with the first equation of system (26), q.e.d.

In subsequent constructions it is convenient to carry out a partial change of sign in the components of the metric tensor

$$g_{\alpha\beta}\to -g_{\alpha\beta},\qquad \alpha,\beta=1,\dots,n.$$

This change is obviously equivalent to the introduction of new components $u^1,\dots,u^n$ of the contravariant velocity $\mathbf{v}$ according to the formulas

$$v^\alpha=u^\alpha,\qquad v_\alpha=-u_\alpha.$$

By distinguishing the variable x^0 in equations (28) by means of the identical change of D_0 to $\partial/\partial x^0$ and making the indicated substitutions connected with the change of sign, we find that the change of sign, (28) leads to

$$\begin{aligned}&\frac{v_0}{|\boldsymbol{v}|}\frac{\partial}{\partial x^0}u_\alpha+v^0u_\alpha\frac{\partial}{\partial x^0}\frac{1}{|\boldsymbol{v}|}+v^0D_\alpha\frac{v_0}{|\boldsymbol{v}|}\\&\qquad+\frac{u^\beta}{|\boldsymbol{v}|}\left(D_\beta u_\alpha-D_\alpha u_\beta\right)+u^\beta\left(u_\alpha D_\beta\frac{1}{|\boldsymbol{v}|}-u_\beta D_\alpha\frac{1}{|\boldsymbol{v}|}\right)=0,\end{aligned}$$

or, after multiplying by $|\boldsymbol{v}|>0$, to

$$\begin{aligned}&v^0\frac{\partial}{\partial x^0}u_\alpha+u^\beta\left(D_\beta u_\alpha-D_\alpha u_\beta\right)+|\boldsymbol{v}|\left(v^0\frac{\partial}{\partial x^0}\frac{1}{|\boldsymbol{v}|}+u^\beta D_\beta\frac{1}{|\boldsymbol{v}|}\right)u_\alpha\\&\qquad+|\boldsymbol{v}|v^0D_\alpha\frac{v_0}{|\boldsymbol{v}|}-|\boldsymbol{v}|u^\beta u_\beta D_\alpha\frac{1}{|\boldsymbol{v}|}=0.\end{aligned}$$

By transforming the last two summands to the form

$$\begin{aligned}|\boldsymbol{v}|v^0D_\alpha\frac{v_0}{|\boldsymbol{v}|}-|\boldsymbol{v}|u^\beta u_\beta D_\alpha\frac{1}{|\boldsymbol{v}|}&=|\boldsymbol{v}|v^0\frac{1}{|\boldsymbol{v}|}D_\alpha v_0+|\boldsymbol{v}|v^0v_0D_\alpha\frac{1}{|\boldsymbol{v}|}+|\boldsymbol{v}|v^\beta v_\beta D_\alpha\frac{1}{|\boldsymbol{v}|}\\&=v^0D_\alpha v_0+|\boldsymbol{v}|^3D_\alpha\frac{1}{|\boldsymbol{v}|}\\&=-v_0D_\alpha v^0+D_\alpha(v_0v^0)-\frac{|\boldsymbol{v}|^3}{|\boldsymbol{v}|^2}D_\alpha|\boldsymbol{v}|\\&=-v_0D_\alpha v^0+D_\alpha\frac{v_0v^0+u_\beta u^\beta}{2},\end{aligned}$$

we obtain

$$\begin{aligned}&v^0\frac{\partial}{\partial x^0}u_\alpha+u^\beta\left(D_\beta u_\alpha-D_\alpha u_\beta\right)+\left(v^0\frac{\partial}{\partial x^0}\ln\frac{1}{|\boldsymbol{v}|}+u^\beta D_\beta\ln\frac{1}{|\boldsymbol{v}|}\right)u_\alpha\\&\qquad+D_\alpha\frac{v_0v^0+u_\beta u^\beta}{2}-v_0D_\alpha v^0=0.\end{aligned}$$

Comparing these equalities with (8), we rewrite them (as the equalities (8) above), using the invariant notation *relative to the group of variables* $x^1, \dots, x^n$,

$$v^0 \frac{\partial}{\partial x^0} \boldsymbol{u} + *(\omega \wedge \boldsymbol{u}) + \lambda \boldsymbol{u} + dq = v_0 \, dv^0. \tag{29}$$

Here $\boldsymbol{u} = u_\alpha \, dx^\alpha$ is the covector field with components $u_1, \dots, u_n$, ω is the vorticity of the field $\boldsymbol{u}$ defined by equality (5),

$$q = \frac{v_0 v^0 + u_\beta u^\beta}{2}$$

is a scalar function analogous to the Bernoulli integral occurring in equation (4) and defined by equality (6). In comparison with (4), equality (29) contains the following additional summands: the *"relaxation" term* $\lambda \boldsymbol{u}$ *with the "friction coefficient"*

$$\lambda = v^0 \frac{\partial}{\partial x^0} \ln \frac{1}{|\boldsymbol{v}|} + u^\beta D_\beta \ln \frac{1}{|\boldsymbol{v}|},$$

and the *free* (*generally speaking, "nongradient"*) *term* $v_0 \, dv^0$.

Now we shall assume that

$$v^0 \equiv c = \text{const} > 0$$

and that the motion under consideration is *slow* in the sense that

$$|u^1|, \dots, |u^n| \ll c$$

("$\ll$" means "much less than"). Relating the variable t with x^0 by the formula $dx^0 = c\,dt$ and assuming that the limits as $c \to \infty$ of the components of the metric tensor g_{ij} exist and that the function

$$\varkappa \left| \frac{\det(g_{ij})}{\det(g_{\alpha\beta})} \right|^{1/2} \to \rho(t, x^1, \dots, x^n), \qquad c \to \infty,$$

(as before, the Greek indices run over $1, \dots, n$, and the Latin ones, over $0, 1, \dots, n$) and that the corresponding limits of partial derivatives of the indicated functions with respect to the variables $t, x^1, \dots, x^n$ also exist, we pass to the limit as $c \to \infty$ in relation (19) (or (24)), after rewriting it in the form

$$D_0(\varkappa H v^0) + D_\alpha(\varkappa H u^\alpha) = 0 \qquad (u^\alpha = v^\alpha).$$

As a result of the passage to the limit, we obtain, as can be easily verified, the coordinate form of relation (3). Assuming additionally the existence of a scalar function $w = w(t, x^1, \dots, x^n)$ such that

$$d(q - \gamma_w) \to 0, \qquad c \to \infty,$$

where γ_w is the Bernoulli integral defined as in (6), we obtain

$$g_{00} \sim \eta^2(t) + \frac{2w}{c^2}, \qquad c \to \infty$$

provided that $\lim_{c\to\infty} g_{00} > 0$. In more details,

$$\begin{aligned} d(q - \gamma_w) &= d\left(\frac{1}{2}g_{00}c^2 + \frac{1}{2}g_{0\alpha}v^\alpha c - w\right) \\ &= c^2 d\left(\frac{1}{2}g_{00} + \frac{1}{2}g_{0\alpha}(v^\alpha/c) - w/c^2\right) \\ &\sim c^2 d\left(g_{00}/2 - w/c^2\right) \to 0, \qquad c \to \infty. \end{aligned}$$

Then the relation

$$\frac{\partial}{\partial t}\boldsymbol{u} + k\boldsymbol{u} + *(\omega \wedge \boldsymbol{u}) + d\gamma_w = 0,$$

where

$$k = \lim_{c\to\infty} \lambda = \frac{d}{dt}\ln\frac{1}{\eta(t)}, \qquad \eta(t) > 0$$

is a limiting relation for (29). Identifying the metric of M with the gravitational field, we conclude that in the case of a *weak gravitational field*, when

$$g_{00} \sim 1 + \frac{2w}{c^2} \qquad (\text{or} \qquad \eta(t) \equiv 1),$$

relation (4) is limiting for (29).

Thus, the validity of the following theorem is proved.

THEOREM 5. *Under our assumptions, relations* (3), (4) *are the limiting cases of relations* (19), (21) *that correspond to slow motions in weak gravitational fields.*

17.6. The concept of the integrating factor. We pass to the analysis of relations (3)–(5). First we describe a construction due to Dezin **[36]** mentioned in the Introduction. A p-form χ defined on the manifold V is called a *factor* of a stationary covector field $\boldsymbol{u}$ if

$$d * (\chi \wedge \boldsymbol{u}) = 0.$$

A *solenoidal field* $\boldsymbol{u}$ is defined as having the factor $\chi \equiv 1$ and an *Euler* field, as having an additional factor ω from (5). Acting on both sides of equality (5) by the operator $*$, we obtain

$$*\omega = \operatorname{sgn}(g) * *d\boldsymbol{u} = \operatorname{sgn}(g)\operatorname{sgn}(g)(-1)^{2(n-2)}d\boldsymbol{u} = d\boldsymbol{u}.$$

Now, combining the requirement that $\boldsymbol{u}$ be solenoidal, the above formula and the condition that the covector field be Euler, we come to the closed invariant system

(30) $$d * \boldsymbol{u} = 0, \qquad d\boldsymbol{u} = *\omega, \qquad d * (\omega \wedge \boldsymbol{u}) = 0,$$

which is an equivalent form of equations (3)–(5) for the case of a stationary flow ($\boldsymbol{u}$ does not depend on time t) of an incompressible fluid ($\rho \equiv \text{const} > 0$).

Indeed, under our assumptions, (3) is equivalent to the first relation from (30), and (5) to the second (as shown above). Relation (4) takes the form

(31) $$*(\omega \wedge \boldsymbol{u}) = -d\gamma.$$

Acting on both sides of identity (31) by the operator d, we obtain the third equality from (30) (since $dd\gamma = 0$).

Conversely, by the third equality from (30), the differential form $*(\omega \wedge \boldsymbol{u})$ is closed and, hence, at least locally, exact ((31) is valid for some scalar function γ).

Now we consider equations (30) in the cylindrical system of coordinates x, y, θ connected with the Cartesian orthogonal coordinates x_1, x_2, x_3 of three-dimensional Euclidean space by the relations

$$x_1 = x, \qquad x_2 = y\cos\theta, \qquad x_3 = y\sin\theta,$$
$$-\infty < x < +\infty, \qquad y > 0, \qquad 0 \leqslant \theta \leqslant 2\pi.$$

We have

$$dx_1 = dx, \qquad dx_2 = \cos\theta\,dy - y\sin\theta\,d\theta, \qquad dx_3 = \sin\theta\,dy + y\cos\theta\,d\theta,$$
$$(dx_1)^2 + (dx_2)^2 + (dx_3)^2 = (dx)^2 + (dy)^2 + y^2(d\theta)^2,$$

and the components of the metric tensor are

$$g_{11} = g_{22} = 1, \qquad g_{33} = y^2, \qquad g_{12} = g_{13} = g_{23} = 0 \qquad (g_{ij} = g_{ji}).$$

An axially symmetric flow is defined in the cylindrical coordinates by the 1-form

$$\boldsymbol{u} = u\,dx + v\,dy + w\,d\theta$$

with components u, v, w of the form

$$u = u(x, y), \qquad v = v(x, y), \qquad w = 0.$$

We have

$$*\boldsymbol{u} = yu\,dy \wedge d\theta - yv\,dx \wedge d\theta + y^{-1}w\,dx \wedge dy = yu\,dy \wedge d\theta - yv\,dx \wedge d\theta,$$
$$d * \boldsymbol{u} = \left(\frac{\partial}{\partial x}(yu) + \frac{\partial}{\partial y}(yv)\right) dx \wedge dy \wedge d\theta,$$

and the first relation of system (30) is written as

$$\frac{\partial}{\partial x}(yu) + \frac{\partial}{\partial y}(yv) = 0.$$

After substituting the fields

$$\boldsymbol{u} = u(x, y)\,dx + v(x, y)\,dy \qquad \text{and} \qquad \omega = \omega_1\,dx + \omega_2\,dy + \omega_3\,d\theta$$

in the second equation of the indicated system, we obtain

$$d\boldsymbol{u} = \left(\frac{\partial v}{\partial x} - \frac{\partial u}{\partial y}\right) dx \wedge dy = *\omega = y^{-1}\omega_3\,dx \wedge dy - y\omega_2\,dx \wedge d\theta + y\omega_1\,dy \wedge d\theta,$$

or

$$\omega_1 = \omega_2 = 0, \qquad y\left(\frac{\partial v}{\partial x} - \frac{\partial u}{\partial y}\right) = \omega_3,$$

which implies $\omega_3 = \omega_3(x, y)$.

The third relation from system (30) is written as

$$d * (\omega \wedge \boldsymbol{u}) = d * (\omega_3\,d\theta \wedge (u\,dx + v\,dy)) = d * (-\omega_3 u\,dx \wedge d\theta - \omega_3 v\,dy \wedge d\theta)$$
$$= \frac{\partial}{\partial y}(y^{-1}\omega_3 v)\,dx \wedge dy + \frac{\partial}{\partial x}(y^{-1}\omega_3 u)\,dx \wedge dy = 0,$$

or

$$\frac{\partial}{\partial x}(y^{-1}\omega_3 u) + \frac{\partial}{\partial y}(y^{-1}\omega_3 v) = 0.$$

17.7. The invariant form of the Navier-Stokes equations. The invariant form of the *Navier-Stokes equations* for an incompressible fluid ($\rho = \text{const} > 0$) is formally obtained by adding to the left-hand side of equality (4) the "viscous" term $-\nu\Delta \boldsymbol{u}$, where

$$\Delta \equiv d\delta + \delta d$$

is a multidimensional analog of the Laplace operator introduced above. The corresponding relation additionally taking into account the influence of external nonpotential mass forces defined by some solenoidal covector field (1-form) $\mathfrak{I} = \mathfrak{I}(t,x)$, is written as

$$\frac{\partial}{\partial t}\boldsymbol{u} + *(\omega \wedge \boldsymbol{u}) + d\gamma - \nu\Delta\boldsymbol{u} = \mathfrak{I} \qquad (d * \mathfrak{I} = 0). \tag{32}$$

Here

$$\gamma = \frac{\langle \boldsymbol{u}, \boldsymbol{u}\rangle^2}{2} + \varphi(t,x) + \frac{p}{\rho}$$

is the Bernoulli integral with the "potential" $\varphi = \varphi(t,x)$ defined by the gradient component of the field of mass forces, an exact form $d\varphi$. The first two equations of system (30) are preserved. The metric of the manifold is assumed to be stationary (otherwise, the summand $\partial * \rho/\partial t = \rho\partial(*1)/\partial t$ is preserved in the equation of continuity (3)).

The necessary (and at least locally sufficient) condition of consistency of equation (32) (the exact form $\partial \boldsymbol{u}/\partial t + *(\omega \wedge \boldsymbol{u}) - \nu\Delta\boldsymbol{u} - \mathfrak{F} = -d\gamma$ is necessarily closed: $dd\gamma = 0$) leads to the *Helmholtz relation* for ω which is obtained by acting on both sides of equality (32) by the operator $\operatorname{sgn}(g) * d$ and by using identity (5) (or the second relation of system (3)) and the definition of the operator δ that is, $(\operatorname{sgn}(g) * d * (\omega \wedge \boldsymbol{u}) = (-1)^n \delta(\omega \wedge \boldsymbol{u}))$. The corresponding complete system playing the role of (30) for the case of an incompressible fluid is written as

$$\begin{gathered}\frac{\partial}{\partial t}\omega + (-1)^n\delta(\omega\wedge\boldsymbol{u}) - \nu\Delta\omega = \mathfrak{F} \qquad (\mathfrak{F} = \operatorname{sgn}(g) * d\mathfrak{I}),\\ d * \boldsymbol{u} = 0, \qquad d\boldsymbol{u} = *\omega\end{gathered} \tag{33}$$

(to obtain this system we additionally use the equalities

$$\begin{aligned}\operatorname{sgn}(g) * d\Delta\boldsymbol{u} &= \operatorname{sgn}(g) * d(d\delta + \delta d)\boldsymbol{u} = \operatorname{sgn}(g) * d\delta d\boldsymbol{u}\\ &= \operatorname{sgn}(g) * d\delta * \omega = *d * (-1)^n d * *\omega = *d * (-1)^n d \operatorname{sgn}(g)\omega\\ &= \operatorname{sgn}(g)(-1)^n * d * d\omega = \delta d\omega = d\delta\omega + \delta d\omega = \Delta\omega\end{aligned}$$

and $\delta\omega = 0$, which follows from the third equality of system (33)).

The indicated form of the main hydrodynamic equations allows us to introduce the notion of a "basic" flow (mentioned in §11.4) in an arbitrary system of coordinates. Thus, the corresponding Stokes system is written as

$$\frac{\partial}{\partial t}\omega - \nu\Delta\omega = \mathfrak{F}, \qquad d * \boldsymbol{u} = 0, \qquad d\boldsymbol{u} = *\omega. \tag{34}$$

The covector fields $\boldsymbol{u}$ satisfying simultaneously (30) and (34) (which are both "Euler" and "Stokes") are by definition *basic flows*. As already mentioned in the Introduction, the majority of the well-known solutions of Navier-Stokes equations investigated with respect to stability are the basic flows considered in a suitable system of curvilinear coordinates. The invariant relations used here enable one to study the indicated flows independently of the choice of a local coordinate system, on one hand, and, on the other hand, to obtain new coordinate representations.

Due to the presence of the second relation in system (33), an $(n-2)$-form ψ such that

$$*\boldsymbol{u} = \operatorname{sgn}(g)\, d\psi \tag{35}$$

is defined (at least locally). For $n = 2$, ψ is a scalar, for $n = 3$, ψ is a vector (a covector field). In the general case, following [**36**], we call ψ the *vector potential* of the field $\boldsymbol{u}$ and assume that

$$d * \psi = 0 \qquad (\text{or } \delta\psi = 0). \tag{36}$$

For $n = 2$, equality (36) is satisfied identically. For $n = 3$ and the Euclidean metric, equalities (35), (36) lead to the classical system

$$\operatorname{rot} \Psi = \mathbf{U}, \qquad \operatorname{div} \Psi = 0,$$

where Ψ and $\mathbf{U}$ are vector fields associated to the forms ψ and $\boldsymbol{u}$, respectively. In terms of the vector potential, the third relation of system (33) takes the form of the multidimensional Poisson equation,

$$-\Delta\psi = \omega. \tag{37}$$

Indeed, we find from (35) that

$$* * \boldsymbol{u} = \operatorname{sgn}(g)(-1)^{n-1}\boldsymbol{u} = \operatorname{sgn}(g) * d\psi \quad \text{or} \quad \boldsymbol{u} = (-1)^{n-1} * d\psi.$$

The substitution of the obtained $\boldsymbol{u}$ in the third relation of system (33) gives

$$d\boldsymbol{u} = (-1)^{n-1} d * d\psi = *\omega \quad \text{or} \quad (-1)^{n-1} * d * d\psi = * * \omega.$$

But

$$(-1)^{n-1} * d * d\psi = -\operatorname{sgn}(g))\delta d\psi = -\operatorname{sgn}(g)(\delta d\psi + d\delta\psi) = -\operatorname{sgn}(g)\Delta\psi,$$
$$* * \omega = \operatorname{sgn}(g)\omega,$$

which implies (37).

Then the system of equations (33) reduces to the form

$$\begin{gathered} \frac{\partial}{\partial t}\omega - \delta(\omega \wedge *d\psi) - \nu\Delta\omega = \mathfrak{F}, \\ -\Delta\psi = \omega, \qquad \delta\psi = 0. \end{gathered} \tag{38}$$

In the study of equations (33), along with the vector potential, it is often useful to use the *stream functions*, i.e., a system of $n-1$ scalar functions $\varphi_1, \dots, \varphi_{n-1}$ connected with $\boldsymbol{u}$ by the equality

$$*\boldsymbol{u} = d\varphi_1 \wedge \ldots \wedge d\varphi_{n-1} \tag{39}$$

guaranteeing that the second relation of system (33)

$$d*\boldsymbol{u} = d(d\varphi_1 \wedge \ldots \wedge d\varphi_{n-1}) = (dd\varphi_1) \wedge \ldots \wedge d\varphi_{n-1} - d\varphi_1 \wedge d(d\varphi_2 \wedge \ldots \wedge d\varphi_{n-1}) = \ldots = 0$$

holds (the rule of repeated differentiation of the exterior product of a pair of differential forms is used here); for $n = 2$, the stream function φ_1 coincides (up to an additive constant) with ψ.

The stream functions $\varphi_1, \dots, \varphi_{n-1}$ are simultaneously the integrals (the so-called *first integrals*) of the corresponding autonomous dynamical system on V generated by the trajectories z_t of the stationary covector field $\boldsymbol{u}$,

$$\begin{aligned}\frac{d}{dt}\varphi_\lambda(z_t) &= u^\alpha D_\alpha \varphi_\lambda = \operatorname{sgn}(g) * (\boldsymbol{u} \wedge *d\varphi_\lambda) = \operatorname{sgn}(g) * (d\varphi_\lambda \wedge *\boldsymbol{u}) \\ &= \operatorname{sgn}(g) * (d\varphi_\lambda \wedge d\varphi_1 \wedge \dots \wedge d\varphi_{n-1}) = 0, \qquad \lambda = 1, \dots, n-1,\end{aligned}$$

(since $d\varphi_\lambda \wedge d\varphi_\lambda = 0$). Locally, the existence of the required set of the first integrals is ensured by the regularity (smoothness and absence of singular points) of the flow $\boldsymbol{u}$ **[12]**. Globally, there may be no complete set. The presence of a smooth trajectory z_t of the flow $\boldsymbol{u}$ everywhere densely filling out some subdomain Q of the manifold V excludes the existence of the required functions in Q (and, hence, in the whole of V).

Indeed, since each stream function φ_λ is constant on a trajectory z_t dense in Q, $\varphi_\lambda \equiv \text{const}$ in Q (as a continuous function) and, hence, $\boldsymbol{u} \equiv 0$ in Q (equality (39)), which contradicts the assumed regularity of $\boldsymbol{u}$ in Q.

The trajectories z_t with the indicated properties are often called *ergodic*. The computer experiments mentioned in the Introduction reveal the presence of ergodic trajectories for the ABC-flows. A rigorous (mathematical) justification of ergodicity of these trajectories is not known to the author.

References

1. G. N. Abramovich, *The theory of a free jet and its applications*, Trudy TSAGI (1936), no. 293. (Russian)
2. S. Agmon, A. Douglis, and L. Nirenberg, *Estimates near the boundary for solutions of elliptic partial differential equations satisfying general boundary conditions.* I, Comm. Pure Appl. Math **12** (1959), 623–727.
3. V. V. Aleksandrov and Yu. D. Shmyglevskiĭ, *On inertia and shear flows*, Dokl. Akad. Nauk SSSR **274** (1984), 280–283; English transl. in Soviet Phys. Dokl. **29** (1984).
4. G. V. Alekseev, *On the uniqueness and smoothness of turbulent flows of an ideal fluid*, Dinamika Sploshn. Sredy **1973**, no. 15, 7–17. (Russian)
5. D. V. Anosov, *Geodesic flows on closed Riemannian manifolds of negative curvature*, Trudy Mat. Inst. Steklov. **90** (1967); English transl. in Proc. Steklov Inst. Math. **1969**.
6. V. I. Arnol′d, *Conditions for non-linear stability of stationary plane curvilinear flows of an ideal fluid*, Dokl. Akad. Nauk SSSR **162** (1965), no. 5, 975–978; English transl. in Soviet Math. Dokl. **6** (1965).
7. ——, *A variational principle for the three-dimensional stationary flows of an ideal fluid*, Prikl. Mat. Mekh. **29** (1965), no. 5, 846–851; English transl. in J. Appl. Math. Mech. **29** (1965).
8. ——, *On an apriori estimate in the theory of hydrodynamical stability*, Izv. Vyssh. Uchebn. Zaved. Mat. **1966**, no. 5, 3–5; English transl. in Amer. Math. Soc. Transl. Ser. 2 **79** **1969**.
9. ——, *Remarks concerning the behaviour of three-dimensional ideal fluids under a small perturbation of the initial velocity field*, Prikl. Mat. Mekh. **36** (1972), no. 5, 255–262; English transl. in J. Appl. Math. Mech. **36** (1972).
10. ——, *The asymptotic Hopf invariant and its applications*, Materials All-Union School on Differential Equations with an Infinite Number of Variables and Dynamical Systems with an Infinite Number of Degrees of Freedom (Dilizhan, 1973), Izdat. Akad. Nauk Armyan. SSR, Erevan, 1974, pp. 229–256; English transl. in Selecta Math. Soviet. **5** (1986).
11. ——, *Mathematical methods of classical mechanics*, "Nauka", Moscow, 1974; English transl., Springer-Verlag, New York, 1978.
12. ——, *Additional chapters of the theory of ordinary differential equations*, "Nauka", Moscow, 1974; English transl., *Geometrical methods in the theory of ordinary differential equations*, Springer-Verlag, New York, 1983.
13. V. I. Arnol′d and L. D. Meshalkin, *A. N. Kolmogorov seminar on selected problems of analysis* (1958–1959), Uspekhi Mat. Nauk **15** (1960), no. 1, 247–250.
14. I. Ya. Bakel′man, A. L. Verner, and B. E. Kantor, *Introduction to differential geometry "in the large"*, "Nauka", Moscow, 1973. (Russian)
15. S. N. Bernstein, *Collected works. Vol.* III: *Differential equations and calculus of variations and geometry* (1903–1947), Izdat. Akad. Nauk SSSR, Moscow, 1960. (Russian)
16. G. Birkhoff and E. H. Zarantonello, *Jets, wakes and cavities*, Academic Press, New York, 1957.
17. M. Sh. Birman and G. E. Skvortsov, *On the square summability of highest derivatives of the solution of the Dirichlet problem in a domain with piecewise smooth boundary*, Izv. Vyssh. Uchebn. Zaved. Mat. **1962**, no. 5, 11–21. (Russian)
18. A. V. Bitsadze, *Some classes of partial differential equations*, "Nauka", Moscow, 1981; English transl., Gordon and Breach, New York, 1988.
19. N. Bourbaki, *Groupes et algèbres de Lie*, Chap. 9, Masson, Paris, 1982.
20. N. N. Bukholtz, *Basic course of theoretical mechanics*, Part 2, GITTL, Moscow, 1939. (Russian)
21. M. M. Vainberg and V. A. Trenogin, *Theory of branching of solutions of nonlinear equations*, "Nauka", Moscow, 1969; English transl., Nordhoff Internat. Publ., Leyden, 1974.
22. M. M. Vainberg, *Variational method and the method of monotone operators in the theory of nonlinear equations*, "Nauka", Moscow, 1972; English transl., Wiley, New York, 1973.

23. H. Weyl, *The method of orthogonal projection in the potential theory*, Duke Math. J. **7** (1940), 411–444.
24. I. N. Vekua, *Generalized analytic functions*, Fizmatgiz, Moscow, 1959; English transl., Pergamon Press, London, and Addison-Wesley, Reading, MA, 1962.
25. I. M. Gelfand, *Lectures on linear algebra*, 4th ed., "Nauka", Moscow, 1971; English transl. of 2nd ed., Interscience, New York, 1961.
26. E. B. Gledzer, F. V. Dolzhanskiĭ, and A. M. Obukhov, *Systems of hydrodynamic type and their application*, "Nauka", Moscow, 1981. (Russian)
27. K. K. Golovkin and O. A. Ladyzhenskaya, *Solutions of non-stationary boundary value problems for Navier-Stokes equations*, Trudy Mat. Inst. Steklov. **59** (1960), 100–114. (Russian)
28. M. A. Gol′dshtik, *Vortex flows*, "Nauka", Moscow, 1981. (Russian)
29. I. S. Gromeka, *Collected works*, Izdat. Akad. Nauk SSSR, Moscow, 1952. (Russian)
30. S. H. Gould, *Variational methods for eigenvalue problems. An introduction to the Weinstein method of intermediate problems*, 2nd ed., Univ. of Toronto Press, Toronto, 1966.
31. N. Dunford and J. T. Schwartz, *Linear operators. Part* I. *General theory*, Pure Appl. Math., vol. 7, Interscience Publ., New York, 1958.
32. A. A. Dezin, *Invariant differential operators and boundary problems*, Trudy Mat. Inst. Steklov. **68** (1962); English transl. in Amer. Math. Soc. Transl. Ser. 2 **41** (1964).
33. ———, *Some models connected with Euler's equations*, Differentsial′nye Uravneniya **6** (1970), no. 1, 17–26; English transl. in Differential Equations **6** (1970).
34. ———, *On a class of vector fields*, Complex Analysis and Its Application, "Nauka", Moscow, 1978, pp. 203–208. (Russian)
35. ———, *General questions of the theory of boundary problems*, "Nauka", Moscow, 1980; English transl., *Partial differential equations*, Springer-Verlag, Berlin, 1987.
36. ———, *Invariant forms and some structural properties of the hydrodynamic Euler equations*, Z. Anal. Anwendungen **2** (1983), 401–409.
37. ———, *Multi-dimensional analysis and discrete models*, "Nauka", Moscow, 1990. (Russian)
38. D. D. Joseph, *Stability of fluid motions.* I, Springer Tracts Nat. Philos., vol. 27, Springer-Verlag, Berlin, 1976.
39. P. A. M. Dirac, *General theory of relativity*, Wiley, New York, 1975.
40. B. A. Dubrovin, S. P. Novikov, and A. T. Fomenko, *Modern geometry — methods and applications. Part* I, "Nauka", Moscow, 1979; English transl., Graduate Texts in Math., vol. 93, Springer-Verlag, New York, 1984.
41. D. N. Zubarev, V. G. Morozov, and O. V. Troshkin, *A bifurcation model of turbulent flow in a channel*, Dokl. Akad. Nauk SSSR **290** (1986), no. 2, 313–317; English transl. in Soviet Phys. Dokl. **31** (1986).
42. D. N. Zubarev, V. G. Morozov, O. V. Troshkin, and A. Yu. Danilenko, *Bifurcation model of simple turbulent flows*, MIAN SSSR, Moscow, 1988. (Russian)
43. Yu. P. Ivanilov and G. N. Yakovlev, *On the bifurcation of fluid flows between rotating cylinders*, Prikl. Mat. Mekh. **30** (1966), 910–916; English transl. in J. Appl. Math. Mech. **30** (1966).
44. L. V. Kantorovich and G. P. Akilov, *Functional analysis in normed spaces*, "Nauka", Moscow, 1959; English transl., 2nd ed., Pergamon Press, Oxford, 1982.
45. A. N. Kolmogorov, *Foundations of the theory of probability*, 2nd ed., "Nauka", Moscow, 1974; English transl. of 1st ed., Chelsea Publishing, New York, 1950; 1956.
46. G. Comte-Bellot, *Écoulement turbulent entre deux parois parallèles*, Publicacions scientifiques et techniques du Ministère de l'Air, no. 419, 1965.
47. N. E. Kochin, I. A. Kibel′, and N. V. Roze, *Theoretical hydromechanics*, vol. I, 6th ed., vol. II, 4th ed., Fizmatgiz, Moscow, 1963; English transl. of vol. I, 5th ed, Interscience, New York, 1964.
48. M. A. Krasnosel′skiĭ, *Topological methods in the theory of non-linear integral equations*, GITTL, Moscow, 1956; English transl., The Macmillan Co., New York, 1964.
49. M. A. Krasnosel′skiĭ, A. I. Perov, A. I. Povolotskiĭ, and P. P. Zabreĭko, *Plane vector fields*, Fizmatgiz, Moscow, 1963; English transl., Academic Press, New York, 1966.
50. S. G. Kreĭn, *On functional properties of operators of vector analysis and hydrodynamics*, Dokl. Akad. Nauk SSSR **93** (1953), no. 6, 969–972. (Russian)
51. A. L. Krylov, *On the stability of the poiseuille flow in a planar channel*, Dokl. Akad. Nauk SSSR **159** (1964), no. 5, 978–981; English transl. in Soviet Math. Dokl. **5** (1964).
52. R. Courant and D. Hilbert, *Methods of mathematical physics*, vol. I, Interscience Publ., New York, 1953.

53. A. F. Kurbatskiĭ and A. T. Onufriev, *Modeling of turbulent transfer in the wake after a cylinder with the use of equations for the third moments*, Zh. Prikl. Mekh. i Tekhn. Fiz. **6** (1979), 99–107; English transl. in J. Appl. Mech. Tech. Phys..
54. M. A. Lavrent′ev and B. V. Shabat, *Methods of the theory of functions of a complex variable*, 4th ed., "Nauka", Moscow, 1973. (Russian)
55. ———, *Problems of hydrodynamics and their mathematical models*, "Nauka", Moscow, 1973; 2nd ed., 1977. (Russian)
56. O. A. Ladyzhenskaya, *The mathematical theory of viscous incompressible fluid*, "Nauka", Moscow, 1970; English transl., Math. Appl., vol. 2, Gordon and Breach, New York, 1963; 2nd ed., 1969.
57. H. Lamb, *Hydrodynamics*, Dover, New York, 1945.
58. L. D. Landau and E. M. Lifshits, *Theoretical physics. Vol* VI. *Fluid dynamics*, 3rd ed., "Nauka", Moscow, 1986; English transl. of 1st ed.,, *Fluid mechanics*, Pergamon Press, London, and Addison-Wesley, Reading, MA, 1959.
59. E. M. Landis, *Second order equations of elliptic and parabolic type*, "Nauka", Moscow, 1971. (Russian)
60. R. L. Liboff, *Introduction to the theory of kinetic equations*, Wiley, New York, 1969.
61. C. C. Lin, *The theory of hydrodinamic stability*, Cambridge Univ. Press, Cambridge, 1955.
62. L. G. Loĭtsyanskiĭ, *Mechanics of liquids and gases*, "Nauka", Moscow, 1987; English transl., Pergrammon Press, New York, 1960.
63. H. A. Lorentz, *Theories and models of ether*, ONTI NKTP SSSR GROLN, Moscow, 1936. (Russian)
64. J. E. Marsden and M. McCracken, *The Hopf bifurcation and its applications*, Appl. Math. Sci., vol. 19, Springer-Verlag, New York, 1976.
65. V. P. Maslov, *Deterministic theory of hydrodynamic turbulence*, Teoret. Mat. Fiz. **65** (1985), no. 3, 448; English transl. in Theor. and Math. Phys. **65** (1986).
66. L. D. Meshalkin and Ya. G. Sinaĭ, *Investigations of the stability of a stationary solution of a system of equations for the plane movement of an incompressible viscous fluid*, Prikl. Mat. Mekh. **25** (1961), no. 6, 1140–1143; English transl. in J. Appl. Math. Mech. **25** (1961).
67. L. M. Milne-Thomson, *Theoretical hydrodynamics*, 4th ed., The Macmillan Co., New York, 1960.
68. A. S. Monin and A. M. Yaglom, *Statistical fluid dynamics*, Part I, "Nauka", Moscow, 1965; English transl., MIT Press, Cambridge, MA, 1971.
69. M. Morse, *Topological methods in the theory of functions of a complex variable*, Ann. of Math. Stud., vol. 15, Princeton Univ. Press, Princeton, NJ, 1947.
70. I. N. Naumova and Yu. D. Shmyglevskiĭ, *On the streamlines of a spiral flow*, Zh. Vychisl. Mat. i Mat. Fiz. **25** (1985), 312–313; English transl. in U.S.S.R. Comput. Math. and Math. Phys. **25** (1985).
71. V. G. Nevzglyadov, *A phenomenological theory of turbulence*, Dokl. Akad. Nauk SSSR **47**, no. 3, 169–173. (Russian)
72. V. V. Nemytskiĭ and V. V. Stepanov, *Qualitative theory of differential equations*, OGIZ, Moscow, 1949; English transl., Princeton Math. Ser., vol. 22, Princeton Univ. Press, Princeton, NJ, 1960.
73. L. Nirenberg, *Topics in nonlinear functional analysis*, Courant Inst. Math. Sci., New York Univ., New York, 1974.
74. A. M. Obukhov, *Turbulence and dynamics of the atmosphere*, Gidrometeoizdat, Leningrad, 1988. (Russian)
75. A. S. Parkhomenko, *What is a line*, GITTL, Moscow, 1954. (Russian)
76. W. Pauli, *Theory of relativity*, Pergamon Press, New York, 1958.
77. I. G. Petrovskiĭ, *Lectures on partial differential equations*, GITTL, Moscow, 1950; English transl., Interscience Publ., New York, 1954.
78. Yu. L. Klimontovich and H. Engel-Herbert, *Averaged stationary turbulent Couette and Poiseuille flows in an incompressible fluid*, Zh. Tekhn. Fiz. **54** (1984), 440–449. (Russian)
79. L. S. Pontryagin, *Foundations of combinatorial topology*, "Nauka", Moscow, 1947; English transl., Graylock Press, Rochester, NY, 1952.
80. ———, *Topological groups*, 4th ed., "Nauka", Moscow, 1984; English transl., 5th ed., Princeton Math. Ser., vol. 2, Princeton Univ. Press, Princeton, NJ, 1958.
81. M. M. Postnikov, *Introduction to the Morse theory*, "Nauka", Moscow, 1971. (Russian)
82. M. I. Rabinovich and M. M. Sushchik, *Regular and chaotic dynamics of structures in fluid flows*, Uspekhi Fiz. Nauk **160** (1990), no. 1, 3–64; English transl. in Soviet Phys. Uspekhi **33** (1990).
83. G. de Rham, *Differentiable manifolds*, Grundlehren Math. Wiss., vol. 266, Springer-Verlag, Berlin, 1984.
84. F. Riesz and B. Sz-Nagy, *Functional analysis*, Ungar, New York, 1955.

85. B. L. Rozhdestvenskiĭ and I. N. Simakin, *Modeling of turbulent flows in a two-dimensional channel*, Zh. Vychisl. Mat. i Mat. Fiz. **25** (1985), no. 1, 96–121; English transl. in U.S.S.R. Comput. Math. and Math. Phys. **25** (1985).
86. D. Ruelle and F. Takens, *On the nature of turbulence*, Comm. Math. Phys. **20** (1971), 167–192.
87. L. I. Sedov, *Mathematical methods for constructing new models of continuous media*, Uspekhi Mat. Nauk **20** (1965), no. 5, 121–180; English transl. in Russian Math. Surveys **20** (1965).
88. S. L. Sobolev, *Applications of functional analysis in mathematical physics*, Izdat. Gos. Leningrad Univ., Leningrad, 1950; English transl., Transl. Math. Monographs, vol. 7, Amer. Math. Soc., Providence, RI, 1963.
89. V. A. Solonnikov, *Estimates for the solutions of the nonstationary linearized system of the Navier-Stokes equations*, Trudy Mat. Inst. Steklov. **70** (1964), 213–317; English transl. in Amer. Math. Soc. Transl. Ser. 2 **75** (1968).
90. ———, *On the differential properties of the solution of the first boundary-value problem for a nonstationary system of Navier-Stokes equations*, Trudy Mat. Inst. Steklov. **73** (1964), 222–291. (Russian)
91. S. Sternberg, *Lectures on differential geometry*, Prentice-Hall, Englewood Cliffs, NJ, 1964.
92. S. Stoïlow, *Leçons sur les principes topologiques de la théorie des fonctions analytiques*, 2nd ed., Gauthier-Villars, Paris, 1956.
93. R. Témam, *Navier-Stokes equations. Theory and numerical analysis*, Stud. Math. Appl., rev. ed., vol. 2, North-Holland, Amsterdam, 1979.
94. J. B. Keller and S. Antman (eds.), *Bifurcation theory and nonlinear eigenvalue problems*, Benjamin, Inc., New York, 1969.
95. O. V. Troshkin, *Some properties of Euler fields*, Differentsial′nye Uravneniya **18** (1982), no. 1, 138–144; English transl. in Differential Equations **18** (1982).
96. ———, *Admissibility of the set of boundary values in a steady-state hydrodynamic problem*, Dokl. Akad. Nauk SSSR **272** (1983), no. 5, 1086–1090; English transl. in Soviet Phys. Dokl. **28** (1983).
97. ———, *Algebraic structure of the two-dimensional stationary Navier-Stokes equations, and nonlocal uniqueness theorems*, Dokl. Akad. Nauk SSSR **298** (1988), no. 6, 1372-1376; English transl. in Soviet Phys. Dokl. **33** (1988).
98. ———, *Topological analysis of the structure of hydrodynamic flows*, Uspekhi Mat. Nauk **43** (1988), no. 4, 129–158; English transl. in Russian Math. Surveys **43** (1988).
99. ———, *Two-dimensional flow problem for the steady Euler equations*, Mat. Sb. **180** (1989), no. 3, 354–374; English transl. in Math. USSR-Sb. **66** (1990).
100. ———, *Propagation of small perturbations in an ideal turbulent medium*, Dokl. Akad. Nauk SSSR **307** (1989), no. 5, 1072–1076; English transl. in Soviet Phys. Dokl. **34** (1989).
101. ———, *To the wave theory of turbulence*, Computing Center, Akad. Nauk SSSR, Moscow, 1989. (Russian)
102. *Turbulence. Principle and primary*, Mir, Moscow, 1980. (Russian)
103. Ho Chih Ming and P. Huerre, *Perturbed free layers*, Annual Review of Fluid Mechanics, Vol. 16, Annual Reviews, Palo Alto, CA, 1984, pp. 365–424.
104. L. T. Chernyĭ, *Relativistic models of continuos media*, "Nauka", Moscow, 1983. (Russian)
105. H. Schlichting, *Entstehung der Turbulenz* (1959), Springer-Verlag, Berlin, 351–450.
106. A. I. Shnirel′man, *Geometry of the group of diffeomorphism and the dynamics of an ideal incompressible fluid*, Mat. Sb. **128** (1985), no. 1, 82–109; English transl. in Math. USSR-Sb. **56** (1987).
107. A. Einstein, *Reasons for formation of curved riverbeds and the so-called Baire law*, Collected Scientific Works, vol. 4, "Nauka", Moscow, 1967, pp. 74–77. (Russian)
108. I. Ekeland and R. Témam, *Convex analysis and variational problems*, Stud. Math. Appl., vol. 1, North-Holland, Amsterdam, 1976.
109. V. I. Yudovich, *A two-dimensional problem of unsteady flow of an ideal incompressible fluid across a given domain*, Mat. Sb. **64** (1964), 562–588; English transl. in Amer. Math. Soc. Transl. Ser. 2 **57** (1966).
110. ———, *An example of the birth of a secondary stationary or periodic flow under the loss of stability of a laminar flow of a viscous incompressible fluid*, Prikl. Mat. Mekh. **29** (1965), 453–467; English transl. in J. Appl. Math. Mech. **29** (1965).
111. ———, *Secondary flows and fluid instability between rotating cylinders*, Prikl. Mat. Mekh. **30** (1966), no. 4, 688–698; English transl. in J. Appl. Math. Mech. **30** (1966).
112. ———, *On the onset of convection*, Prikl. Mat. Mekh. **30** (1966), no. 6, 1000–1005; English transl. in J. Appl. Math. Mech. **30** (1966).
113. ———, *On the bifurcation of rotational fluid flows*, Dokl. Akad. Nauk SSSR **169** (1966), 306–309; English transl. in Soviet Phys. Dokl. **11** (1966).

114. D. G. Andrews, *On the existence of nonzonal flows satisfying sufficient conditions for stability*, Geophys. Astrophys. Fluid Dynamics **28** (1984), 243.
115. V. Arnold, *Sur la courbure de Riemann des groupes de difféomorphismes*, C. R. Acad. Sci. Paris **260** (1965), 5668–5671.
116. ———, *Sur la topologie de écoulements stationnaires des fluides parfaits*, C. R. Acad. Sci. Paris **261** (1965), 17–20.
117. ———, *Sur la géometrie différentielle des groupes de Lie de dimension infinie et ses applications à l'hydrodynamique de fluides parfaits*, Ann. Inst. Fourier (Grenoble) **16** (1966), 319–361.
118. ———, *Sur un principe variationnel pour les écoulements stationnaires des liquides parfaits et ses applications aux problèmes de stabilité non linéaires*, J. Mécanique **5** (1966), 29–43.
119. G. K. Batchelor, *An introduction to fluid mechanics*, Cambridge Univ. Press, Cambridge, 1967.
120. J. T. Beale, T. Kato, and A. Majda, *Remarks on the breakdown of smooth solutions for* $3 - D$ *Euler equations*, Comm. Math. Phys. **94** (1984), 61–66.
121. M. Berger, *An eigenvalue problem for nonlinear elliptic partial differential equations*, Trans. Amer. Math. Soc. **120** (1965), 145–184.
122. P. Bradshaw, D. H. Ferris, and N. P. Atwell, *Calculation of boundary-layer development using the turbulent energy equation*, J. Fluid Mech. **28** (1967), 593–616.
123. G. L. Brown and A. Roshko, *Density effects and large structure in turbulent mixing layers*, J. Fluid Mech. **64** (1974), 775–816.
124. L. Caffarelli, R. Kohn, and L. Nirenberg, *Partial regularity of suitable weak solutions of the Navier-Stokes equations*, Comm. Pure Appl. Math. **35** (1982), 771–831.
125. G. F. Carnevale and T. G. Shepherd, *On the interpretation of Andrew's theorem*, Geophys. Astrophys. Fluid Dynamics **51** (1990), 1–17.
126. P. Cartier, *Spectre de l'équation de Schrödinger, application à la stabilité de la matière*, Lecture Notes in Math., vol. 677, Springer-Verlag, Berlin, 1978, pp. 88–104.
127. Chern Shuh-Jye and J. E. Marsden, *A note on symmetry and stability for fluid flows*, Geophys. Astrophys. Fluid Dynamics **51** (1990), 19–26.
128. S. Childress, *New solutions to the kinematic dynamo problem*, J. Math. Phys. **11** (1970), 3063–3076.
129. D. Coles, *Transition in circular Couette flow*, J. Fluid Mech. **21** (1965), 385–425.
130. P. E. Dimotakis and G. L. Brown, *The mixing layer at high Reynolds number*, J. Fluid Mech. **78** (1976), 535–560.
131. C. L. Dolph, *Non-linear integral equations of the Hammerstein type*, Proc. Nat. Acad. Sci. U.S.A. **31** (1945), 60–65.
132. ———, *Non-linear integral equations of the Hammerstein type.*, Trans. Amer. Math. Soc. **66** (1949), 289–307.
133. T. Dombre, U. Frisch, J. M. Green, M. Hénon, A. Mehr, and A. M. Soward, *Chaotic streamlines in the ABC flows*, J. Fluid Mech. **167** (1986), 353–391.
134. C. P. Donaldson, *Calculation of turbulent shear flows for atmospheric and vortex motions*, AIAA J. **10** (1972), 4–12.
135. H. L. Dryden, *Recent advances in the mechanics of boundary layer flow*, Adv. Appl. Mech., vol. 1, Academic Press, Boston, MA, 1948, pp. 1–40.
136. M. L. Dubreil-Jacotin, *Sur les théorémes d'existence relatifs aux ondes permanentes périodiques à deux dimensions dans les liquides heterogenes*, J. Math. Pures Appl. **13** (1937), 217–291.
137. D. G. Ebin and J. Marsden, *Groups of diffeomorphisms and the notion of an incompressible fluid*, Ann. of Math. (2) **92** (1970), 102–163.
138. C. Foias, U. Frisch, and R. Temam, *Existence de solutions* C^∞ *des equations d'Euler*, C. R. Acad. Sci. Paris Sér. A **280** (1975), 505–508.
139. L. E. Fraenkel, *On corner eddies in plane inviscid shear flow*, J. Fluid Mech. **11** (1961), 400–406.
140. L. E. Fraenkel and M. S. Burger, *A global theory of steady vortex rings in an ideal fluid*, Acta Math. **132** (1974), 13–51.
141. A. Hammerstein, *Nichtlinear Integralgleichungen nebst Anwendungen*, Acta Math. **54** (1930), 117–176.
142. M. Hénon, *Sur la topologie les lignes de courant dans un cas particulier*, C. R. Acad. Sci. Paris Sér. A **262** (1966), 312–314.
143. J. Heywood, *On classical solutions of the non-stationary Navier-Stokes equations in two and three dimensions*, Fluid Dynamics Trans. **10** (1980), 177–203.
144. ———, *The Navier-Stokes equations: on the existence, regularity, and decay of solutions*, Indiana Univ. Math. J. **29** (1980), 639–681.

145. E. Hopf, *A mathematical example displaying the features of turbulence*, Comm. Pure Appl. Math. **1** (1948), 303–322.
146. ______, *Über die Anfangswertaufgabe für die hydrodynamischen Grundgleichungen*, Math. Nachr. **4** (1951), 213–231.
147. C. O. Horgan and W. E. Olmstead, *Stability and uniqueness for a turbulence model of Burgers*, Quart. Appl. Math. **36** (1978), 121–127.
148. T. Kato, *On classical solutions of the two-dimensional nonstationary Euler equation*, Arch. Rational Mech. Anal. **25** (1967), 188–200.
149. Y. Kobashi, Proc. 2nd Japan Nat. Cong. Appl. Mech., 1962, p. 223.
150. M. T. Landahl, *A wave-guide model for turbulent shear flow*, J. Fluid Mech. **29** (1967), 441–459.
151. E. M. Landesman and A. C. Lazer, *Linear eigenvalues and a nonlinear boundary value problem*, Pacific J. Math. **33** (1970), 311–328.
152. B. E. Launder, *Phenomenological modeling: Present...and future?*, Whither Turbulence? Turbulence at Crossroads (J. L. Lumley, ed.), Lecture Notes in Phys., vol. 357, Springer-Verlag, Berlin, 1990, pp. 439–535.
153. E. M. Landesman, A. C. Lazer, and D. R. Meyers, *On saddle point problems in the calculus of variations, the Ritz algorithm, and monotone convergence*, J. Math. Anal. Appl. **52** (1975), 594–614.
154. J. Leray, *Etude de diverses equations integrales nonlineares*, J. Math. Pures Appl. **12** (1933), 1–82.
155. ______, *Essai sur les mouvements plans d'un liquide visqueux que limitent des parois*, J. Math. Pures Appl. **13** (1934), 331–418.
156. ______, *Sur le mouvent d'un liquide visqueux emplissant l'espace*, Acta Math. **63** (1934), 193–248.
157. M. J. Lighthill, *On sound generated aerodynamically.* I. *General theory*, Proc. Roy. Soc. London Ser. A **211** (1952), 564–587.
158. G. L. Mellor and H. J. Herring, *Survey of the main turbulent field closure models*, AIAA J. **11** (1973), 590–599.
159. L. M. Milne-Thomson, *On the steady motion in two dimensions of inviscid incompressible fluid*, Zanichelli, Bologna, 1969.
160. H. K. Moffatt, *Magnetostatic equilibria and analogous Euler flows of arbitrarily complex topology.* I. *Fundamentals*, J. Fluid Mech. **159** (1985), 359–378.
161. ______, *Magnetostatic equilibria and analogous Euler flows of arbitrarily complex topology.* II. *Stability considerations*, J. Fluid Mech. **166** (1986), 359–378.
162. ______, *On the existence of localized rotational disturbances which propagate without change of structure in an inviscid fluid*, J. Fluid Mech. **173** (1986), 289–302.
163. E. L. Mollo-Christensen, *Jet noise and shera-flow instability seen from an experimenter's viewpoint*, ASME J. Appl. Mech. **89E** (1967), 1.
164. J. J. Moreau, *Une méthode de "cinematique fonctionelle" en hydrodynamique*, C. R. Acad. Sci. Paris **249** (1959), 2156–2158.
165. M. Morse, *The calculus of variations in the large*, Amer. Math. Soc. Colloq. Publ., vol. 18, Amer. Math. Soc., New York, 1934.
166. ______, *The topology of pseude-harmonic functions*, Duke Math. J. **13** (1946), 21–42.
167. M. Morse and M. Heins, *Topological methods in the theory of functions of a single complex variable.* I, II, Ann. of Math. (2) **46** (1945), 600–666.
168. ______, *Topological methods in the theory of functions of a single complex variable.* III, Ann. of Math. (2) **47** (1946), 233–273.
169. L. Nirenberg, *On nonlinear elliptic partial differential equations and Hölder continuity*, Comm. Pure Appl. Math. **6** (1953), 103–150.
170. A. Pumir and E. D. Siggia, *Vortex dynamics and the existence of solutions to the Navier-Stokes equations*, Phys. Fluids **30** (1987), 1606–1626.
171. P. H. Rabinowitz, *Existence and nonuniqueness of rectangular solutions of the Bénard problem*, Arch. Rational Mech. Anal. **29** (1968), 32–57.
172. O. Reynolds, *On the dynamical theory of incompressible viscous fluids and the determination of the criterion*, Phil. Trans. Roy. Soc. London Ser. A **186** (1894), 123.
173. J. Rotta, *Statistische Theorie nichthomogener Turbulenz.* I, Z. Phys. **129** (1951), 547–572.
174. V. Scheffer, *Turbulence and Hausdorff dimension*, Turbulence and Navier-Stokes Equations, Lecture Notes in Math., vol. 565, Springer-Verlag, Berlin, 1976, pp. 174–183.
175. ______, *Partial regularity of solutions to the Navier-Stokes equations*, Pacific J. Math. **66** (1976), 535–552.
176. ______, *Hausdorff measure and the Navier-Stokes equations*, Comm. Math. Phys. **55** (1977), 97–112.

177. ______, *The Navier-Stokes equations in space dimension four*, Comm. Math. Phys. **61** (1978), 41–68.
178. ______, *The Navier-Stokes equations on a bounded domain*, Comm. Math. Phys. **73** (1980), 1–42.
179. D. Serre, *A propos des invariants de diverses équations d'évolution*, Research Notes in Math. **70** (1982), 326–336.
180. ______, *Invariants et dégénérescence symplectique de l'équation d'Euler des fluides parfaits incompressibles*, C. R. Acad. Sci. Paris Sér. I Math. **298** (1984), 349–352.
181. R. Témam, *Local existence of C^∞ solutions of the Euler equations of incompressible perfect fluids*, Turbulence in Navier-Stokes Equations, Lecture Notes in Math., vol. 565, Springer-Verlag, Berlin, 1976, pp. 184–194.
182. W.Thomson, *On the propagation of laminar motion through a turbulently moving inviscid liquid*, Phil. Mag. **24** (1887), 342.
183. O. V. Troshkin, *On wave properties of an incompressible turbulent fluid*, Phys. A **168** (1990), 881–899.
184. W. Velte, *Stabilität und Verzweigung stationärer Lösungen der Navier-Stokesschen Gleihungen bein Taylorproblem*, Arch. Rational Mech. Anal. **22** (1966), 1–14.
185. F. B. Weissler, *Single point blow-up for a semilinear initial value problem*, J. Differential Equations **55** (1984), 204–224.
186. J. L. Lumley (ed.), *Whither turbulence? Turbulence at crossroads*, Proc. Workshop, Cornell University, Ithaca, NY, 1989, Lecture Notes in Phys., vol. 357, Springer-Verlag, Berlin, 1990.
187. N. M. Wigley, *Asymptotic expansions at a corner of solutions of mixed boundary value problems*, J. Math. Mech. **13** (1964), 549–576.
188. ______, *Development of the mapping function at a corner*, Pacific J. Math. **15** (1965), 1435–1461.
189. ______, *Mixed boundary value problems in plane domains with corners*, Math. Z. **115** (1970), 33–52.
190. W. Wolibner, *On théoréme sur l'existence du mouvement plan d'un fluide parfait, homogéne, incompressible, pendant un temps infiniment longue*, Math. Z. **37** (1933), 698–726.
191. V. C. Patel and M. R. Head, *Some observations on skin friction and velocity profiles in fully developed pipe and channel flows*, J. Fluid Mech. **38** (1969), 181–201.
192. G. S. Beavers and E. M. Sparrow, *Low Reynolds number turbulent flow in large aspect ratio rectangular ducts*, J. Basic Engn. **93** (1971), 296–299.